Electronics in Textiles and Clothing

Design, Products and Applications

Electronics in Textiles and Clothing

Design, Products and Applications

L. Ashok Kumar
C. Vigneswaran

CRC Press
Taylor & Francis Group
Boca Raton   London   New York

CRC Press is an imprint of the
Taylor & Francis Group, an **informa** business

CRC Press
Taylor & Francis Group
6000 Broken Sound Parkway NW, Suite 300
Boca Raton, FL 33487-2742

First issued in paperback 2020

© 2016 by Taylor & Francis Group, LLC
CRC Press is an imprint of Taylor & Francis Group, an Informa business

No claim to original U.S. Government works

Version Date: 20150917

ISBN 13: 978-0-367-57537-3 (pbk)
ISBN 13: 978-1-4987-1550-8 (hbk)

Visit the Taylor & Francis Web site at
http://www.taylorandfrancis.com

and the CRC Press Web site at
http://www.crcpress.com

Contents

List of Figures

Preface

Electronics in Textiles and Clothing mainly deals with the fundamentals of electronics and their applications in textiles and clothing product development. With an increasing awareness about e-textiles, researchers and scientists today are looking for ways to integrate textile materials with electronics for communication/signal transferring applications. Following this, many researchers are working toward developing wearable electronic products for industrial applications based on functional properties and the needs of the end users.

The book is divided into seven chapters. Chapter 1 discusses the application of wearable electronics and outlines the textile fibres used in wearable electronics for the development of sensing fabrics and their functional end uses. This chapter also briefly discusses various wearable sensors and technologies that can be integrated into clothing and the usage of body sensor network, its wearability and technical challenges. Chapter 2 discusses the different types of yarn and fabric production techniques, modifications made to conventional textile machines to develop fabrics using specialty yarns and the problems faced during the production processes and their solutions. Chapter 3 describes novel sensors developed and integrated in sensorized garments for monitoring vital signs and body gesture/posture. The main advantage of these systems is that they can be worn for long periods of time without discomfort. Several issues deriving from the employment of this new technology that has constrained the realization of these unobtrusive devices have been addressed. Moreover, it has been pointed out that the use of these sensorized garments is a valid alternative to existing instrumentation applicable in several health-care areas. Finally, results on the performances of such sensing systems are briefly reported. The work continues toward potential commercial exploitation. This chapter also presents one of the most challenging aspects of creating wearable electronic circuits and integration with wearable electronic fabrics. The design of circuits, selection of controllers for different fabrics, types of printed circuit boards, programming methods for controlling various signals and the testing methodologies involved have been discussed. Chapter 4 discusses the development of wearable electronics products using cotton-wrapped nichrome yarn, POF yarns–based nichrome fabric for communication applications, illumination and teleintimation. This chapter also discusses specialty garments manufactured using nichrome wires and POF fabrics and the plug-in methods for integrating miniaturized electronic circuits in the garment. Chapter 5 discusses the software developed to track soldiers based on the GPS/GPRS technology. The purpose of body sensor networks (BSN) is to provide an integrated hardware and software platform for facilitating future development of pervasive monitoring systems. Chapter 6 deals with the development of flexible solar tents utilized for power generation and aimed at supplying power to the load, which can be utilized for the electronic equipment and wearable electronic products used by soldiers during war. Chapter 7 discusses the development of fabric integrated wearable electronic products for use in wearable mobihealth care systems, smart clothes for ambulatory remote monitoring, electronic jerkins, heating gloves and pneumatic gloves. The design and integration of electronic circuits in the latter three products are briefed in this chapter.

The main focus of this book is the development of wearable electronic products based on various functional needs such as defence, medical, health monitoring and security. This book also describes the future scope of e-textiles using conductive materials for medical, health-care textile product development and safety aspects. This book will be very useful to undergraduate and postgraduate students of electrical and electronic engineering, instrumentation engineering, biomedical engineering, textile technology and apparel technology; researchers working in academic and industrial R&D; and colleges and universities offering textile technology and electronic engineering programs. In short, this book presents an original approach in emphasizing the interface between electronics and textile materials dealing with diverse methods and techniques used in industrial practices.

We hope that the book will serve as a guideline to researchers, scientists and industrialists, as well as be a useful source of information to students in the important field of developing wearable textile products through e-textiles.

Dr. L. Ashok Kumar
Professor
Department of Electrical and Electronic Engineering
PSG College of Technology
Coimbatore, India

Dr. C. Vigneswaran
Associate Professor
Department of Fashion Technology
PSG College of Technology
Coimbatore, India

Acknowledgements

The authors are always thankful to the Almighty for their perseverance and achievements. The authors are grateful to L. Gopalakrishnan, managing trustee, PSG Institutions, and Dr. R. Rudramoorthy, principal, PSG College of Technology, Coimbatore, India, for their wholehearted cooperation and encouragement in this successful endeavour.

I (Ashok Kumar) acknowledge those people who helped me in completing this book. This book would not have come to its completion without the help of my students, department staff and institute and especially my project staff. I sincerely thank the management and our principal, guru and mentor Dr. Rudramoorthy for providing a wonderful platform to perform. I am thankful to all my students, who are doing their project and research work with me. But the writing of this book was mainly possible because of the support of my family members, parents and sisters. Most importantly, I am very grateful to my wife, Y. Uma Maheswari, for her constant support during the writing, without her this book would not have been possible. I express my special gratitude to my daughter, A. K. Sangamithra. Her smiling face and support have helped immensely in overcoming backlogs and completing this book. My apologies to her; I took a significant amount of my playtime with her to write this book.

Dr. C. Vigneswaran acknowledges his wife, V. Kanchanamala, and daughters for their constant support and encouragement.

The authors wish to thank all their friends and colleagues who have been with them in all their endeavours with their excellent, unforgettable help and assistance in the successful execution of this work.

Authors

Dr. L. Ashok Kumar is a professor in the Department of Electrical and Electronics Engineering, PSG College of Technology, Coimbatore, India. He has completed his graduate program in electrical and electronics engineering. He did his postgraduation with electrical machines as his major. He also holds an MBA with a specialization in HRD. He completed his PhD in wearable electronics. He has been a teaching faculty at PSG College of Technology since 2000. After completing his undergraduate degree, Dr. Kumar joined as project engineer in Servalakshmi Paper and Boards Pvt. Ltd. (now renamed ITC PSPD, Kovai), Coimbatore. He has been actively involved in numerous government-sponsored research projects funded by DRDO, DSIR, CSIR, DST, L-RAMP, Villgro and AICTE. He has published many technical papers in reputed national and international journals and conferences. He visited many countries for presenting his research work and industry consultancy activities. He was selected as a visiting professor at the University of Arkansas, Fayetteville, Arkansas. He was trained in Germany in the installation of grid-connected and stand-alone solar PV systems. He is a registered charted engineer from IE (India). Dr. Kumar is a recipient of many national awards from IE, ISTE, STA, NFED, etc. He is a member of various national and international technical bodies, including IE, ISTE, IETE, TSI, BMSI, ISSS, SESI, SSI, CSI and TAI. He has received several awards, namely, Young Engineer Award, Dr. Triguna Charan Sen Prize, P.K. Das Memorial Best Faculty Award, Prof. K Arumugam National Award, Best Researcher Award and Outstanding Educator and Best Researcher Award. His research areas include wearable electronics, solar PV and wind energy systems, textile control engineering, smart grid and power electronics and drives.

Dr. C. Vigneswaran is an associate professor in the Department of Fashion Technology, PSG College of Technology, Coimbatore, India, since 2004. He completed his MTech (textile technology) from PSG College of Technology in wearable electronics and PhD in textile technology from Anna University, Chennai, India, in bioprocessing of organic cotton textiles. He stood state first in diploma in man-made fibre technology during 1992–1995. He received the Best Student Award from Lee Educational Trust, UK, during the 79th World Textile Conference organized by The Textile Institute, UK, in 1998; Precitex Award for best student in spinning from Precitex Rubber Industries, Mumbai, during 1998–1999; Dr. Triguna Charan Sen Prize Award from The Institution of Engineers (India), Kolkata, for the Best Technical Research Paper in Textile Discipline for 2008–2009 during the 23rd Indian Engineering Congress, Warangal, on December 12, 2008; "Bharat Shiksha Ratan Award" in 2010 from The Global Society for Health and Educational Trust,

New Delhi, for his excellence in the academic field in textiles; and recently (2012–2013) he received the Young Engineer Award from The Institution of Engineers (India) in the textile engineering discipline. Dr. Vigneswaran has published 24 technical papers in national and international journals and presented eight technical papers in national and international technical conferences. His areas of specialization are fibre science, yarn manufacturing, garment manufacturing technology, apparel machinery and equipment and apparel costing. He has a perfect blend of industrial and teaching experience for more than 12 years.

1

Wearable Electronics

1.1 Introduction

Electronics in textiles is an emerging interdisciplinary field of research that brings together specialists in information technology, microsystems, materials and textiles. The focus of this new area is on developing the enabling technologies and fabrication techniques for the economical production of flexible, conformably and optionally, large area textile-based information systems that are expected to have unique applications for both civilian and military arenas. The synergistic relationship between computing and textiles can also be characterized by specific challenges and needs that must be addressed in the context of electronic textile–based computing.

The *wearable electronic textiles* arena has generated a massive demand and significant opportunities stem from increased consumer demand for lightweight electronic devices integrated with clothing. Technological advances have enabled electronics to become smaller and more powerful. Many industry experts are seeking to take this one step further, by integrating electronics into textiles to increase user mobility and comfort. This involves understanding available technologies, and how they can be used to develop interactive electronic textiles. These researches provide a better understanding of the current work and focus on the wearable electronic textiles arena and may potentially open the area to new concepts and ideas. When a research study can provide information to assist industry with further understanding of a new emerging area, it proves to be worthwhile and relevant.

A significant invention in the field of textiles is the Jacquard head by *J.M. Jacquard*; he was the first to use binary information processor for development of user interface mode. At any given point, the thread in a woven fabric can be in one of the two states or positions: on the face or on the back of the fabric. The cards were punched or cut according to the required fabric design. A hole in the card signified that the thread would appear on the face of the fabric, while a blank meant that the end would be left down and appear on the back of the fabric. The Jacquard head was used on the weaving loom or machine for rising and lowering the warp threads to form desired patterns based on the lifting plan or programme embedded in the cards.

Thus, the Jacquard mechanism set the stage for modern-day binary information processing. The *second industrial revolution* is also known as the information processing revolution (Jayaraman 1990). In fact, when Intel introduced its Pentium class of microprocessors, one of the advertisements had a *fabric of chips* emerging from a weaving machine; this eloquently captured the essence of chip making, a true blending of art and science, much like the design and production of textiles. Today, more than two centuries later, researchers are finding themselves looking at the possibility of integrating textiles and electronics, but in a different way.

Humans are used to wearing clothes from the day they are born and, in general, no special *training* is required to wear them, that is to use the interface. In fact, it is probably the most universal of human–computer interfaces and is one that humans need, use, have familiarity with, and which can be easily customized (Marculescu 2003). Moreover, humans enjoy clothing and this universal interface of clothing can be *tailored* to fit the individual's preferences, needs and tastes including body dimensions, budgets, occasions and moods. Textiles can also be designed to accommodate the constraints imposed by the ambient environment in which the user interacts, that is different climates or operating requirements. In addition to these two dimensions of functionality (or protection) and aesthetics, if *intelligence* can be embedded or integrated into textiles as a third dimension, it would lead to the realization of clothing as a personalized and flexible wearable information infrastructure (Park and Jayaraman 2001a).

1.2 Textiles and Information Processing

A well-designed information processing system should facilitate the access of information anytime, at any place by anyone – the three As. The *ultimate* information-processing system should not only provide for large bandwidths, but also have the ability to see, feel, think and act. In other words, the system should be totally *customizable* and be in tune with the human. Of course, clothing is probably the only element that is *always there* and in complete harmony with the individual. Also, textiles provide the ultimate flexibility in system design by virtue of the broad range of fibres, yarns, fabrics and manufacturing techniques that can be deployed to create products for desired end-use applications. Moreover, textile fabrics provide *large* surface areas that may be needed for *hosting* the large numbers of sensors and processors that might be needed for deployment over large terrains, for example a battlefield. The opportunities to build in redundancies for fault tolerance make textiles an *ideal* platform for information processing. An *intelligent* garment differs from a traditional garment primarily in this capacity to receive input from its wearer or from its environment, and use that input to activate or change the state of an associated technology. A garment may be more or less intelligent based on the number of sensors and their ability to detect context, synthesize input or cause change. The field of intelligent clothing is characterized by the fusion of two distinct fields: *functional clothing design* and *portable technology*. These two fields have developed separately, but both are necessary for the successful design of intelligent clothing.

1.3 Chronology of Wearable Electronics

The concept of wearable computing (*wearables*) emerged in the mid-1990s at a time when carrying an *always-on* computer combined with a head-mounted display and control interface first became a practical possibility. In July 1996, a workshop 'Wearables in 2005' was sponsored by the *U.S. Defense Advanced Research Projects Agency*. This was attended by industrial, university and military visionaries to work on the common theme of delivering computing to the individual. They defined wearable computing as 'data gathering and disseminating devices which enable the user to operate more efficiently'. These devices are carried or worn by the user during normal execution of his/her tasks. One of the first advocates and adopters of this form of computer usage further defined wearable computing and arrived at three fundamental properties (Mann 1994). First, a wearable computer is worn, not carried, in such a way as it can be regarded as being part of the user; second it is user controllable, not necessarily involving conscious thought or effort and, finally, it operates in real time – it is always active (though it may have a sleep mode) and be able to interact with the user at any time (Mann 1997b).

Using these definitions, it was possible to retrospectively recognize early applications of wearable computing. These included the shoe mounted roulette wheel prediction system by Thorp and Shannon, first implemented in 1961, subsequently successfully developed and used by the 'Eudemons' and by 1983, commercialized by Keith Taft and others (Thorp 1998). Dive computers, which first became common in the 1980s, are also worn, are user controllable and have sensors which operate in real time. In the art world, Stelarc has experimented with body sensors and actuators (Stelarc 1997), and many artists have developed unusual musical controllers. These wearable interfaces were used sensor systems affixed to the body or clothing of a performer measuring movement and/or body functions such as heart rate or skin resistance. The interfaces connected to audio equipment such as midi devices and sound synthesizers, sometimes also worn on the body. Joe Paradiso at M.I.T. has taken a particular interest in these and had created his own devices, providing a comprehensive overview of this field in an IEEE Spectrum article 'New Ways to Play: Electronic Music Interfaces' (Paradiso 1997; Paradiso and Feldmeirer 2005).

Many other inventive backroom constructors also produced wearable systems, most notably Mann with his WearCam and WearComp devices. Originally starting by building a wearable 'photographer's assistant', he developed a series of wearables from the 1970s to the present day featuring body-mounted cameras and lighting, head-mounted displays, audio interfaces and many of the other features commonly associated with wearable computing (Mann 1997b). As the development of the wearable computer was originally inspired by the availability of battery powered head-mounted displays, it has been closely linked to this technology. An overview of the challenges presented by these displays was summarized by Duchamp in 1991. These were the hassle of the head gear, low-resolution, eye fatigue and the requirement for dim lighting conditions (Duchamp et al. 1991).

The health and well-being of service personnel also require special attention. The sensate liner developed at Georgia Institute of Technology was designed specifically to monitor the vital signs of combat casualties, as well as automatically detect and characterize a wound in real time using bullet entry detection (Lind et al. 1997). Further health-monitoring applications are presented in the following section. The applications described previously have used position-sensing technology to assist in a variety of tasks. The knowledge of where the user is located clearly provides the basis for many wearable designs. Wearables can also be designed to monitor well-being and activity – the how and what of the user. This form of context sensing has been put to use in wearable computers for medical and health applications and has met with more success than in any other field. Body invasive devices, such as heart pacemakers, have become commonplace. However, as these devices are generally not user controllable, they do not fall under the definition of wearable computers. Wearables have the potential to monitor health to assist with improving performance, for example sports; prevention and detection of illness through diagnosis, and even treatment, though this usually involves some invasive procedure. Examples of treatment by a wearable are insulin pump therapy for diabetics (Doyle et al. 2004) and a brain implant to facilitate communication with speech-incapable patients (Bakay et al. 1999).

Textiles are ubiquitous in our society, and provide the ideal base or support for wearable monitoring electronics. Because textile garments or accessories can be worn close to the skin, integrating electronics within the textiles gives benefits unrivalled by other systems. Measurements of data requiring close contact with the skin are possible, with minimal effects on the comfort of the wearer, and with minimal disruption to his/her day-to-day activities. To date, a number of wearable electronic textiles have been developed. Many are multi-layer systems, consisting of at least an internal layer (in contact with the skin) and an external one, with connected electronics and circuitry.

The main advantage with textiles is their flexibility, which relates to some extent to wearing comfort. Knitted fabrics have the advantage of being stretchable and deformable to some extent, and have been used where the fabrics need to be close to the body, or close fitting, such as for leotards or other sportswear. Comparatively, woven fabrics provide more dimensional stability, and are more suitable where large movements of the body are not an essential factor to consider.

1.4 Application Groups

The main important stakeholders for wearable electronics in textiles are end users and a short overview of the major application of wearable electronics is given as under:

1.4.1 Retailer Support

A successful embedded electronics for retailer support should cover the needs in logistics, such as stock control, quality insurance control and anti-theft protection. For this purpose, there are several integrated RF indent-tags already in use. These tags consist mainly of an RF antenna used to transfer energy to the indent-tag and to establish a communication link between the tag and the control equipment. The most primitive tags consist of a simple antenna-capacitor resonance circuit. The more complex ones contain simple micro-controllers and non-volatile memories.

1.4.2 Service Support

Service support is more important in developed countries like Europe and United States. Here consumer tends to sort clothes into different categories before washing. Therefore, the washing machines are able to treat the clothes in very different ways. Naturally, it happens rarely that a black sock is hiding in a white shirt. This results in coloured shine on the originally white clothes. The consumer can avoid such accidents if the washing machine recognizes type, number and required optimal treatment of the clothes to be washed. In case of conflict, an error message is generated.

1.4.3 User Convenience

Built-in electronics may control and support more advanced textile functionalities like temperature, moisture, etc. For that purpose, secure heating and cooling elements are necessary.

1.4.4 User Interfacing

User interfacing enables the interfacing between the user and electronic belongings or external (out of body) networks and terminals. Sound, gesture and temperature may control such interfaces. Suitable components are microphones, loudspeaker, textile keyboards, flexible displays and more complex devices.

1.4.5 Appliance Networking

The networking between different electronic modules is a key feature necessary for the application of user interfaces. One scenario might be the connection of a cellular phone ported in the right pocket, with a personal digital assistant (PDA) in the left pocket, Bluetooth interface in the trousers and the user interface integrated into the coat.

1.4.6 Networking with External Networks

The networking with external networks can be done in real time or by *batch processing*, online and offline. Another offline method might be the use of exchangeable storage elements like multimedia cards, etc. Online external networking might be possible as a direct communication link established by *Bluetooth, DECT, GSM* or *UMTS RF* connections. They require a suitable *TWRX RF* part and a base-band processor handling all the necessary protocol work. Broadcasting applications are also viable. They might rely on standards like *DVB-T* or *DAB*. Here the system requires a tuner, demodulation stage and decompression and decryption unit. Common to all applications is the need for a power supply that enables relative long stand-by times and that can be recharged or refuelled with high customer convenience. Wearable computers have the potential to enhance the day-to-day activities of the user. Applications that support many specialized activities have been explored and demonstrated by Ockerman and Pritchett (1998). The Metronaut mobile computer is being able to sense information and locate position via bar code reader, provide wide range communications via two-way pager, supporting applications like navigation, messaging, and scheduling and still consuming less than 1 W of power and weighing less than 1 lb (Smailagic and Siewiorek 1994). Posture and gesture analysis, together with the monitoring of body kinematics, is a field of increasing interest in bioengineering and several connected disciplines. In the paper by Lorussi et al. (2004a), some typical features of distributed sensing systems, as well as a methodology to read signals from such systems are described.

1.5 Need for Wearable Electronics

The need for truly wearable body sensors is clear: while invasive, restrictive, uncomfortable sensors are the only choice, viable applications will remain restricted to those arenas (medical, military) where the body sensing function is crucial to the subject's health or well-being. But the trade-offs for sensing accuracy must also be overcome to preserve the full range of possible applications. An understanding of the requirements for mainstream wear ability is necessary to further wearable body-sensing research in this direction. The paper by Tröster et al. (2003) reflects the state-of-the-art in electronic textiles and presents first implementations. Depending on the application, the electronic functionality can be fully integrated or a modular approach can be chosen, where clothing provides a kind of *platform* for several possible modules. In order to achieve these objectives, a close interdisciplinary cooperation in the fields of electrical, textile and clothing technology is necessary.

They are expected to measure and react in an intelligent way. They can monitor a person and his or her environment, and interpret such data. In cases where health problems or environmental threats are detected, a smart suit can send out a warning or even protect instantaneously. Consequently, they will become an important tool for prevention. In the long term, a smart suit can assist in the rehabilitation process by supplying of drugs, activation of muscles and many more. Weber et al. (2002a) presented various technology components to enable the integration of electronics into textiles. Key elements are a packaging and interconnect technology for deep textile integration of electronics. An interconnect and packaging technology is demonstrated using a polyester narrow fabric with several warp threads replaced by copper wires which are coated with silver and polyester.

1.5.1 Health and Wellness Monitoring

As the world population is aging and health-care costs are increasing, several countries are promoting *aging in place* programmes which allow older adults and individuals with chronic conditions to remain in the home environment while they are remotely monitored for safety and for the purpose of facilitating the implementation of clinical interventions. Monitoring activities performed by older adults and individuals with chronic conditions participating in *aging in place* programmes has been considered a matter of paramount importance. Accordingly, extensive research efforts have been made to assess the accuracy of wearable sensors in classifying activities of daily living (ADL). Mathie et al. (2004) have studied the feasibility of using accelerometers to identify the performance of ADL by older adults monitored in the home environment. Sazonov et al. (2009) have developed an in-shoe pressure and acceleration sensor system that was used to classify activities including sitting, standing and walking with the ability of detecting whether subjects were simultaneously performing arm reaching movements. Giansanti et al. (2008) have developed an accelerometer-based device designed for step counting in patients with Parkinson's disease (PD). Aziz et al. (2007) have used wearable sensors to monitor the recovery of patients after abdominal surgery.

Several research works have been conducted: activity monitoring for wellness applications has great potential to increase exercise compliance in populations at risk. For example wearable technology has been used to monitor physical activities in obese individuals and to facilitate the implementation of clinical interventions based on encouraging an active and healthy lifestyle. Long-term monitoring of physiological data can lead to improvements in the diagnosis and treatment, for instance, of cardiovascular diseases. Commercially available technology provides one with the ability to achieve long-term monitoring of heart rate, blood pressure, oxygen saturation, respiratory rate, body temperature and galvanic skin response (GSR). Clinical studies are currently carried out to evaluate and validate the performance of wearable sensor platforms to monitor physiological data over long periods of time and improve the clinical management of patients, for instance, with congestive heart failure. Several ongoing studies are focused on clinically assessing wearable systems developed as part of major research projects. For instance LiveNet, a system developed at the MIT Media Laboratory that measures 3-D acceleration, ECG, EMG and galvanic skin conductance, is under evaluation for monitoring Parkinsonian symptoms and detecting epileptic seizures. LifeGuard is a custom data logger designed to monitor health status of individuals in extreme environments (space and terrestrial). The system has undergone testing in hostile environments with good results. As part of the FP5 programme of the European Commission, a project named AMON resulted in the development of a wrist-worn device capable of monitoring blood pressure, skin temperature, blood oxygen saturation and ECG. The device was developed to monitor high-risk patients with cardiorespiratory problems.

1.5.2 Safety Monitoring

A number of devices have been developed for safety monitoring applications, such as detecting falls and relaying alarm messages to a caregiver or an emergency response team. The Life Alert Classic by Life Alert Emergency Response Inc. and the AlertOne medical alert system are examples of commercially available devices designed for safety monitoring. These devices are simple emergency response devices consisting of a pendant or watch with a push button. By pressing the button, one has the ability to wirelessly relay an alarm message to operators located in a remote call centre. Other systems integrate sensors into the body-worn unit. For instance the Wellcore system employs advanced microprocessors and accelerometers to monitor the body's position. The system detects falls as distinct events from normal movements, and automatically relays a message to the designated response centre or nurse call station. Another device in this category is the MyHalo™ by Halo Monitoring™. The system is worn as a chest strap and detects falls, while it monitors heart rate, skin temperature, sleep/wake patterns and activity levels. The BrickHouse system is equipped with an automatic fall detector and a manual panic button. Finally, among the numerous commercially available systems, it is worth to mention ITTM EasyWorls, a system based on a mobile phone that is equipped with balance sensors which trigger automatic dialling of SOS numbers if the system detects a sudden impact. Reliable detection of falls via wearable sensors has been achieved by many research groups.

Researchers at CSEM have developed an automatic fall detection system in the form of a wristwatch. The device implements functions such as wireless communication, automatic fall detection, manual alarm triggering, data storage and a simple user interface. Even though the wrist is a challenging sensor location to detect a fall event, researchers working on the project achieved 90% sensitivity and 97% specificity in the detection of simulated falls. Bourke et al. (2008) have adapted an alternative approach and used a tri-axial accelerometer embedded in a custom-designed vest to detect falls. Bianchi et al. (2010) used instead a barometric pressure sensor as a surrogate measure of altitude to discriminate real fall events from normal ADL. When tested among a cohort of 20 young healthy volunteers, the proposed method demonstrated considerable improvements in sensitivity and specificity compared to an existing accelerometer-based technique. Finally, among the numerous systems developed by researchers to detect falls, it is worth to mention that Lanz and Messelodi (2009) have developed Smart-Fall, a system that relies on an accelerometer embedded in a cane to detect falls. The authors argued that canes are assistive devices that people widely use to overcome problems associated with balance disorders and therefore, embedding the system in the cane is a very appealing solution to achieve unobtrusive monitoring while assuring safety to individuals.

Safety monitoring applications typically require detection of emergency events. The sensing technology used for such applications must be extremely robust and reliable. A great deal work has been done towards developing wearable systems to monitor individuals working in hostile environments in response to emergency situations. The Proe-TEX project, carried out as part of the FP6 programme of the European Commission, is an example of such work. The project resulted in the development of a new generation of smart garments to monitor emergency-disaster personnel (Figure 1.1). These garments enable the detection of health status parameters of the users and environmental variables such as external temperature, presence of toxic gases and heat flux passing through the garments. Extensive testing of the garments is being carried out both in laboratories, specialized in physiological measures, and in simulated fire-fighting scenarios. Advances achieved in the aforementioned projects could be used to design robust systems for home health monitoring to be deployed to detect emergency events such as falls and seizures.

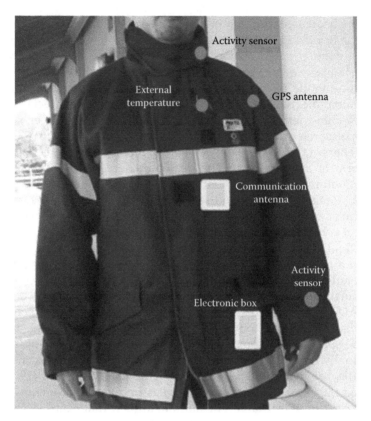

FIGURE 1.1
Smart garments integrate sensors, communication, processing and power management directly into the garment to continuously monitor emergency responders. (Courtesy of Smartex, Italy.)

1.5.3 Home Rehabilitation

An emerging area of application of wearable technology is the use of wearable sensors to facilitate the implementation of home-based rehabilitation interventions. Systems that aim to facilitate the implementation of rehabilitation exercise programmes often leverage the combination of sensing technology and interactive gaming or virtual reality (VR) environments. For example, Rehabilitation Engineering Research Center (RERC) at the University of Southern California is building on VR gaming to address compliance and motivation challenges. VR simulation technology using specialized interface devices has been applied to improve motor skills in subjects undergoing rehabilitation to address functional deficits including reaching, hand function and walking. It has been proposed that such VR-based activities could be delivered in the home via a telerehabilitation approach to support patients' increased access to rehabilitation and preventive exercise programming. When this is put in an interactive game-based context, the potential exists to enhance the engagement and motivation needed to drive neuroplastic changes that underlie motor process maintenance and improvement. However, home-based systems need to be affordable and easy to deploy and maintain, while still providing the interactional fidelity required to produce the meaningful motor activity required to foster rehabilitative aims and promote transfer to real world activities. An example of such systems is the Valedo system by Hocoma AG. The Valedo system is a medical back training device, which improves patient's compliance and allows one to achieve increased motivation by real-time augmented feedback based on trunk movements. It transfers trunk movements from two wireless sensors into a motivating game environment and guides the patient through exercises specifically designed for low back pain therapy. To facilitate challenging the patient and achieving efficient training, the exercises can be adjusted according to the patient's specific needs. Several other systems are currently under development. For instance GE Healthcare is developing a wireless medical monitoring system that is expected to allow one to gather physiological and movement data, thus facilitating rehabilitation interventions in the home setting. Another example of home-based rehabilitation technology is the Stroke Rehabilitation Exerciser (SRE) developed by Philips Research. The Stroke Rehab Exerciser coaches the patient through a sequence of exercises for motor retraining, which are prescribed by the physiotherapist and uploaded to a patient unit. A wireless inertial sensor system records the patient's movements, analyzes the data for deviations from a personal movement target and provides feedback to the patient and the therapist. Major efforts have been made by European groups to develop systems suitable for home-based interventions that rely on wearable technology. A project that was part of the myHeart initiative led to the development of a sensorized garment–based system to facilitate rehabilitation interventions in the home setting. The system allows patients to increase the amount of motor exercise they can perform independently, providing them with a real-time feedback based on wearable sensors embedded in the garment across the upper limb and trunk. After the feedback phase, data is stored in a central location for review and statistics. Workstations can be installed either at home or at the hospital to support patients, regardless of their location. Two other major initiatives in the field include the research programmes set in place by TRIL and CLARITY Centers in Dublin, Ireland. The TRIL centre brings together industry and academia to conduct research studies in older adults and examine how technology can enable health and social care.

1.5.4 Assessment of Treatment Efficacy

A quantitative way of assessing treatment efficacy can be a valuable tool for clinicians in disease management. By knowing what happens between outpatient's visits, treatment interventions can be fine-tuned to the needs of individual patients. Another important application could be for use in randomized clinical trials. By gathering accurate and objective measures of symptoms, one could reduce the number of subjects and the duration of treatment required to observe an effect in a trial of a new therapy. In patients with PD, careful medication titration, based on detailed information about symptom response to medication intake, can significantly improve the patient's quality of life. Medication titration in patients with late stage PD is often challenging as fluctuations in a patient's motor symptom manifest over several hours and hence cannot be observed in a typical outpatient appointment (often lasting no more than 30 min). Patient diaries are unreliable due to perceptual bias and inaccurate reporting about motor status. The aforementioned issues limit the ability of physicians to optimally adjust medication dosage and to test new compounds for the treatment of PD. The use of a sensor-based system to monitor PD symptoms is a promising approach to improve the clinical management of patients in the late stages of the disease. Major PD symptoms have typical motor characteristics, which can be captured using motion sensors such as accelerometers. Alberto and Bonomi (2011) used a portable tri-axial accelerometer placed on the shoulder to monitor severity of dyskinesias in PD patients. Dyskinesia is a side effect of medication intake, and they can cause significant discomfort to patients. Manson and colleagues showed that there is a correlation between accelerometer output and severity of dyskinesia in patients with PD. The ability to estimate the severity of symptoms via processing sensor data recorded during ADL is important for practical applications. Thielgen et al. (2004) have showed that accelerometers can be used to automatically quantify tremor severity scores via 24 h ambulatory home monitoring in patients with PD. Gait impairments such as shuffling and freezing are characteristics of PD. They have explored the correlation between gait parameters and motor scores in patients with PD. The authors used a biaxial accelerometer mounted on the lower back to measure gait features such as stride frequency, step symmetry and stride regularity. Strong correlation was found between walking regularity and motor scores capturing the severity of PD symptoms.

Movement sensors can also be used to automate clinical testing procedures. Salarian et al. (2004) and Weiss et al. (2010) have proposed instrumented versions of the timed up-and-go test for identifying gait impairments due to PD. They have shown that instrumented tests lead to an improved sensitivity to gait impairments compared to observation methods. Besides, sensor-based methods can also be extended to long-term home monitoring. Based on this body of work, an ambulatory gait analysis system, based on wearable accelerometers, for patients with PD, has been proposed by Salarian et al. (2010). Intensive long-term rehabilitation poststroke is an important factor in ensuring motor function recovery. Tracking changes in motor function can be used as a feedback tool for guiding the rehabilitation process. Uswatte et al. (2005) have shown that accelerometer data can provide objective information about real-world arm activity in stroke survivors. In their study, 169 stroke survivors undergoing constraint-induced movement therapy wore an accelerometer on both wrists for a period of 3 days. The results indicated good patient compliance and showed that by simply taking the ratio of activity recorded on impaired and unimpaired arm using accelerometers, one can gather clinically relevant information about upper extremity motor status. Prajapati et al. (2011) have performed a similar study for the lower extremities. The authors used two wireless accelerometers placed on each leg to monitor walking in stroke survivors. Results showed that the system was able to monitor the quantity, symmetry and major biomechanical characteristics of walking. Finally, Patel et al. (2010) have showed that, using accelerometers placed on the arm, it is possible to derive accurate estimates of upper extremity functional ability.

1.6 Textile Fibres Used for Wearable Electronics Applications

Functional apparel design has historically sought to meet user needs with material properties and design features, but now, many of these needs can be met with more flexibility and control using powered technology. At the same time, engineers in technology fields are turning their attention to increasingly portable devices and encountering the design challenges that are the province of functional apparel designers. Currently, there are few functioning intelligent garments in the commercial arena despite the availability of facilitating technology. Existing garments are primarily designed for protective functions, including protection from extreme cold, physiological monitoring for emergency conditions and wearer GPS information for emergency intervention (Rantanen and HaoNnnikaoNinen 2005; Sensatex Inc. 2002).

Continuous miniaturization of electronic components has made it possible to create smaller and smaller electrical devices which can be worn and carried all the time. Together with developing fibre and textile technologies, this has enabled the creation of truly usable smart clothes that resemble clothes more than wearable computing equipment. These intelligent clothes are worn like ordinary clothing and provide help in various situations according to the application area. Rantanen and HaoNnnikaoNinen (2005) describes the design and implementation of a survival smart clothing prototype for the arctic environment. With the advancements in technology and science, electronics has been getting smaller and smaller, which enables researchers and scientists to weave the electronics and interconnections into the fabric. This new technology enables to establish a new concept called *Electronic Textiles* (*e-Textiles*), which allows cost-effective and efficient solutions for various applications.

The blossoming research field of electronic textiles seeks to integrate ubiquitous electronic and computational elements into fabric. This section concerns one of the most challenging aspects of the design and construction of e-textile prototypes: namely, engineering the attachment of traditional hardware components to textiles. Three new techniques for attaching off-the-shelf electrical hardware to e-textiles were presented: (a) the design of fabric PCBs or iron-on circuits to attach electronics directly to a fabric substrate; (b) the use of electronic sequins to create wearable displays and other artefacts and (c) the use of socket buttons to facilitate connecting pluggable devices to textiles.

Buechley and Eisenberg (2007) have focused on using easily obtained materials and developing user-friendly techniques to develop methods that will make e-textile technology available to crafters, students and hobbyists. Locher and Troster (2007) describe transmission lines structures screen-printed on fabrics. Textile structures can be divided into (a) woven, (b) non-woven or pressed fabric and (c) knitted fabrics. Woven fabrics are usually quite dense. Non-woven fabrics are less dense, weaker and lack a regular Structure. Knitted fabrics have a low density and are elastic. In each category, a conducting fabric can be constructed either by incorporating conducting threads, or by plating after textile has been formed.

> *Woven conductor fabrics*: The early conducting thread woven fabrics were so poor that they could not be used in a resonant antenna. Their performance will probably improve with further development.

> *Knitted fabrics*: The interest in knitted cloth stems from their application as a stretchable fabric in constructing a laminated garment. Knitted fabrics can be constructed from pure copper. However, using just metal threads tends to damage the knitting machines, and most samples have been constructed with at least 50% nylon. To date, no antennas constructed with knitted fabrics have had satisfactory performance characteristics; the reasons for the poor performance of the knitted fabrics have not yet been verified. Possible causes include: additional inductance due to using fine wire threads, poor contact between the knitted wires and too little conducting surface in the material (compared with plain woven sheets). Future research will concentrate on using finer knits.

Wireless communication and wearable computers coupled with clothing forms a new approach to wearable computing (Troster et al. 2003; Yao et al. 2005). This so-called *smart clothing* has become a potential alternative for a wide range of personal applications, including safety and entertainment as well as applications requiring privacy. The basis for smart clothes is ordinary clothing, which is augmented with electrical or non-electrical components. Also the fabric of the clothing may itself be intelligent. Based on these, an intelligent garment can better fulfil its primary function as clothing, and also give some added value to a user. Electronically implemented intelligence in smart clothing includes electrical components such as processors, sensors and communication equipment which help the clothing to adapt to the changing environment and the user's needs. On the other hand, non-electrical functions may also provide necessary tools for the user to survive in uncommon situations. Mizell (1999) introduced the terms *Tool Model* and *Clothing Model* to describe the different usage models of wearable systems. Steve Mann has carried out extensive work in the field of making computer systems wearable (Mann 1997a).

Probably the best-known example of smart clothing is a textile keyboard and a synthesizer embedded into a denim jacket (Post and Orth 1997). Intelligence in the form of electrical components has also been embedded into other pieces of clothing, for example gloves (Perng et al. 1999), ties (Schmidt et al. 1999), suspenders (Gorlick 1999), undergarments (Mann 1996) and footwear (Paradiso and Feldmeier 2005). Electro Textiles Company Limited has adopted another view in the development of smart clothes. They started by introducing new functionality into textile material without compromising other clothing-like properties such as washability and flexibility. The Defence Clothing and Textile Agency has done research on material development for smart clothing. They have studied shape memory materials (Russell et al. 2000), active ventilation, heat transfer through garments and reactive waterproof materials (Elton 2000). Leckenwalter et al. (2000) have developed many new materials based on polytetrafluoroethylene (PTFE) technology and has concentrated on waterproof, windproof and breathable fabrics.

Du Pont developed the plastic optical fibre in 1964, and Mitsubishi Rayon later introduced the technology for it as a commercial product. Tampere University of Technology studies POF manufacturing methods in the Technical Textiles laboratory of fibre material science. Many electronic functions can be built into textiles based on low-tension electrical connection. However, limitations related to electromagnetic disturbances and moisture are evident. In contrast to the electric connections, light transmission is not affected by the defects of physical surroundings. The light wave guides may be made of silica glass, which are very efficient for data transmission, especially at long distances. In textile integration, POF is a reasonable choice because of its high mechanical flexibility and durability.

1.7 Sensing Fabrics

The sensing fabrics for monitoring physiological and biomechanical have been developed in the implementation of wearable systems for personalized health care and their evolution in time for e-textile solutions

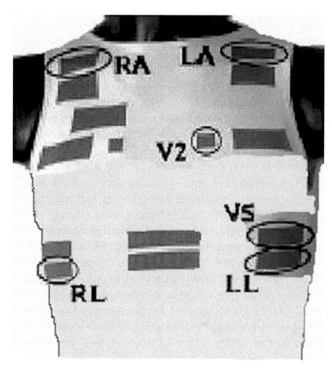

FIGURE 1.2
Sensing fabric. (From Pacelli, M. et al., Sensing fabrics for monitoring physiological and bio-mechanical variables: E-textile solutions, *Proceedings of the Third IEEE-EMBS*, 2006, pp. 1–4.)

(Figure 1.2). In the last few years, monitoring systems based on multi-functional instrumented garments are playing an innovative role in the development of more human-oriented monitoring devices. Smart fabrics allow the monitoring of patients over extended period, in a natural con-test, in biomedicine, as well as in several health-focused disciplines, such as bio-monitoring, rehabilitation, telemedicine, tele-assistance, ergonom-ics and sport medicine. The innovation in this field originated with the development of a new generation of textile sensors, combining electron-ics and informatics novelties, leading to the integration of multiple, smart functions into textiles-based sensing interfaces, aiming to the reduction of any impediment. The wearable instrumented garments are based on conductive and piezoresistive fabric that was developed to work as tex-tile sensors, where the mechanical and thermal properties are tuned with those of textile materials. The innovation has exploited different technologies ranging from flat and circular knitting to woven process, as well as the use of a cut and sews approach to manufacture sensing elastic fabric on which pies resistive sensors are printed, according to an engi-neered body map.

1.7.1 Wearable Sensor on Knitted Fabrics

A common textile process like flat-knitting technology (*Steger SA4, Switzerland*) allows the implementation of fabric where defined yarns are confined to insulated domains, realizing multi-layered structures where the conductive surface is sandwiched between two insulated standard textile surfaces. Sensors, electrodes and connections can be fully integrated into the fabric and produced in one single step, by combining conductive and non-conductive yarns. The electrical properties of fabric are due to the interaction between the fibres inside the yarn and the interaction between the single loops inside the fabric. The whole textile structure has to be considered as a complicated array of electrical impedances. Most sensors that are used for the detection of vital signs and users movements need to be in close contact with the body (*like a second skin*). The use of seamless knitting dedicated machine can provide elastic, adherent and comfortable garments with these inherent properties.

1.7.2 Fabric Electrodes Realized by Flat-Knitting Technology

The conductive fabric electrodes, used in the frame of Wealthy European project (*IST-2001-37778*), have been knitted by means of flat-knitting machine (*Steiger SA4, Switzerland*). They have been realized with commercial stainless steel threads twisted around a standard continuous viscose (*or cotton*) textile yarn. The quality of bioelectrical signals gathered in dynamic condition can be improved by coupling fabric electrodes with a hydrogel membrane, purchased by ST&D Ltd. (*Belfast, UK*). The membrane reduces the contact resistance between the skin and the electrode, and increases the stability of the contact with its adhesive properties, without affecting the comfort. The pH of the chosen membrane avoids skin irritation.

1.7.3 Fabric Electrodes Realized by Seamless Knitting Technology

Seamless sensorized garment have been processed with Santoni knitting machines, using stainless steel yarns produced by Bekintex to realize the electrodes and the Meryl® Skin life purchased by Nylstar as basal yarn. Garments with different level of complexity, in term of sensors distribution, have been developed during the course of the European project My Heart (*IST-2002-507816*). Seamless electrodes have been tested and used without the hydrogel membrane as the goal of the project was to interface the sensors directly with the skin.

1.7.4 Bioelectrical Signals through Fabric Electrodes

Fabric electrodes realized by flat-knitting machinery have been used to monitor electrocardiogram (ECG), and respiratory activity by means of impedance pneumography (IP). Fabric electrodes realized by flat-knitting machinery

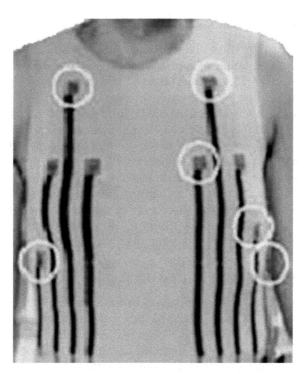

FIGURE 1.3
Wearable sensors on knitted fabrics. (From Pacelli, M. et al., Sensing fabrics for monitoring physiological and biomechanical variables: E-textile solutions, *Proceedings of the Third IEEE-EMBS*, 2006, pp. 1–4.)

(Figure 1.3). Circled ones are for ECG acquisition. ECG signals have been acquired by using an acquisition card (*National Instruments PCI-6036E*). Signals have been conditioned by means of GRASS TELEFACTOR model 15LT device, with gain set to 1000. A shirt has been manufactured by seamless machine in order to gather ECG and IP signal. Fabric electrodes realized by seamless knitting technology: electrode positions for ECG acquisition. Signals from the seamless system have been conditioned by using a device developed by CSEM (CSEM 2008) and developed within the WEALTHY project. To improve the electrical signal quality in dynamic conditions, the electrodes have been wet.

1.7.5 Knitted Piezoresistive Fabric Realized by Seamless Knitting Technology

Piezoresistive fabric sensors have been manufactured with Santoni seamless machines using intarsia technique. The used conductive yarn (*Belltron 9R1 is produced by Kanebo Ltd., and it is coupled with an elastomer Lycra®*).

1.7.6 Printed Piezoresistive Fabric Manufactured by Serigraphy Technology

Coated fabric sensors have been manufactured with a conductive silicone, not designed for application in textile field. The viscosity of this material has been reduced to allow the use of an industrial coating process. Print screening technology has been evaluated as textile approach to coat the elastic substrate with the conductive elastomer. This approach allows to print on the fabric, the desired sensors topography as well as to solve the connection issue by using the same materials for both the functions: sensorial and circuital.

1.7.7 Fabric Strain Sensors

The properties of piezoresistive fabric sensors, a protocol of mechanical characterization, have been implemented. The fabric strain sensors have been subjected to predetermined mechanical stimuli imposed by a PC controlled system. Corresponding variations of electrical resistance have been collected through voltage divider, gathered by an acquisition card (*National Instruments PCIMIO-16E-4*) with sampling rate of 64 Hz. Several samples of strain sensors have been subjected to different uniaxial mechanical stimuli following signals such as step and trapezium, both of them with variable strain amplitudes, and sinusoidal cycles with variable strain amplitudes at selected frequencies. Uniaxial mechanical stimuli have been applied along the length of both kinds of piezoresistive sensors. The mechanical characterization aims to study the electrical response of fabric strain sensor as a function of the external mechanical stimuli.

1.7.8 Biomechanical Signals through Fabric Strain Sensors

Knitted system for the acquisition of abdominal and thoracic respiratory activity, in order to evaluate the performances of knitted piezoresistive fabric sensors for biomechanical monitoring, the knitted piezoresistive sensors have been tested to detect both the respiration signal as a function of thorax movement and the elbow bends. A seamless t-shirt with fabric strain sensors has been manufactured to keep a tab on the respiratory signal. Strain fabric sensors signals have been acquired using a voltage divider to convert resistance to voltage, gathered by an acquisition card (*National Instruments PCI 6036*) with a sampling rate of 1000 Hz. The respiration signals acquired with textile sensors have been compared with a respiratory effort transducer: model SSL5B, contained in the BIOPAC® MP30 system. The knitted piezoresistive fabric sensors are sensitive to changes in thoracic or abdominal circumference that occur during the respiratory activity. The elbow bends signal

FIGURE 1.4
Garments for biomechanical monitoring. (From Pacelli, M. et al., Sensing fabrics for monitoring physiological and biomechanical variables: E-textile solutions, *Proceedings of the Third IEEE-EMBS*, 2006, pp. 1–4.)

detected by knitted piezoresistive fabric sensors that has been compared with a commercial movement tracking system (*electrogoniometer by Biometric®*). The piezoresistive sensors' performances allow the detection of movement index, while the printed piezoresistive sensors have been showed to be more efficient in the realization of wearable kinaesthetic systems for gesture and posture monitoring. The realization of a garment devoted to the analysis of human movements is an application of printed piezoresistive fabric (Figure 1.4).

From design and manufacturing perspective, a special process has been set up to realize biomechanical monitoring garments, where sensors and connections are realized with the same conductive materials. This choice has been adopted in order to solve a critical issue related to the connectivity between the coated strain sensors and the electronics, keeping the elasticity and wearability of garment. For this reason, printing and cabling was realized before the cut and sew phase, by using a hybrid solution: coating on the sewing done with conductive flexible yarns that are used as conductive cables.

1.8 Wearable Sensors

Clothing is a basic need of a human being besides food and shelter. About 6000 years ago, man started to replace the inflexible animal skin with manufactured textiles. The body protection functionality has been enlarged by aesthetic attributes. Beyond their protective and aesthetic functions, clothes as our second skin have the potential to acquire an additional functionality as a personalized and flexible information platform (Park and Jayaraman 2001b). Wearable computing will change the mobile computing landscape, provided that the unobtrusive integration of electronics in our daily outfit becomes feasible. Today people are actively involved in their health and increasing their activity level to prevent health risk factors. Applications in the medical field are highly accepted by customers because of personal benefits provided like extending active lifetime by disease prevention or disease management based on objective measures. Vital signs like ECG, EEG or respiration monitored almost continuously can give information about the cardiovascular system of a patient. A recent approach is to investigate concepts in which sensors are integrated into clothing for monitoring purposes: measuring signals of movements, temperature, GSR and patient's activities. To pick up electrical signals by intelligent textiles, one essential component of a system design is using an appropriate electrode design. Permanent monitoring, which will become possible with textile sensors, opens up new perspectives for traditional parameters too. Permanent monitoring supported by self-learning devices will allow the set-up of personal profiles for each individual, so that conditions deviating from the normal state can be traced as soon as possible. There are a variety of sensing technologies that could be envisioned for context recognition: for instance audio, video, photography, acceleration, light, air and body temperature, heat flux, humidity, pressure, heart rate, strain gauges, GSR, ECGs or electromyograms. Streaming video is overly expensive on bandwidth and many physiology sensors require skin contact or special outfits. There is clearly a trade-off between informative and unobtrusive sensing.

The advanced medical monitor system (AMON) is a next-generation wearable medical monitoring computer that has been developed by the European Union (Amon 2001). It provides complex monitoring, data analysis and communication capabilities in a single wrist worn unit. AMON has been conceived as a clinical device for high-risk patients requiring constant monitoring, logging and analysis of their vital signs. Belardinelii et al. (1998) and Jovanov et al. (2000) presented a wrist worn medical monitoring computer designed to free high-risk patient from the constraints of stationary monitoring equipment. The system combines complex medical monitoring, data analysis and communication capabilities in a truly wearable watch like form. The paper summarizes the functionality, architecture and implementation of the system. Mühlsteff (2004) describes and characterizes dry electrodes based on silicone rubber concerning their usability for functional clothes regarding skin-electrode impedance, warm-up time, motion artefact and robustness. Health monitoring applications were initially explored for military purposes with the objective of remotely determining the physical status of troops in the field. The personnel status monitor was designed to predict when a soldier is either injured or fatigued using a wide range of sensors, processing boards and a wristwatch display (Satava 1997). A simpler low-cost, lightweight, non-invasive and adaptable system employed a single neck-mounted acoustic sensor to listen to the sounds of blood flow, respiration and the voice, while minimizing ambient sound (Siuru 1997). The sensor can collect information related to the functions of the heart, lungs and digestive tract or it can detect changes in voice or sleep patterns, other activities and mobility. Extensive testing with soldiers and fire-fighters has demonstrated the effectiveness of this design to help understand the interrelations among physiology, the task at hand and the surrounding environment. More recently, health-monitoring wearables have become commercially available in the form of the Body media product range. This is based on an armband design with sensors for detecting movement, heat flux, skin temperature, near-body temperature and GSR. Data can be either viewed in real time via a wireless link, or downloaded for analysis using the Internet.

Meanwhile academic research continues with health-monitoring wearables such as the University of Birmingham's Sensvest for monitoring sports activity (Knight et al. 2005a); the WEALTHY Wearable Health Care System which seeks to improve the comfort of wearable systems by integrating sensors with the fabric of the users clothes and the grid-enabled system which can display live data, historical data or perform data mining developed by the University of Nottingham (Crowe et al. 2004). Providing assistance for people with special needs has become an important role for wearables. Many systems have been explored to provide the visually impaired with guidance. Early examples of this were developed at the University of California, Santa Barbara, using GPS (Loomis 1985). Evolving from a bulky backpack design, the current system weighs only a few pounds and is worn in a pack slung over the shoulder. A radically different approach was taken by the University of Bristol (Campbell et al. 1995) in which real-time video from a body worn camera produced images in which areas such as pavements were classified, identified and presented to the visually impaired user as registered colour-coded areas on a head mounted display.

Interaction and communication in this field can also be assisted by wearables, for instance with the deaf using M.I.T.'s American Sign Language Recognizer (Starner and Paradiso 2004), and the forwarding of images from body worn cameras at accident scenes to hospitals from medics while talking to the waiting doctors using British Telecom's CamNet system. The continuing challenge of medical wearables is the achievement of interfaces which can be worn, and will operate reliably, without the conscious involvement of the user. Perhaps more than in any other wearable computing field, it is important that the wearable augments and assists daily life and does not interfere with the normal functions, especially for users who may have special needs. Computational fashion in MIT with creations features biosensor and movement sensors controlling displays in the garments structure. There is undoubted appeal for such fashion items and products are starting to become available commercially. The first physiological parameter measured using intelligent textiles is the heart activity. Heart signals are one of the basic parameters in health care. The heart is a muscle controlled by the brain through electrical impulses (Clark 1998). Since the body is a vessel filled with aqueous electrolyte, these signals can be detected in all of its parts. The generated difference in electrical potential can be measured on different spots on the skin. This signal is the ECG. The ECG of a normal person at rest has a fundamental frequency of about 1.2 Hz (Golden 1973). Priniotakis et al. (2007) have demonstrated that electrochemical impedance spectroscopy is a useful technique to measure the properties of textile electrodes in an accurate and repeatable way.

Several wireless ECG monitoring systems have been proposed (Fensli et al. 2005; Otto et al. 1998; Proulx et al. 2006; Shih et al. 2004). Shnayder et al. (2005) present a description and evaluation of a wireless version of a system based on these innovative ECG sensors. Experimental results show that the wireless interface will add minimal size and weight to the system while providing reliable operation. The development of an active implantable device for measuring ECG is presented in Riistama et al. (2007). A number of wearable physiological monitoring systems have been developed. To measure ECG and blood pressure, the subject's attention is required. An unobtrusive and wearable, multi-parameter ambulatory physiological monitoring system for space and terrestrial application, termed LifeGuard (Mundt et al. 2005), has been developed. The system uses the conventional electrodes to acquire ECG and interfaces a separate non-invasive cuff-based blood pressure monitor and wires from the sensors to the data acquisition hardware that are routed around the subject. The U.S. Army Medical Research and Material Command has developed the Warfighter Physiological Status Monitoring (WPSM) system (Hoyt et al. 2002), which is intended to be a next-generation combat uniform featuring a configurable array of miniaturized sensors. These sensors acquire various physiological data and transmit them to a central hub device so as to be stored, analyzed using life sign decision support algorithms (Savell et al. 2004) and sent to command communication networks when desired. The project CodeBlue from Harvard University is a wireless infrastructure, which is intended for deployment in emergency medical care, integrating low power, wireless vital sensors, PDA and PC (Lorincz et al. 2004). Vivometrics™, a U.S.-based company, has developed 'Life Shirt' to acquire a number of physiological parameters (Carter et al. 2004; Wilhelm et al. 2006). This system uses the conventional electrodes to acquire ECG, and there is no online analysis of data. A wireless sensor network (WSN) for smart electronics shirt has been developed which allows the monitoring of individual biomedical data and is transmitted for further processing using a wireless link (Carmo et al. 2005). A novel wearable cardiorespiratory signal sensor device for monitoring sleep condition at home using a belt-type with conductive fabric sheets and a PVDF film is developed (Choi and Jiang 2006). A wearable system with sensing bio-clothes for monitoring five leads of ECG signal and body gesture/posture without transmission is reported (Pacelli et al. 2006; Paradiso et al. 2005). Ambulatory health status monitoring consisting of multiple sensor nodes that monitor motion and heart activity by a personal server is reported (Otto et al. 2006). Smart sensors for continuous detection of vital physiological parameters such as ECG, heart rate and blood pressure and physical gait signals with Bluetooth wireless transmission for determination of fall to prevent injury have been developed (Choon et al. 2006).

Though the aforementioned systems use new innovative sensors, they lack in terms of the number of vital parameters being monitored. A number of wearable physiological monitoring systems have been put into practical use for health monitoring of the wearer in hospital and real life situations and their performances have been reported. The Armband Sense Wear (BodyMedia Inc., Pittsburgh, PA) wearable body monitor has been used to study bodily movement and energy expenditure in normal subjects and chronic obstructive pulmonary disease (COPD) subjects. Pandian et al. (2008) describes the wearable physiological monitoring system, which uses an array of sensors connected to a central processing unit with firmware for continuously monitoring physiological signals. The data collected can be correlated to produce an overall picture of the wearer's health.

Smart textiles with wireless communication to monitor the bio-signals are increasing in the last few years. For the purpose of acquiring the exact signals, commercial electrodes are usually used, especially in the ECG application (Healey and Logan 2005). In addition, fabric-based electrodes are also studied (Scilingo et al. 2005). In the study by Shen et al. (2006), a wearable fabric-based sensing band is developed. Combining with the textile structure design and the processing algorithm, any mechanical signals resulting from exercise can be eliminated and the ECG signals are cleared from motion artefacts.

Locher et al. (2004) presented a smart fabric system with temperature-sensing capability. The woven structure of the hybrid fabric itself represents an array of temperature sensors that can be utilized to measure the temperature profile of a surface (De Vries et al. 1993). Currently, many research labs concentrate on the side of signal transmission (Jung et al. 2003a,b). Temperature measurement is one option, but fabrics have the capacity to integrate other sensing functionality. Lorussi et al. (2004b) work on fabric-based strain sensors in order to track body postures. Companies such as Softswitch, Eleksen and ITP are investigating the applications of fabric pressure sensors.

Physiological measures of interest in rehabilitation include heart rate, respiratory rate, blood pressure, blood oxygen saturation and muscle activity. Parameters extracted from such measures can provide indicators of health status and have tremendous diagnostic value. Until recently, continuous monitoring of physiological parameters was possible only in the hospital setting. But today, with developments in the field of wearable technology, the possibility of accurate, continuous, real-time monitoring of physiological signals is a reality. Integrating physiological monitoring in a wearable system often requires ingenious designs and novel sensor locations. For example Asada et al. (2003) presented a ring sensor design for measuring blood oxygen saturation (SpO_2) and heart rate. The ring sensor was completely self-contained. Worn on the base of the finger (like a ring), the integrated techniques for motion artefact reduction can be designed to improve measurement accuracy. Applications of the ring sensor ranged from the diagnosis of hypertension to the management of congestive heart failure. A self-contained wearable cuff-less photoplethysmographic (PPG)-based blood pressure monitor was subsequently developed by the same research group (Shaltis et al. 2006). The sensor integrated a novel height sensor based on two MEMS accelerometers for measuring the hydrostatic pressure offset of the PPG sensor relative to the heart. The mean arterial blood pressure was derived from the PPG sensor output amplitude by taking into account the height of the sensor relative to the heart. Another example of ingenious design is the system developed by Corbishley et al. (2008) to measure respiratory rate using a miniaturized wearable acoustic sensor (microphone). The microphone was placed on the neck to record acoustic signals associated with breathing, which were band-pass filtered to obtain the signal modulation envelope.

In recent years, physiological monitoring has benefited significantly from developments in the field of flexible circuits and the integration of sensing technology into wearable items. An ear-worn, flexible, low-power PPG sensor for heart rate monitoring was introduced by Patterson et al. (2009). The sensor is suitable for long-term monitoring due to its location and unobtrusive design. Although systems of this type have shown promising results, additional work appears to be necessary to achieve motion artefact reduction. Proper attenuation of motion artefacts is essential to the deployment of wearable sensors. Some of the problems due to motion artefacts could be minimized by integrating sensors into tight-fitting garments. A comparative analysis of different wearable systems for monitoring respiratory function was presented by Lanata et al. (2010). The analysis showed that piezoelectric pneumography performs better than spirometry. Nonetheless, further advances in signal-processing techniques to mitigate motion artefacts are needed. Biochemical sensors have recently gained a great deal of interest among researchers in the field of wearable technology. These types of sensors can be used to monitor the biochemistry as well as levels of chemical compounds in the atmosphere (e.g. to facilitate monitoring people working in hazardous environments). From a design point of view, biochemical sensors are perhaps the most complex as they often require collection, analysis and disposal of body fluids.

Advances in the field of wearable biochemical sensors have been slow, but research has recently picked up pace due to the development of micro- and nano-fabrication technologies. For example, Dudde et al. (2006) have developed a minimally invasive wearable closed-loop quasi-continuous drug infusion system that measures blood glucose levels and infuses insulin automatically. The glucose monitor consists of a novel silicon sensor that continuously measures glucose levels using a micro-perfusion technique, and continuous infusion of insulin is achieved by a modified advanced insulin pump. The device has integrated Bluetooth communication capability for displaying and logging data and receiving commands from a PDA device. An array of biochemical sensors has been developed as part of the BIOTEX project, supported by the European Commission. Specifically, the BIOTEX project deals with the integration of biochemical sensors into textiles for monitoring body fluids. Within this project, researchers have developed a textile-based fluid-collecting system and sensors for in vitro and in vivo testing of pH, sodium and conductivity from body sweat. By in vitro and in vivo testing of the wearable system, the system can be used for real-time analysis of sweat during physical activity. As part of another project called ProeTEX, Curone et al. (2010) have developed a wearable sensorized garment for fire-fighters, which integrates a CO_2 sensor with sensors to measure movement, environmental and body temperature, position, blood oxygen saturation, heart rate and respiration rate. The ProeTEX system can warn the fire-fighters of a potentially dangerous environment and also provide information about their well-being to the control centre. The systems developed in the projects mentioned previously could be relied upon to design robust e-textile-based wearable systems for remote health-monitoring applications. The biochip contains an integrated biosensor array for detecting multiple parameters and uses a passive micro-fluidic manipulation system instead of active micro-fluidic pumps. Finally, applications in rehabilitation of remote monitoring systems relying on wearable sensors have largely relied upon inertial sensors for movement detection and tracking. Inertial sensors include accelerometers and gyroscopes. Often, magnetometers are used in conjunction with them to improve motion tracking. Today, movement sensors are inexpensive, small and require very little power, making them highly attractive for patient-monitoring applications. Ambient sensors are examples of instrumented environments; these include sensors and motion detectors on doors that detect opening of, for instance, a medicine cabinet, refrigerator or the home front door. This approach has the characteristic of being totally unobtrusive and of avoiding the problem of misplacing or damaging wearable devices. *Smart home* technology that includes ambient and environmental sensors has been incorporated in a variety of rehabilitation related applications.

1.8.1 Electrode Technologies

1.8.1.1 Dry Electrodes

In contrast to wet Ag/AgCl electrodes, dry electrodes are designed to operate without an explicit electrolyte. Instead, it is usually supplied by moisture on the skin (i.e. *sweat*). Numerous variations of dry electrodes exist, ranging from simple stainless steel discs to micro-fabricated silicon structures with built-in amplifier circuitry. Employing dry contact sensors is somewhat more challenging in practice than traditional techniques largely due to the increased skin-electrode impedance, although the impedance can be quite comparable to wet electrodes after a few minutes due to sweat and moisture build-up. Successful designs use either an active electrode circuit to buffer the signal before driving any cabling or alternatively penetrate the skin to achieve low contact impedance. In its simplest form, a dry electrode can be built from any conductive material in contact with the skin, such as a flat metal disc and is well known in the literature. As an example, Valchinov et al. (2004) present a modern variation of this design. Performance and signal quality of these simple electrodes can be as good as wet electrodes, especially if an amplifier is onboard. Dry electrodes work well for quick measurements (such as exercise machines), but suffer from usability problems for normal clinical applications. Standard wet electrodes usually include an adhesive material to fix the electrode in proper locations, and a hydrogel or wet-foam to both lower the skin impedance, and buffer the electrode against mechanical motion. Adding an adhesive material to place these dry electrodes in the proper clinical locations for continuous use eliminates many of its comfort/convenience advantages. Nevertheless, the simplicity and durability of metal dry electrodes make it highly useful for applications like ECG event monitors where short, infrequent use over long periods of time is expected. Flexible versions of the dry electrode based on rubber, fabric or foam are also possible and more appealing from both a comfort and usability standpoint. Softer materials have the advantage of conforming easily against the skin, increasing comfort and contact area. Gruetzmann et al. (2007) demonstrated a foam electrode, which exhibited excellent stability with increased resistance to motion artefact versus the wet and rigid dry Ag/AgCl electrode.

The high-resistance layer of the skin, the stratum corneum, is typically abraded or hydrated to achieve a lower resistance and better electrode contact. It is also possible to penetrate the 10–40 mm layer with micro-fabricated needles. Bypassing the stratum corneum can achieve a contact as good as, if not better, than a standard Ag/AgCl electrode without the need for any skin preparation or gel. To date, preliminary data has been available for EEG applications of this electrode. However, long-term studies on the hygiene comfort and safety of this technology are unavailable. The authors have observed irritation and slight pain when using these electrodes. It is certainly conceivable that they must be single use, and necessarily be packaged pre-sterilized. For EEG, recording signals reliably through thick layers of hair remains one of the key challenges. One technique, using dry sensors that do not require scalp preparation, involves the use of thin fingers that can penetrate through hair, first described in a patent by Gevins et al. in 1990. Several research groups have demonstrated this technique successfully. Matthews et al. (2007) have presented one well-characterized version of this sensor, and show that the EEG signal obtained can be largely comparable to wet electrodes, for stationary subjects. However, the high skin-contact impedance results in a much larger motion artefact with the dry sensors. Fiedler et al. (2015) have published a TiN-based fingered dry electrode that reported an impedance of 14–55 kW/finger versus around 10 MW/finger. The final type of dry electrode, first demonstrated by Richardson in 1969, does not require ohmic contact at all. In Richardson's original design, a simple aluminium disc was anodized to form a large blocking capacitor in series with the skin. Signals were capacitive coupled to the input of an FET buffer amplifier and subsequently connected to standard instrumentation (Richardson et al. 1969).

Taheri et al. (1994) have expanded on their design by fabricating an insulated electrode on a silicon substrate which integrated a buffer amplifier.

It was also designed to have multiple, redundant sensing sites along with a simple algorithm to select the channels that are most likely to have a good contact. The combination of a good dielectric material with physical skin contact means that the coupling capacitance for insulated electrodes is relatively large: from 300 pF to several nanofarads. As a result, designing a bias network with low noise and frequency response for clinical grade signals is very feasible with a standard high-impedance input FET amplifier. In most respects, the usage and performance of insulated electrodes is quite similar to dry electrodes in practice. Some limited data exists that suggest capacitive coupled electrodes suffer from less skin-motion artefact noise than dry electrodes. More detailed studies need to be conducted to determine what advantage, if any, can be achieved by inserting a layer of insulation between the skin and electrode. From an electrical perspective, the high capacitance of the thin insulation layer is effective at signal frequencies and has no effect on the signal quality vis-a-vis dry electrodes. One obvious downside, however, is that the insulated nature of the electrode precludes a frequency response down to DC, which may be important for certain applications.

1.8.1.2 Non-Contact, Capacitive Electrodes

Wet and dry electrodes both require direct physical skin contact to operate. The final type of sensor, the non-contact electrode, can sense signals with an explicit gap between the sensor and body. This enables the sensor to operate without a special dielectric layer and through insulation like hair, clothing or air. Non-contact electrodes have been typically described simply as coupling signals through a small capacitance (10 s pF). In reality, however, there is typically an important resistive element (>100 MW) as well, since the typical insulation (fabric) will also have a non-negligible resistance. As shown previously, signal coupling through non-contact electrodes can be actually dominated by the resistive part of the source impedance which can cause a large input voltage noise. Designing an amplifier to acquire signals from such high source impedance is quite challenging. Typical design problems include achieving a high enough input impedance and a stable bias network that does introduce excessive noise. Finally, very high impedance nodes are susceptible to any stray interference and motion-induced artefacts. Nevertheless, Prance et al. (1994) have demonstrated a working non-contact system with an array of 25 ECG sensors that were designed to acquire signals with 3 mm spacing from the body. A low-leakage biasing circuit using a bootstrapped reverse diode, combined with positive feedback to neutralize the parasitic input capacitance, was used to achieve extremely high impedance, reported at (10^{16} Ω, 10^{-17} F). However, it is not clear how these measurements were made or over what bandwidth. In addition, the effective input impedance with neutralization is a complex function of both the coupling capacitance and frequency.

Prance et al. (2000a,b) have studied an improved version based on the INA116 electrometer instrumentation amplifier from Burr-Brown (Texas

Instruments) with a lower noise floor. It again utilizes positive feedback for neutralization of the input capacitance. While the specifics were not published, it can be inferred that the process is far from perfect, as it requires manual calibration and different devices do not match well. The ability to sense biopotential signals through insulation has resulted in ingenious implementations ranging from sensors mounted on beds, chairs and even toilet seats. In general, the signal quality ranges from poor to quite good, as long as proper shielding and subject grounding techniques are utilized. Kim et al. (2000) make an important contribution in this field by extending the analysis for the driven-right-leg scheme for capacitive applications. In particular, he shows that an active ground, even capacitive coupled, is highly effective at reducing line noise. It is worthwhile to note that the active ground connection can be capacitive as well for a system that is truly non-contact. A few other key publications in this field have mentioned the need for least dry contact to ensure proper operation. This extra degree of common-mode rejection is especially useful in light of the input impedance problem. Unfortunately, specific key circuit and construction details for non-contact sensors have generally not been available in the literature. In particular, the critical information relating to inputs biasing capacitance neutralization and circuit reference/grounding that allow someone to duplicate the sensor and experiments has been scarce.

A complete design for a non-contact, wireless ECG/EEG system can improve and summarizes upon their previous designs. These non-contact sensor designs are very simple and robust, manufactured completely on a standard PCB with inexpensive and commonly available components (*chip resistors, capacitors and the National LMP7723 and LMP2232*). The critical input node was left floating, and it was found that the input can reliably self-bias purely through the device's internal ESD protection structure. Since no extra conductive devices were added to the input, the circuit achieved the optimal noise performance of the amplifier. The DC offset was simply removed with a passive high-pass filter before the second, differential gain stage. The sensor performed well in laboratory environments and 60 Hz noise was virtually absent through the use of proper shielding, an active ground and a fully isolated, wireless system. Several authors have demonstrated performance comparable to clinical adhesive electrodes, through a t-shirt, with a moving subject for ECG. The caveat, however, was that this required a tight vest and chest band to secure the non-contact electrodes in place. This highlights the key, unresolved, problem with non-contact electrodes – susceptibility to motion induced artefacts. For non-contact electrodes, artefacts tend to be dominated by three sources. First, the high-impedance, capacitive-coupled, input node of the electrode exhibits a large time settling time constant. Second, displacements in the electrode-to-skin distance can cause artefacts. Finally, friction between the electrode and insulation (*fabric, hair*, etc.) can cause large voltage excursion at the sensitive input. Typically non-contact electrodes exhibit poor settling times due to the high-pass characteristic at the electrode.

Faster recovery is possible by shifting the corner frequency of the high-pass filter, but at a cost of distorting the signal waveform. Achieving a good frequency response without the settling time problem remains an unsolved challenge. All known non-contact sensor designs deliberately limit the high-pass corner frequency to at least around 0.5–1 Hz, which introduces appreciable distortion in the ECG waveform. The clinical usefulness of this distorted ECG versus the standard trace is not known by the authors and needs further consideration. Simple models have been devised to model and solve the displacement artefact for capacitive ECG sensors, but rely on precise knowledge of the coupling capacitance. Thus, while effective in simulations and controlled bench experiments; it is yet to be reliably demonstrated on actual live recordings. On the other hand, there is no known solution to friction-induced artefacts. As it stands, there is no real impediment to building fully functional non-contact sensors from standard off-the-shelf amplifiers, and the actual implementation can be as simple as a dry electrode, with proper component selection. For actual usage, the non-contact electrode's susceptibility to motion artefacts, friction and thermal noise is problematic.

The relative utility of dry-contact and non-contact electrodes, in contrast with the more established and widespread wet-contact electrodes, is inextricably tied to novel system applications or tools that it can enable. Enabling systems application domains for two main clinical needs are cardiac and neurological monitoring. Examples of systems in their applications environments for clinical ECG and EEG use are illustrated in Figure 1.5.

1.8.1.3 Skin-Electrode Modelling

The noise properties of biopotential electrodes are complex and the result of multiple different mechanisms. It is the sum of the intrinsic thermal noise, excess

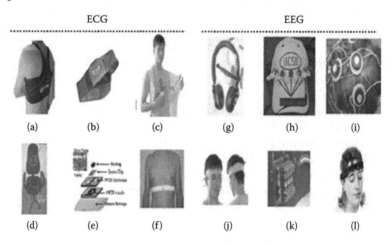

FIGURE 1.5
Dry and non-contact electrode systems. ECG – (a) chest harness, (b) polar heart strap, (c) non-contact vest, (d) chair, (e) wireless band-aid and (f) dry chest strap. EEG – (g) Neurosky single-channel headset, (h) dry MEMs cap, (i) fingered dry EEG harness, (j) dry/non-contact EEG headband, (k) dry active electrode ENOBIO wireless dry sensor and (l) EEG headband.

noise from chemical reactions at the interface, biological noise from the body and external noise from movement artefacts. In addition, the properties of the electrode and front-end electronics have a strong influence on how well a system can reject external interference such as 50/60 Hz noise. While literature exists on the noise properties of electrodes, especially for standard wet Ag/AgCl types, there is no unified model that can fully predict the performance of the interface. A model of a generic skin–electrode interface is shown in Figure 1.6. Electrically, it can be thought of as a series connection of parallel RC, each element representing a specific layer in the interface. Typically one high-impedance layer dominates the noise properties of the electrode (e.g. *dry skin, fabric*) and influences its overall properties from electrode offset to immunity to long-term drift.

Electrical activity from within the body is coupled via different layers (e.g. *skin, hydrogel, Ag/AgCl*), each with an associated resistance and capacitance to the input of an amplifier (Table 1.1). Typically, the properties of one layer will dominate the

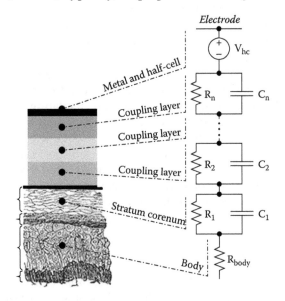

FIGURE 1.6
Electrode-skin model. (From Chi Ed.H. et al., Social network activity, In *Proceedings of the SIGCHI Conference on Human Factors in Computing Systems*, San Jose, CA, 3(1), 2010.)

TABLE 1.1

Measured Material Electrical Properties

Wet Ag/AgCl	660 kΩ / 36 nF
HASL	6 MΩ / 5.2 nF
Latex	100 MΩ / 280 pF
Silver cloth	40 MΩ / 900 pF
Solder mask	1.5 GΩ / 400 pF
Cotton	1 GΩ / 40 pF

behaviour of the interface, including noise and offset (*Vhc*). The equivalent circuit diagram is provided with the noise-generating sources for the electrode. The model is simplified by assuming that the skin–electrode interface is represented by a single R–C element representing the dominant layer (Chi et al. 2010).

The research provides several useful options for implementing future body area physiological sensors and provides a simple test protocol to measure and compare the noise of new electrode materials. Several conductive and non-conductive materials were tested. Although excess interface and biological noise render it impossible to fully predict the behaviour of electrodes from analytical models, empirical data suggests that overall noise levels are strongly correlated with electrode impedance within the frequencies of interest, not just the DC resistance level. For example solder mask and cotton have extremely high resistivity, (1 GW), orders of magnitude higher than a standard Ag/AgCl electrode. Nevertheless, solder mask was able to provide relatively low-noise signals due to its high capacitance. Cotton, however, was noisy within ECG/EEG frequencies due to its low capacitance. For dry and non-contact electrodes where it is impossible to achieve a low contact resistance, it may be possible to achieve a high signal quality by maximizing both the electrode's resistance and capacitance. Excellent signals, particularly for ECG, may be obtained with simple materials including bare lead-free PCB. Contrary to what may be expected from their insulating nature, latex and solder mask provided signals that were quite comparable to a dry metal electrode. Furthermore, acceptable signals may be obtained with comfortable materials such as cotton. Interestingly, silver cloth performed quite poorly and is no better than regular cotton fabric. It is expected that non-traditional electrode materials will become increasingly important for medical and non-medical use, especially given the importance of subject comfort for wireless, wearable sensors.

1.8.2 Body Sensor Network

Body sensor networks (BSNs) comprise wearable or implanted sensors that collect, process and communicate physiological information from the body. With health-care costs soaring, we need a more efficient and cost-effective approach to medical diagnosis, treatment and care. BSNs could revolutionize health care by offering miniaturized, unobtrusive, nearly continuous monitors to provide unprecedented levels of medical observation while simultaneously reducing the need for visits to the doctor. BSNs provide feedback to users that can illuminate potential health concerns earlier, encourage healthier lifestyles and improve personal wellbeing. There is also the potential for using the collected data to provide real-time assessments of an individual's condition and need that could trigger a real-time assistance mechanism, such as a balance assistive device or a wearable defibrillator. If adopted widely, this technology could also upgrade well patient monitoring, help at home for the aged people, report on medication effectiveness and cut the costs of health care across the board. Li and Tan (2005) have studied the BSN-based context-aware QRS detection. The algorithm uses the context information provided by the BSN to improve the QRS detection performance by dynamically selecting the leads with best SNR and taking advantage of the best features of two complementary detection algorithms. The accelerometer data from the BSN are used to classify the patients' daily activity and provide the context information. The classification results indicate both the type of the activities and their corresponding intensity, which is related to the signal/noise ratio of the ECG recordings. Activity intensity is first fed to lead selector to eliminate the leads with low SNR, and then is fed to a selector for selecting a proper QRS detector according to the noise level. MIT-BIH noise stress test database is used to evaluate the algorithms. Recent significant progresses in wireless sensing/monitoring and wearable/implantable biosensors have enabled BSNs, a promising personal and ubiquitous health-care candidate solution. BSNs are capable of sensing, communicating and processing different physiological parameters through biosensor nodes and helping physicians to make clinical decisions. A BSN usually consists of several implantable or wearable biosensors, such as ECG, EEG, glucose sensors, accelerometers, blood pressure and oxygen saturation (SpO_2) sensors, temperature sensors and even ingestible camera pills. These sensors continuously monitor vital signs and report data to a powerful external device, such as a PDA, cell phone or bedside monitor station. A novel BSN-based energy efficient QRS detection scheme has been developed. The algorithm uses the context information provided by the BSN to improve the QRS detection performance by dynamically selecting the leads with the best SNR and taking advantage of the best features of two complementary detection algorithms. An energy-efficient BSN-MAC is adopted to guarantee the low-power wireless communication during the context information collection. In BSN-MAC, adopt a cross-layer design strategy considering the information obtained from the sensors to guide the medium access. The network coordinator (NC) and the sensors interact to achieve efficient power management. The coordinator controls the communication by varying the superframe structure. The sensors provide real-time feedback to BSN coordinator with application-specific and sensor-specific information. The two candidate algorithms are chosen to be relatively simple on purpose, because they impose lower computation requirements on the hardware, which is helpful to conserve the precious energy in BSNs and prolong the overall network life. Although the two candidate algorithms used in this chapter may not be the best choice, this work demonstrates that using multi-channel ECGs and context information obtained from BSNs can improve the performance of existing QRS detectors by dynamically selecting the leads with the best SNR and switching between two complementary detection algorithms.

1.8.2.1 Anatomy of a Body Sensor Network

A BSN consists of one or more sensor nodes that form a communication network. Usually, one of the nodes acts as a base station, which aggregates information from the body-distributed sensor nodes and ultimately conveys it across existing networks to other stakeholders like the wearer's caretakers and physicians. Smart phones are already supporting many useful health related apps, and future generation smart phones are the obvious choice to serve double duty as a BSN base station. The base station thus plays a different role from other nodes in the system, and it has more resources in terms of bigger size, available energy, longer range radios, and more memory and computing power. The sensor nodes themselves each contain a set of sensors for particular physiological data, processing hardware for computation and a wireless radio for communication. Some BSN nodes may replace the radio with a transceiver that communicates across the body, using the surface of the skin as a communication channel. Based on the application requirements, the sensed data may be streamed wirelessly or stored locally (*typically in a flash memory*) for later transmission or download.

1.8.2.2 BSNs versus Wireless Sensor Networks

While BSNs hold these basic components in common with generic WSNs, which have provided a popular research topic for years, there are several important distinguishing features for BSNs. First, the body itself prevents the reuse of the same node across many roles due to the specific features of measuring physiological data. For example an ECG sensor that measures the electrical activity of the heart works well in the torso area, but it cannot work on the wrist; a gait sensor probably uses accelerometers and gyroscopes and is best located on the legs; the sensor on the head that measures brain activity (electroencephalogram [EEG]) probably cannot also measure respiration rate. Second, the proximity of the base station and the unique nature of each node's sensing modality makes node-to-node communication largely unnecessary. Instead of the ad hoc network arrangements common in WSNs, BSNs tend to use a star-hub topology with each sensor node communicating only with the base station (as the hub). Third, the number of nodes in a BSN (*<10 for most systems*) is smaller than in common WSNs (e.g. *>1000s for large-scale environmental monitoring*), but the per node data rates are typically much higher, with ECG or EEG data being inherently more dynamic than environmental temperature data. Finally, the wearable nature of the BSN nodes places unique constraints on their specification and design to make them acceptable for widespread use but also provides more regular node access for maintenance and re-charging.

1.8.2.3 Wearability

Specifically, BSN nodes will not be adopted if they are too inconvenient, uncomfortable or unsightly. The level of tolerable chunkiness for a BSN node is proportional to its importance to the user; to encourage widespread use of BSNs, system designers must shrink the size of the nodes to the sub-cubic centimetre level and give them conformal, wearable shapes that disappear to the point that they are forgettable by the wearer. Such small form factors place a severe limit on the amount of energy that the node can store locally (e.g. *small batteries*). This creates a challenge to provide the necessary functionality for the desired application within a tight energy budget. Some BSN nodes could use batteries that are recharged regularly, for example if the user removes them at night. Other nodes might need to be used continuously for longer periods, although as noted earlier, unlike some WSNs, they are unlikely to be worn for much more than a week or so before removal for some activities (e.g. *bathing, sleeping*). In an effort to further reduce the node size and to extend their usable lifetime, some BSN researchers are pursuing custom solutions that supplement or replace the battery by harvesting energy from the body environment (e.g. *body heat, vibration, solar*) (Calhoun et al. 2012). Eliminating the battery through energy harvesting would shrink the node, make it cheaper and potentially improve the possibility of integrating it into items people already wear like clothing, shoes or jewellery.

1.8.2.4 Technical Challenges

While the BSN field has made good progress in developing systems to sense data and get it where it needs to go, additional research is necessary to convert that data into information relevant to the target application. This requires advancements not only in signal processing and data mining but also in interdisciplinary collaborations that involve application domain expertise. In the long run, these technologies must be demonstrated to improve patient outcomes, enhance quality of life and reduce health-care costs – and the demonstrations must be through real deployments on real people. Other technical challenges facing BSNs include security, privacy and safety – all of which are essential given the target medical applications and are made challenging given the severe resource constraints and dynamic nature of the system and environment. Highly important research thrusts also include incorporating additional sensing modalities, providing real-time guarantees (*especially for applications involving actuation*), achieving ultra-low-power sensing/processing/communication/networking and realizing usability across a diverse group of users. BSNs enable a diverse and powerful new direction in medical care. While technical and societal challenges remain, promising initial forays in BSN development show the potentially revolutionary possibilities for this technology.

1.8.2.5 Practical Design Considerations

Broadly speaking, two approaches have been taken to resolve the issue of electrode-skin contact impedance for low-noise, low-artefact biopotential sensing. The traditional solution has been to simply abrade the skin to obtain a very low contact resistance (*5–10 kW*). At the other extreme, one common practice has been to employ an amplifier with such high input impedance that the skin-electrode impedance becomes negligible. For wet electrodes, neither extreme is necessary, but the problem of contact impedance becomes a much more pressing problem for dry and non-contact sensors, for which maximizing input impedance is the only viable alternative. Achieving truly non-conductive, non-contact sensing, however, is difficult in practice. Fully accounting for the electrical coupling between the skin and the electrode, and its effect on noise, is generally quite complex, because of the different layers of coupling involved through skin and the coupling medium. Low resistance layers generate no appreciable thermal noise. High resistivity layers may generate large thermal noise voltages, but these voltages get shunted away as long as the impedance of the parallel capacitance is sufficiently low over the frequencies of interest. At the most basic level, the coupling impedance can be described as a single resistance in series with a parallel conductance–capacitance combination. In practice, all electrode types couple signals both resistively and capacitive in the frequencies of interest for biopotential signals. The interplay between electrode conductance and capacitance is one of the critical factors determining the limits on noise performance. Also, the success in reducing noise by increasing coupling resistance depends on the impedance level of the coupling capacitance, which strongly depends on frequency. For low capacitive coupling (*at large distance*), higher electrode resistances translate directly into increased noise levels, both intrinsically due to thermal noise and induced by motion and friction artefacts. According to increasing the coupling resistance only lowers noise for values of resistance Rc larger than $1/\omega Cc$. This value becomes exceedingly large for increasing electrode distances. For this reason, the most demanding applications where close proximity to the skin cannot be warranted, like research EEG over haired skull, still require wet electrodes. In summary, nearly all aspects of the performance of an electrode are critically limited by the physical properties of the interface between skin and the electrode, rather than amplifier goodness criteria.

1.8.3 Wireless Body Area Sensor Network: Health Monitoring

Recent technological advances in sensors, low-power microelectronics and miniaturization and wireless networking enabled the design and proliferation of WSNs capable of autonomously monitoring and controlling environments. One of the most promising applications of sensor networks is for human health monitoring. A number of tiny wireless sensors, strategically placed on the human body, create a wireless body area network (WBAN) that can monitor various vital signs, providing real-time feedback to the user and medical personnel. The WBANs promise to revolutionize health monitoring. However, designers of such systems face a number of challenging tasks, as they need to address often quite conflicting requirements for size, operating time, precision and reliability. In this chapter, we present hardware and software architecture of a working WSN system for ambulatory health status monitoring. The system consists of multiple sensor nodes that monitor body motion and heart activity, a NC and a personal server running on a PDA or a personal computer. Recent technological advances in wireless networking, microelectronics integration and miniaturization, sensors and the Internet allow us to fundamentally modernize and change the way health-care services are deployed and delivered. Focus on prevention and early detection of disease or optimal maintenance of chronic conditions promise to augment existing health-care systems that are mostly structured and optimized for reacting to crisis and managing illness rather than wellness. The anticipated change and emerging new services are well-timed to help cope with the imminent crisis in the health-care systems caused by current economic, social and demographic trends. The overall health-care expenditures in the United States reached $1.8 trillion in 2004, though almost 45 million Americans do not have health insurance. On the other hand, many companies have already been plagued by high-rising costs of health-care liabilities. With current trends in health-care costs, it is projected that the total health-care expenditures will reach almost 20% of the gross domestic product (GDP) in less than 10 years from now, threatening the well-being of the entire economy. The demographic trends are indicating two significant phenomena: an aging population due to increased life expectancy and baby boomers' demographic peak. Life expectancy has significantly increased from 49 years in 1901 to 77.6 years in 2003. According to the U.S. Bureau of the Census, 308 *System Architecture of WBAN for Ubiquitous Health Monitoring*, the number of elderly over age 65 is expected to double from 35 million to nearly 70 million by 2025 when the youngest baby boomers retire. This trend is global, so the worldwide population over age 65 is expected to more than double from 357 million in 1990 to 761 million in 2025. These statistics underscore the need for more scalable and more affordable health-care solutions. Wearable systems for continuous health monitoring are a key technology in helping the transition to more proactive and affordable health care. They allow an individual to closely monitor changes in her or his vital signs and provide feedback to help maintain an optimal health status. If integrated into a telemedical system, these systems can even alert medical personnel when life-threatening changes occur. In addition, the wearable systems can be used for health monitoring of patients in ambulatory settings. For example they can be used as a part of a diagnostic procedure, optimal maintenance of a chronic condition, a supervised recovery from an acute event or surgical procedure, to monitor adherence to treatment guidelines (e.g. *regular cardiovascular exercise*), or to monitor effects of drug therapy. During the last few years, there has been a significant increase in the number and variety of wearable health-monitoring devices, ranging from simple pulse monitors, activity monitors and portable Holter monitors, to sophisticated and expensive implantable sensors. However, wider acceptance of the existing systems is still limited by the following important restrictions. Traditionally, personal medical monitoring systems, such as Holter monitors, have been used only to collect data. Data processing and analysis are performed offline, making such devices impractical for continual monitoring and early detection of medical disorders. Systems with multiple sensors for physical rehabilitation often feature unwieldy wires between the sensors and the monitoring system. These wires may limit the patient's activity and level of comfort and thus negatively influence the measured results. In addition, individual sensors often operate as stand-alone systems and usually do not offer flexibility and integration with third-party devices.

One of the most promising approaches in building wearable health monitoring systems utilizes emerging WBANs. A WBAN consists of multiple sensor nodes, each capable of sampling, processing and communicating one or more vital signs (*heart rate, blood pressure, oxygen saturation, activity*) or environmental parameters (*location, temperature, humidity, light*). Typically, these sensors are placed strategically on the human body as tiny patches or hidden in users' clothes, allowing ubiquitous health monitoring in their native environment for extended periods of time. A number of recent research efforts focus on wearable systems for health monitoring. Researchers at the MIT Media Lab have developed MIThril, a wearable computing platform compatible with both custom and off-the-shelf sensors. The MIThril includes ECG, skin temperature and GSR sensors. In addition, they demonstrated step and gait analysis using three-axis accelerometers, rate gyros and pressure sensors. MIThril is being used to research human behaviour recognition and to create context-aware computing interfaces. CodeBlue, a Harvard University research project, is also focused on developing WSNs for medical applications. They have developed wireless pulse oximeter sensors, wireless ECG sensors and tri-axial accelerometer motion sensors. Using these sensors, they have demonstrated the formation of ad hoc networks. The sensors, when outfitted on patients in hospitals or disaster environments, use the ad hoc networks to transmit vital signs to health-care givers, facilitating automatic vital sign collection and real-time triage. In this chapter, we describe a general WBAN architecture and how it can be integrated into a broader telemedical system. To explore feasibility of the proposed system and address open issues, we have designed a prototype WBAN that consists of a personal server, implemented on a PDA (PDA) or personal computer (PC), and physiological sensors, implemented using off-the shelf sensor platforms and custom-built sensor boards. The WBAN includes several motion sensors that monitor the user's overall activity and an ECG sensor for monitoring heart activity.

1.8.3.1 System Architecture

The proposed wireless body area sensor network for health monitoring integrated into a broader multitier telemedicine system is illustrated in Figure 1.7. The telemedical system spans a network comprised of individual health-monitoring systems that connect through the Internet to a medical

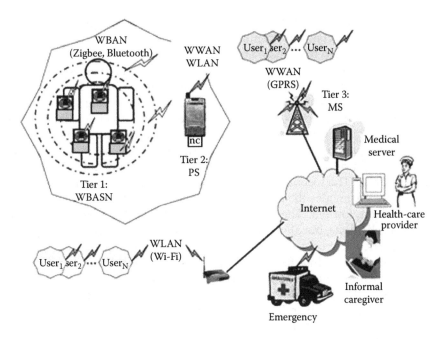

FIGURE 1.7
Health monitoring system network architecture. (From Otto, C. et al., *J. Mobile Multimedia*, 1(4), 307, 2006.)

server tier that resides at the top of this hierarchy. The top tier, centred on a medical server, is optimized to service hundreds or thousands of individual users, and encompasses a complex network of interconnected services, medical personnel and health-care professionals. Each user wears a number of sensor nodes that are strategically placed on her body. The primary functions of these sensor nodes are to unobtrusively sample vital signs and transfer the relevant data to a personal server through wireless personal network implemented using ZigBee (*802.15.4*) or Bluetooth (*802.15.1*). The personal server, implemented on a PDA, cell phone or home personal computer, sets up and controls the WBAN; provides graphical or audio interface to the user and transfers the information about health status to the medical server through the Internet or mobile telephone networks (e.g. *GPRS, 3G*). The medical server keeps electronic medical records of registered users and provides various services to the users, medical personnel and informal caregivers. It is the responsibility of the medical server to authenticate users, accept health-monitoring session uploads, format and insert this session data into corresponding medical records, analyze the data patterns, recognize serious health anomalies in order to contact emergency care givers and forward new instructions to the users, such as physician prescribed exercises. The patient's physician can access the data from his/her office via the Internet and examine it to ensure the patient is within expected health metrics (*heart rate, blood pressure, activity*), ensure that the patient is responding to a given treatment or that a patient has been performing the given exercises. A server agent may inspect the uploaded data and create an alert in the case of a potential medical condition. The large amount of data collected through these services can also be utilized for knowledge discovery through data mining. Integration of the collected data into research databases and quantitative analysis of conditions and patterns could prove invaluable to researchers trying to link symptoms and diagnoses with historical changes in health status, physiological data or other parameters (e.g. *gender, age, weight*). In a similar way, this infrastructure could significantly contribute to monitoring and studying of drug therapy effects. The second tier is the personal server that interfaces WBAN sensor nodes, provides the graphical user interface and communicates with services at the top tier. The personal server is typically implemented on a PDA or a cell phone, but alternatively can run on a home personal computer. This is particularly convenient for in-home monitoring of elderly patients. The personal server interfaces the WBAN nodes through a NC that implements ZigBee or Bluetooth connectivity. To communicate to the medical server, the personal server employs mobile telephone networks (*2G, GPRS, 3G*) or WLANs to reach an Internet access point.

The interface to the WBAN includes the network configuration and management. The network configuration encompasses the following tasks: sensor node registration (*type and number of sensors*), initialization (e.g. *specify sampling frequency and mode of operation*), customization (e.g. *run user specific calibration or user-specific signal processing procedure upload*) and set-up of a secure communication (*key exchange*). Once the WBAN network is configured, the personal server manages the network, taking care of channel sharing, time synchronization, data retrieval and processing and fusion of the data. Based on synergy of information from multiple medical sensors, the PS application should determine the user's state and his or her health status and provide feedback through a user-friendly and intuitive graphical or audio user interface. The personal server holds patient authentication information and is configured with the medical server IP address in order to interface the medical services. If the communication channel to the medical server is available, the PS establishes a secure communication to the medical server and sends reports that can be integrated into the user's medical record. However, if a link between the PS and the medical server is not available, the PS should be able to store the data locally and initiate data uploads when a link becomes available. This organization allows full mobility of users with secure and near-real time health information uploads. A pivotal part of the telemedical system is tier 1 – wireless body area sensor network. It comprises a number of intelligent nodes, each capable of sensing, sampling, processing and communicating of physiological signals. For example an ECG sensor can be used for monitoring heart activity, an EMG sensor for monitoring muscle activity, an EEG sensor for monitoring brain electrical activity, a blood pressure sensor for monitoring blood pressure, a tilt sensor for monitoring trunk position and a breathing sensor for monitoring respiration, while the motion sensors can be used to discriminate the user's status and estimate her or his level of activity.

Each sensor node receives initialization commands and responds to queries from the personal server. WBAN nodes must satisfy requirements for minimal weight, miniature form-factor, low-power consumption to permit prolonged ubiquitous monitoring; seamless integration into a WBAN, standards-based interface protocols, and patient-specific calibration, tuning, and customization (Figure 1.8). The wireless network nodes can be implemented as tiny patches or incorporated into clothes or shoes. The network nodes continuously collect and process raw information, store them locally and send processed event notifications to the personal server. The type and nature of a health-care application will determine the frequency of relevant events (*sampling, processing, storing* and *communicating*). Ideally, sensors periodically transmit their status and events, therefore significantly reducing power consumption and extending battery life. When local analysis of data is inconclusive or indicates an emergency situation, the upper level in the hierarchy can issue a request to transfer raw signals to the next tier of the network. Patient

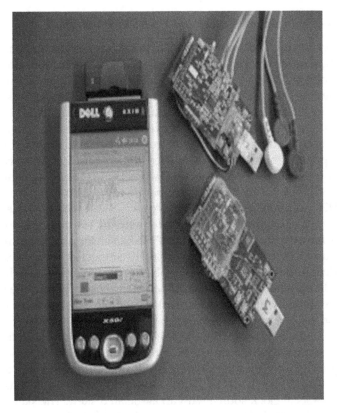

FIGURE 1.8
Prototype WBAN. (From Otto, C. et al., *J. Mobile Multimedia*, 1(4), 307, 2006.)

privacy, an outstanding issue and a requirement by law, must be addressed at all tiers in the health-care system. Data transfers between a user's personal server and the medical server require encryption of all sensitive information related to the personal health. Before possible integration of the data into research databases, all records must be stripped of all information that can tie it to a particular user. The limited range of wireless communications partially addresses security within WBAN; however, the messages can be encrypted using either software or hardware techniques. Some wireless sensor platforms have already provided a low-power hardware encryption solution for ZigBee communications.

Personal server with NC, ECG sensor with electrodes and a motion sensor are shown in Figure 1.8. Similarly, the integrated ECG and tilt sensor (*eActiS*) consists of the Tmote sky platform and an intelligent signal-processing module (*ISPM*). The ISPM is similar to the ActiS IAS board, but includes single-channel bio-amplifier for three-lead ECG/EMG. Electrodes are connected and placed on the chest for monitoring heart activity. When the sensor is worn on the chest, it also serves as an upper body tilt sensor.

1.8.3.2 Communication Protocol and Time Synchronization

Communication protocol was designed to minimize resources and uphold the spirit of the ZigBee star network topology. All communications are between a sensor node and the NC. Each communication superframe is divided into 50 ms timeslots used for message transmissions. Each sensor uses its corresponding timeslot to transmit sensor data, command acknowledgements and event messages. The first timeslot, however, belongs to the NC and is used for transmitting configuration commands from the personal server. The NC also transmits periodic beacon messages used to synchronize the start of super frames. This organization also serves as practical collision avoidance, making more efficient use of the available bandwidth when compared to using only CC2420 Collision Sense Multiple Access (*CSMA*) scheme. Time synchronization is crucial in providing a means for network communication protocol, as well as for event correlation. In the WBAN, sensor nodes are distributed about the user's body and are only wirelessly connected; the sensors operate for extended periods, sampling and analyzing physiological data. It becomes necessary to correlate detected events between sensors within the sampling interval. A time stamp mechanism can be employed; however, without a global time reference, the timestamp has no meaning outside of the scope of a sensor node. Synchronizing the session start times and utilizing a local time reference is not sufficient for the problem at hand. Even if two sensors could precisely agree on the start of a health-monitoring session, a running local time would only work for short session durations. Each sensor in the WBAN has a local clock source with an associated skew. The skew is a measure of the difference in frequency between the local clock source and an ideal clock source. As a result of skew, any elapsed time based on a local clock, over time, will differ between any two sensors. This error is cumulative. Consider two 32 KHz crystals differing by a typical 50 ppm (*parts per million*). Over the course of a few hours, the sensor clocks can differ by more than several hundred milliseconds. For an accurate health-care monitoring system and correlating step and gait analysis between two sensors, this is unacceptable, while the optimization of communication channel sharing would be impossible.

1.8.3.3 Graphical User Interface

The user interface to both address the needs of the research prototype WBAN and support a deployed WBAN system are important. The user interface must provide seamless control of the WBAN, implementing all the necessary control over the WBAN. In designing the user interface, five design goals that the PS must support was identified:

- Node identification
- Configuration of sensors
- Calibration of sensors
- Graphical presentation of events and alerts
- Visual real-time data capture (*oscilloscope-type function*)

Any control or feedback capability provided to the user interface must be implemented using the WBAN communication protocol. The protocol provides the tools enabling control of the WBAN and defines what can and cannot be accomplished. Developed protocol was modified to keep a simple set of WBAN message types and still implement complex user interface functions and application flexibility.

1.8.3.4 Sensor Node Identification, Association and Calibration

Sensor node identification requires a method for uniquely identifying a single sensor node to associate the node with a specific function during a health-monitoring session. For example a motion sensor placed on the arm performs an entirely different function from that of a motion sensor placed on the leg. Because two motion sensors are otherwise indistinguishable, it is necessary to identify which sensor should function as an arm motion sensor and which sensor should function as a leg motion sensor. In order to make node identification user friendly and intuitive, a scheme taking advantage of the inherent motion-sensing capabilities of each sensor was developed. Let the user arbitrarily place a motion sensor on his arm or leg, and then try to identify and associate the sensor with the proper function through a series of easy to follow instructions. Form instructs the user to *move the arm sensor* or in a denser WBAN, *move the left arm sensor*. This interface is more intuitive from a user's perspective, but was implemented using only WBAN protocol event messages already implemented for event processing. While the user is moving the sensor, the PS broadcasts an *ACTIS_EVENT_MASKMSG*, requesting all sensors to report activity level estimations. Based on the largest activity estimate returned, the PS can identify which sensor the user is moving and associate it with the appropriate function. The same message can also be used for configuration. Although the user interface presents these as two distinct functions, they are implemented using the same message. Event mask messages are also used to determine the degree of signal processing and the specific events of interest during a given health-monitoring session. This approach allowed us to minimize the complexity of the communication protocol and still provide rich feature set to the application designer and user. The personal server and ActiS nodes support two types of calibration. The first type is a sensor calibration; its purpose is to accommodate sensor-to-sensor variations and the exact nature of the calibration is sensor dependent. This is typically a one-time calibration and is not expected to be a long-term function of the user interface, but certainly necessary for sensor preparation. The second type of calibration is a session calibration, required immediately prior to starting a new monitoring session to calibrate the sensor in the context of its current environment. For example activity sensors on the leg might need an initial calibration of default orientation on the body.

1.8.3.5 Event Processing

The personal server is solely responsible for collecting data and events from the WBAN. Each sensor node in the network is sampling, collecting and processing data. Depending on the type of sensor and the degree of processing specified at configuration, a variety of events will be reported to the personal server. An event log is created by aggregating event messages from all the sensors in the WBAN; the log must then be inserted into a session archive file. The personal server must recognize events as they are received and make decisions based on the severity of the event. Normally R-peak or heartbeat events do not create alerts, and are only logged in the event log. However, the personal server will recognize when the corresponding heart rate exceeds predetermined threshold values. The personal server can alert the user that his or her heart rate has exceeded the target range. In addition to the regular status report in each superframe, the following events are currently supported:

- STEP (*includes timestamp, step length, maximum forces*)
- RPEAK – detection of heart beats using recognition of the R phase; the system generates a precise timestamp and time interval between the current and a previous heart beat or R-R interval (*RRINT*)
- Sensor error (*such as unexpected sensor reset*)
- Force threshold exceeded – above normal accelerations (*potential fall condition*)

- User's activity – AEE (*one-second integration of the 3D motion vector*)
- Triggered user's activity – generates an event if the AEE exceeds a specified threshold

WBAN systems that monitor vital signs promise ubiquitous, yet affordable health monitoring. WBAN systems will allow a dramatic shift in the way people think about and manage their health in the same fashion the Internet has changed the way people communicate to each other and search for information (Figure 1.9). This shift towards more proactive preventive health care will not only improve the quality of life, but also reduce health-care costs. The proliferation of wireless devices and recent advances in miniature sensors prove the technical feasibility of a ubiquitous heath-monitoring system. However, WBAN designers face a number of challenges in an effort to improve user's compliance that depends on the ease of use, size, reliability and security. In order to address some of these challenging tasks, it designed a WBAN prototype that includes accelerometer-based motion sensors, an ECG sensor and a pocket PC-based personal server. Hardware architecture leverages off-the-shelf commodity sensor platforms. Similarly, software architecture builds upon TinyOS, a widely used open-source operating system for embedded sensor networks.

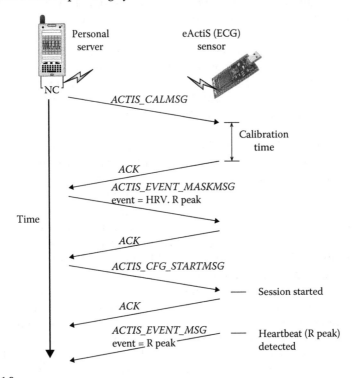

FIGURE 1.9
This figure illustrates a typical session and how WBAN protocol messages are used to calibrate sensors, start a session and receive events. (From Aziz, O. et al., *Surg. Innov.*, 14(2), 83, 2007.)

1.9 Biomonitoring Devices

Multi-functional electroactive fibres and fabrics will give the traditional textile industry a new additional value, the possibility of making daily life healthier, safer and more comfortable, bringing technological advances closer to the public through the use of easy-to-use interfaces between human measuring devices and actuators. The fabrication of such multi-functional interactive fabrics represents a potentially important tool for promoting progress, sustainable development and competitiveness in several disciplines such as health monitoring, rehabilitation, ergonomics, disability compensation, sport medicine, telemedicine and teleoperation.

De Rossi et al. (2003) have developed potential use of smart materials in the realization of sensing strain fabrics and of actuating systems. In particular, the early stage implementation and preliminary testing of fabric-based wearable interfaces are illustrated with reference to a functionalized shirt capable of recording several human vital signs and wearable motion-capture systems. The implementation of truly wearable, instrumented garments, which are capable of recording biomechanical variables, is crucial in several fields of application, from multimedia to rehabilitation, from sport to artistic fields. De Rossi et al. (2003) discussed about wearable devices which can read and record the vital signs and movements of a subject wearing the system. Monitoring of multiple vital signs based on mobile telephony and internet led to creation of solutions that will optimize lifetime health of a person, thereby enhancing quality of life, permit independent living as long as possible, provide real-time support and advice as an electronic *health companion*, and reduce the overall cost of health and medical care during one's entire life span.

Knight et al. (2005b) described the use of accelerometers in wearable systems for a number of applications. In particular, systems for the detection of activity status, assessment of performance and for teaching were explored. Also they suggested that trace of accelerometer data can be used as a referent model to improve technique for skilled performance. This may be best achieved if the trace can easily be dissected into the constituent actions of the whole activity. Anastopoulou (2004) suggest that for future teaching uses, movements for investigation should be less rapid, and less specific, to allow open-ended investigation by the student. Dario and De Rossi (1985) describes the development and applications of an artificial, flexible, force-sensitive skin, that can measure the spatial distribution and magnitude of forces perpendicular to the sensing area. Various sensor structures have been developed for this purpose based on piezoresistive, piezoelectric, opto-electronic or capacitive force-sensing technologies on silicon, printed circuit boards or flexible substrates (Lumelsky et al. 2001; Mei et al. 1999; Minbang and Heping 1991; Snyder and Clair 1978). Dario et al. (1985) describes a novel force-sensing mechanism along with the smart skin fabrication process, the accompanying electronics and representative applications. The novel force-sensing mechanism of contact piezoresistance has been used to create a versatile, flexible, smart skin Berry et al. (2004). Guler and Ertugrul (2003) describe monitoring vital body signs by placing the electronic devices into a garment in a compact way.

1.10 Interfacing Circuits and Garments

The integration of digital components into clothing is becoming an increasingly important segment in wearable computing research. The first indications for this trend are the incorporation of existing mobile technologies, such as PDAs or mobile phones, into jackets via flexible textile circuits. Post and Orth (2000) introduces a functioning prototype of such a flexible network that not only allows communication between wearable components, but is also able to supply power to them. They propose an arrangement of layered textiles as opposed to the more traditional routed circuitry layout, which results in a novel approach towards the concept of a flexible clothing network.

Bonderover and Wagner (2004) have presented a new type of electronic circuits: e-textiles. Such circuits are made by weaving together specialized component fibres. The fibres have electronic components integrated onto them. Various circuits can be formed by weaving these fibres in various patterns. Connections between fibres are made using contact pads, which are free to move against each other, thus keeping the fabric flexible. Use of amorphous silicon thin-film transistors as the active devices and present their fabrication on fibre as well as an inverter circuit woven from the fibres. They made an inverter circuit by weaving component fibres to make a fabric. To keep the fabric flexible, electrical connections between fibres are made by pressure contacts. Also they have demonstrated that an e-textile with a specific electronic function can be woven from generic component fibres. Fuller exploitation of this concept awaits advances in technology for making component fibres.

Wearable technology has the potential of allowing clinicians to track patient status over extended periods of time. This capability could have a significant impact on clinical applications in cardiology, orthopaedics, neurology and other clinical fields. The application of wearable technology would lead to significant improvements in patient management in the field of rehabilitation medicine. In the recent past, researchers have focused their work on two types of wearable systems: (1) wireless BSNs and (2) special garments with embedded sensors (Gibbs and Asada 2005). Wearable systems for continuous monitoring of patient status could positively affect the quality of patient management in rehabilitation hospitals.

E-Textiles are envisioned to be useful for applications in human monitoring, wearable computing and large area sensor networks. Various examples and techniques for creation of e-textile systems are introduced in Marculescu et al. (2002). Fibre form components offer intrinsic integration into textiles for added concealment and comfort: fibre components include batteries (Kim et al. 2000) and acoustic sensors (Piezo Film Sensors 2003). Piezoelectric material was used (Edmison et al. 2002) to sense the movement of the fingers while being integrated in a glove. The Georgia Tech Wearable Motherboard (GTWM) (Firoozbakhsh et al. 2000) is used in a combat situation to assess the seriousness of a soldier's wound and communicate this data. The interconnections in this fabric will determine the location of a bullet penetration. The GTWM can be used as a personal area network (PAN) where the devices carried by the user can interact and share data. The sensor algorithms that run on e-textiles, for example the acoustic beam-forming algorithm discussed in Riley et al. (2002), require information from sensors at specific physical locations, yet the flexibility of fabric means that the relative and absolute locations of elements on the fabric will change. Given this, the communication scheme must support location-specific rather than node-specific communication and addressing. Note that the implementation should take advantage of the fact that the physical connections on the fabric are inherently a two-dimensional grid that, absent faults, does not change even as the locations of elements change. Finally, standalone e-textiles must be power-efficient, requiring computational elements to sleep whenever possible. Communication must account for the cost of activating and routing around such nodes.

The Token Grid network presented in Todd (1992) will be modified for use as the interconnection network. Other types of networks were examined for feasibility, but do not necessarily map well to e-textiles. For example in an e-textile system, the number of nodes to be connected is not known a priori and that number is expected to change throughout the lifetime of the system. The node degree increases linearly with an increase in the dimension of a hypercube (Hwang 1993); this poor scalability factor renders architectures similar to the hypercube unsuitable for e-textiles. Tree-type architectures rely heavily on specific nodes for connections between different branches, which does not map well to the faulty environments of e-textile applications. The scalability and fault tolerance of the token grid are the primary attractive features.

1.10.1 Communication between Circuits and Garments

There are certain professions in which one may have to be highly active and are prone to dangers, such as a soldier fighting the enemy, fire fighters, law enforcement personnel, miners, deep-sea divers and astronauts in space. Considering the mobility and vulnerability, it is very important to monitor the health status and geo-location of the personnel to ensure the safety and effective completion of the assigned job. Hence, there has been a significant shift in the development of the clothing by way of introduction of the sensors into the clothing worn by the personnel. In conventional monitoring systems, there are too many hampering wires, which are meant for acquiring physiological signals, and the system is too bulky to be used for wearable applications (Martin et al. 2000).

The uses of existing technologies and demands for new functions constantly increase, contributing opportunities for further development and incorporation of new technologies. The people have become highly dependent on various forms of technology: in the office, in our cars and at home. The need for continuous communication and access to information has led to the development of portable technologies, devices that have connected us to an electronic world of data. The portability of technology has increased significantly in the last decade, with the development of sleek palm-sized cell phones from the bulky cordless phones of the 1980s, and pocket-sized PDAs from computers that occupied most of desktop and originally occupied whole rooms. These advances are the result of miniaturization of electronics. Computing power has been reduced to the nano-scale, allowing devices to occupy a fraction of the space they once did. The next logical step is wearability. Wearable technology is more than an easily carried device. Clothes are not carried, they are worn. They are more intimately connected to us than hand-held devices, so they are not treated as a separate item. Thus, they are not thought of as a burden or encumbrance that we must remember to carry and to use.

Wearable technology is specifically intended to be worn close to the body for extended periods of time. Thus, it must take into account factors related to human–machine interaction: issues of comfort, mobility, usability and aesthetics. A specific category of wearable technology is emerging: intelligent clothing. Intelligent or *smart* clothing has augmented functions that are derived from new technologies. It is a sub-group of wearable technology, characterized by its emphasis on the integration of technology into ordinary garments, so that the technology becomes part of a garment or garment system. The functions that it fills are generally those traditionally performed by clothing and garment systems, functions that can be augmented to operate more powerfully with technology. The user interface is subtle and relies on sensor input from both user and environment to automate tasks. *Intelligent clothing* is designed to be unobtrusive. It is also designed to reduce the intentional demands on the user through automation of functions.

Future clothing may have a variety of consumer electronics products built into the garments. Physical locations of people and objects are the most widely used context information in context-aware applications. To enable location-aware applications in indoor environments, many indoor location systems such as Active Badge (Want et al. 1992), Active Bat (Harter and Hopper 1994), Cricket (Priyantha et al. 2000), Smart Floor (Orr and Abowd 2000) and RADAR (Bahl and Padmanabhan 2000) were proposed in the past decade. However, there is no widespread adoption of such systems in everyday environments. The main obstacle is the level of system infrastructural support required in the deployment including hardware, installation, calibration, maintenance, etc. Significantly reducing the needed system infrastructure serves as primary motivation to design and prototype a new footstep location system.

Yeh et al. (2007) adapted dead reckoning to track human footsteps from a starting point, such as the entrance of an indoor facility. A wearable location tracker is an important advantage over infrastructure-based indoor location systems. There is a growing body of research on sensor network data management. CodeBlue is a wireless communications infrastructure for medical care applications. It is based on publish/subscribe data delivery where sensors worn by patients publish streams of vital signs and geographic locations to which PDAs or PCs accessed by medical personnel can subscribe. Secure and ad hoc communication, prioritization of critical data and effective allocation of emergency personnel in case of mass casualty events are major emphases of this project. Challenging sensor network application which can highly benefit from various data management strategies has been studied.

A potential research direction involves treating sensor readings as continuous waveforms with integrity constraints. If recent heart rate readings suggest that the heart rate could not have gone beyond normal threshold since the last reading, then there is no need to receive a new heart rate report. WPSM-IC is currently concerned with dismounted warriors and the management of their PANs. The goal at this point is to create a summary of the soldier's physiological state at the hub. In the future, this state information would be disseminated to other battlefield units. The information that is uploaded beyond the individual soldier would be used for some form of remote triage. The remote triage problem, of course, comes with its own technical challenges.

Recent technological advances in wireless networking, microelectronics integration and miniaturization, sensors and the Internet allow us to fundamentally modernize and change the way health-care services are deployed and delivered. Focus on prevention and early detection of disease or optimal maintenance of chronic conditions promise to augment existing health-care systems that are mostly structured and optimized for reacting to crisis and managing illness rather than wellness (Dishman 2004).

Wearable systems for continuous health monitoring are a key technology in helping the transition to more proactive and affordable health care. They allow an individual to closely monitor changes in her or his vital signs and provide feedback to help maintain an optimal health status. If integrated into a telemedical system, these systems can even alert medical personnel when life-threatening changes occur. In addition, the wearable systems can be used for health monitoring of patients in ambulatory settings (Istepanian et al. 2004). For example they can be used as a part of a diagnostic procedure, optimal maintenance of a chronic condition, a supervised recovery from an acute event or surgical procedure, to monitor adherence to treatment guidelines (e.g. regular cardiovascular exercise) or to monitor effects of drug therapy.

During the last few years, there has been a significant increase in the number and variety of wearable health-monitoring devices, ranging from simple pulse monitors, activity monitors and portable Holter monitors, to sophisticated and expensive implantable sensors. However, wider acceptance of the existing systems is still limited by the following important restrictions. Traditionally, personal medical monitoring systems, such as Holter monitors, have been used only to collect data. Data processing and analysis are performed offline, making such devices impractical for continual monitoring and early detection of medical disorders. Systems with multiple sensors for physical rehabilitation often feature unwieldy wires between the sensors and the monitoring system. These wires may limit the patient's activity and level of comfort and thus negatively influence the measured results. In addition, individual sensors often operate as stand-alone systems and usually do not offer flexibility and integration with third-party devices. Finally, the existing systems are rarely made affordable.

One of the most promising approaches in building wearable health-monitoring systems utilizes emerging WBANs (Jovanov et al. 2000). A WBAN consists of multiple sensor nodes, each capable of sampling, processing and communicating one or more vital signs (heart rate, blood pressure, oxygen saturation, activity) or environmental parameters (location, temperature, humidity, light). Typically, these sensors are placed strategically on the human body as tiny patches or hidden in users' clothes, allowing ubiquitous health monitoring in their native environment for extended periods of time.

A number of recent research efforts focus on wearable systems for health monitoring. Researchers at the MIT Media Lab have developed MIThril, a wearable computing platform compatible with both custom and off-the-shelf sensors. The MIThril includes ECG, skin temperature and GSR sensors. In addition, they demonstrated step and gait analysis using three-axis accelerometers, rate gyros and pressure sensors (Pentland 2004). MIThril is being used to research human behaviour recognition and to create context-aware computing interfaces (DeVaul et al. 2003).

Current wearable applications still have several problems to be worked out before mass market. Many of them are related to the maintenance of the device, for example machine washing, recharging of batteries and customer service. Bringing a function close to the body can often be of more service than merely delivering hands-free operating, but interfacing the wearable devices can be complicated. If the wearable device requires both input and output and needs to be operated on-the-go, an integrated wearable user interface may cause more problems than it solves. First of all the usability may suffer if, for example, text input, or any multi-key input, must be used; constructing a soft washable keypad is possible (Post and Orth 1997), but efficient typing may be difficult for lack of suitable rigid surfaces on the body against which to press the keys. Secondly, for the same reasons, high-resolution display output is not yet an option as the flexible displays still really are not flexible and tough enough to withstand regular garment wear (Moore 2002). Woven optical fibre displays (Gould 2003) are softer in feel and lighter, but so far do not offer needed resolution or brightness. Thirdly, hard objects larger than a button or a zipper in soft textile are likely to damage the fabric in machine wash. Constructing them waterproof and rigid enough to be able to take a washing cycle is expensive and time consuming. Finding a perfect location for the display and a keypad is hard from a usability and ergonomic point of view. A shirt with a display cannot show output if the user is wearing a jacket over the shirt. Even further, displays as well as most other output devices consume much energy in relation to sensor electronics and, combined with wireless data transfer, the overall energy consumption may require larger batteries that add weight and bulk. A good alternative to integrated user interfaces in wearable technology would be to use a wirelessly connected mobile tool, preferably one with good input/output capability, one that is widely available and customizable for different target groups and tasks and one that nearly everyone is already familiar using and carrying along with them. By making the mobile phone a part of the system, the toughest problems of added manufacturing (*and purchasing*) costs and the maintenance of the garment-integrated electronics could be solved.

1.10.2 Patient Monitoring: Networks of Wireless Intelligent Sensors

A wearable device for monitoring multiple physiological signals (polysomnograph) usually includes multiple wires connecting sensors and the monitoring device. In order to integrate information from intelligent sensors, all devices must be connected to a PAN. This system organization is unsuitable for longer and continuous monitoring, particularly during the normal activity. For instance monitoring of athletes and computer-assisted rehabilitation commonly involve unwieldy wires to arms and legs that restrain normal activity. Propose a wireless PAN of intelligent sensors as a system architecture of choice, and present a new design of wireless PAN with physiological sensors for medical applications. Intelligent wireless sensors perform data acquisition and limited processing. Individual sensors monitor specific physiological signals (*such as EEG, ECG, GSR,* etc.) and communicate with each other and the personal server. Personal server integrates information from different sensors and communicates with the rest of telemedical system as a standard mobile unit. Present prototype implementation of wireless intelligent sensor (WISE) is based on a very low-power consumption microcontroller and a DSP-based personal server. In future, one can expect all components of WISE getting integrated in a single chip for use in a variety of new medical applications and sophisticated human–computer interfaces. Existing growth of wireless infrastructure will allow a range of new telemedical applications that will significantly improve the quality of health care.

A collection of wearable medical sensors could communicate using PAN or body network, which can be even integrated into user's clothes. Intelligent monitor connects to a specialized medical service only in the event of a medical emergency or if an episode requires intervention. The user or doctor or both could formulate triggers that cause even more data to be collected, additional sensors to be enabled or medical personnel to be contacted. In the case of wireless monitoring systems, security and reliability are particularly important issue. Security can be preserved using the data encryption, balancing strength of encryption with power (both in terms of Watts and MIPS), etc. It is important to emphasize that, in the case of medical monitoring applications, simply wearing the device may disclose to the user's employer/insurer/acquaintances that the user is suffering from a medical condition. Consequently, the wearable monitoring device has to be as unobtrusive as possible, to preserve patient's privacy. Proposed concept of wireless network of WISE sensors would efficiently hide individual sensors and their connection with the personal server. Intelligent sensors would cover only limited range (in the order of 10 ft) and therefore require very low-power consumption for communication.

1.10.3 Telemedical System

Recent advances in microcontroller and sensor technology, including low-power consumption and good performance-to-cost ratio, made possible the manufacture of a whole range of new applications using distributed sensor networks. The concept of intelligent wireless sensors would be an excellent solution for a number of biomedical and monitoring applications, particularly in telemedical environment. WISE sensors as a basic building block of future systems have been developed. Block diagram of the PAN system architecture in a telemedical environment is presented in Figure 1.10. WISE sensors perform local data acquisition and simple signal processing tasks like filtering. PAN is client server network with single

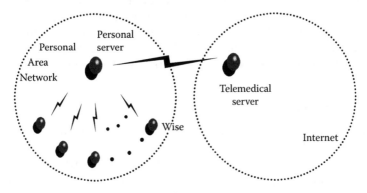

FIGURE 1.10
Block diagram of wireless personal area network in a telemedical environment. (From Otto, C. et al., *J. Mobile Multimedia*, 1(4), 307, 2006.)

server (*PERSONAL SERVER*) and multiple clients (*WISE*). In this system, personal server is a DSP board that executes the following tasks:

- PAN control and supervision
- User interface
- Telemedical server communication using standard cellular link
- Archive of events and signals; compact flash memory card used as a secondary memory

One possible application of a health monitor in telemedical environment is mapping and identity of patient. Multiple physiological signals (such as ECG, EEG, GSR and limb movement) are monitored using intelligent sensors and their state integrated using a low-power DSP-based personal server.

1.10.4 Wireless Intelligent Sensor Wise

The core of this WISE consists of Texas Instruments' microcontroller MPS430F149 that is responsible for A/D data acquisition and processing, analog signal conditioning, the LINX wireless transceiver module TC-916-SC operating at 916 MHz. The controller features 16-bit RISC architecture, ultra-low-power consumption (*400 A in active mode, less than 1 A in standby mode*), 60 KB flash memory, 2 KB RAM and a small 64-pin Quad Flat Pack (QFP) package. Current prototype uses either custom-developed biomedical amplifiers on board or an off-the-shelf two-channel bio-amplifier TETMD A110-1/2 from Teledyne for signal conditioning. It is a battery-powered, compact, ultra-low-power, analog signal-processing amplifier and filter. The signals from the bio-amplifier are converted to digital signals using internal 8-channel, 12-bit analog to digital converter on microcontroller. Additional analog channels are used to monitor battery voltage, wireless link quality and other external analog inputs. Therefore, WISE is capable of reporting the battery status and generating low-battery warnings to the higher system levels. Currently developing a new version (v1.1) will include PCB antenna and will exclude some of the functionality of the old board to decrease the size. Existing boards include both serial and wireless connection to the rest of the network to make extensive testing easier.

1.10.5 Applications

Traditionally, medical monitors were limited to data acquisition, typically implemented as Holters. Holters are used for 24–48 h monitoring of ECG, EEG or polysomnography (EEG, EOG, EMG, EKG, heart rate, breathing, body position, snoring, etc.) and recording on cassette tape or flash memory. Recorded signals are then analyzed off-line using dedicated diagnostic systems. Increased intelligence and low-power consumption of new generations of microcontrollers/DSPs make possible a whole range of intelligent monitor applications. Wireless PAN organization and sensor miniaturization will further enable new applications in this field. Further need for privacy protection and acceptance of implantable sensors and devices requires the introduction of entirely wireless personal network.

Possible applications of wireless PANs in telemedical environment include

- Intelligent portable health monitors, like ECG and ischemia or epilepsy monitoring
- Intelligent control of medication delivery using wireless sensing, dosing and compliance monitoring
- Breathing monitor, for polysomnography, sleeping disorders, stress monitoring
- Activity monitoring using accelerometer MEMS devices
- Aids for disabled individuals
- Computer-assisted rehabilitation
- Battlefield soldier monitoring

A patient wears and/or uses a sensor with a wireless communications link, which enables it to receive commands and transmit physiological and other data (measured value, time of the measurement) to a remote database. The patient also wears and/or uses a medication dosing device/recorder with a wireless link that transmits the dosing history (dose administered and time) to the remote database. A remote intelligent control system determines when a new measurement is needed and then uses the wireless link to signal the sensor, which either automatically makes the measurement or requests that the patient does so. The new dosage schedule is received by the dosing device/recorder and is either automatically administered, or the patient uses the dosing device/recorder to administer the dosage. If the communication link were not available to deliver the new dosage schedule so that the schedule can be obeyed, the system would recalculate a new dosage schedule and try to transmit again. Supervisory medical personnel access the database to monitor the patient measurements and dosing. A supervisory algorithm also monitors operation of the system (including communications link viability) and alerts medical personnel as needed. Several new rehabilitation therapies would make use of devices used by the patient in their home environment during routine activity. Monitoring of compliance of the patient in using the device and assessing therapy could be accomplished without hindering the patient if wireless communication technologies were used.

1.10.6 Software Development

The decreasing portion of money the consumers spend for clothing increases the textile manufacturers' demand for new differentiation potentials. Integrating electronic intelligence into clothes is one possibility. Several institutions have presented research studies on wearable electronics based on already existing consumer devices like cellular phones, MP3-players, etc. Electronic textiles or e-textiles are a new emerging interdisciplinary field of research which brings together specialists in information technology, micro-systems, materials and textiles. The focus of this new area is on develop-ing the enabling technologies and fabrication techniques for the economical manufacture of large-area, flexible, conformable information systems that are expected to have unique applications for both consumer electronics and military industry.

E-textiles offer new challenges for designers and CAD tool developers due to their unique requirements, cutting across from the system to the device and technology.

- The need for a new model of computation intended to support widely distributed applications, with highly unreliable behaviour, but with stringent constraints on longevity of the system.
- Re-configurability and adaptability with low computational over-head. E-textiles rely on simple computing elements embedded into fabric or directly into active yarns. As operating conditions change (*environmental, battery lifetime*, etc.), the system has to adapt and reconfigure on-the-fly to achieve a better functionality.
- Device and technology challenges imposed by embedding simple computational elements into fabric, by building yarns with compu-tational capabilities, or by the need of unconventional power sources and their manufacturing in filament or yarn form.

There have been a handful of attempts to design and build prototype com-putational textiles. In particular, if medical assistance can be provided to an injured soldier within the so-called *golden hour,* mortality can be signifi-cantly reduced. Teleintimation garment can serve as a monitoring system for soldiers that are capable of alerting the medical triage unit when a sol-dier is shot, along with information on the soldier's condition, character-izing the extent of injury and the soldier's vital signs. This garment can be used for the continuous monitoring of the vital signs of athletes to help them track and enhance their performance. In team sports, the coach can track the vital signs and the performance of the player on the field and make desired changes in the players on the field depending on the condi-tion of the player. The knowledge to be gained from medical experiments in space will lead to new discoveries and the advancement of the understand-ing of space.

Monitoring the vital signs of those engaged in critical or hazardous activi-ties, such as pilots, miners, sailors and nuclear engineers, is also possible. Special-purpose sensors that can detect the presence of hazardous materials can be integrated into the garment and enhance the occupational safety of the individuals. Combining the teleintimation garment with a GPS (*Global Positioning System*) and monitoring the well-being of public safety officials (*firefighters, police officers*, etc.), their location and vital signs at all times is also possible, thereby increasing the safety and ability of these personnel to operate in remote and challenging conditions.

1.10.7 Power System

After Alessandro Volta invented the battery in 1799, predating Michael Faraday's dynamo by 32 years, batteries provided the world's first practical electricity source until the wiring of cities in the late 1800s relegated batteries to mobile applications. Despite vacuum tube electronics' weight and large associated battery, people living in the early 1900s lugged such enormous *portable* radios to picnics and other events off the power grid. As electronics became smaller and required less power, batteries could grow smaller, enabling today's wireless and mobile applications explosion. Although economical batteries are a prime agent behind this expansion, they also limit its penetration; ubiquitous computing dream of wireless sensors everywhere is accompanied by the nightmare of battery replacement and disposal.

Exploiting renewable energy resources in the device's environment, however, offers a power source limited by the device's physical survival rather than an adjunct energy store. Energy harvesting's true legacy dates to the water wheel and windmill, and credible approaches that scavenge energy from waste heat or vibration have been around for many decades. Nonetheless, the field has encountered renewed interest as low-power electronics, wireless standards and miniaturization try to flood the world with sensor networks and mobile devices.

Energy recovery from wasted or unused power has been a topic of discussion in recent times. Unused power exists in various forms such as industrial machines, human activity, vehicles, structures and environment sources. Among these, some of the promising sources for recovering energy are periodic vibrations generated by rotating machinery or engines. Primarily, the selection of the energy harvester as compared to other alternatives such as battery depends on two main factors: cost-effectiveness and reliability. In recent years, several energy-harvesting approaches have been proposed using solar, thermoelectric, electromagnetic, piezoelectric and capacitive schemes which can be simply classified in two categories: (1) power harvesting for sensor networks using MEMS/thin/thick film approach, and (2) power harvesting for electronic devices using bulk approach (Snyder 1985). The small portable FM radio or walkman requires a small power level of 30 and 60 mW. These magnitudes of power are possible to be continuously harvested from human and industrial activity (Choon et al. 2006). Priya (2007) provides a comprehensive coverage of the recent developments in the area of piezoelectric energy harvesting using low profile transducers and provides the results for various energy-harvesting prototype devices.

Power source is one of the key elements for pervasive sensing. It often dominates the size and lifetime of the sensors. Thus far, battery remains the main source of energy for sensor nodes. In order to provide a constant energy supply, power scavenging is an actively pursued research topic. A number of power-scavenging sources have currently been proposed, which include motion, vibration, air flow, temperature difference, ambient electromagnetic fields, light and infra-red radiation. For instance Mitcheson et al. (2004) developed a vibration-based generator designed for wearable/ implantable devices.

With the need for an alternative source of power come many solutions that although seem plausible on paper, but in reality do not operate as required. The human body experiences many different motions that could in theory be converted into electrical power Mitcheson (2004). But most of this power is not at a usable level. Studies have shown that out of all the energy-dissipated areas on the body, the foot experiences the largest amount. The energy/power that is dissipated while walking/running is large enough to be converted to a usable electrical level. The work by Kendall and Palin (1998) has focused on implementing piezoelectric materials embedded in the sole of shoes, and although this work has shown a lot of promise, it does have its limitations. The performance of electromagnetic generators designed for integration in a person's shoes is described by Carroll and Duffy (2005).

There has been a steady increase in the technology worn and carried by the individual soldier. Today's soldier carries a wide array of electronic devices such as computers, communications equipment, enhanced sensory devices, weapon systems, etc., all requiring portable power. The increasing demand for power required by these electronic devices, coupled with a requirement to function as a distinct military system, has significantly increased the weight of the load carried by the individual soldier. The army is aware of the situation and has identified problem areas of battery weight, space and power density for the current rechargeable power sources carried by the soldier. One area identified for further study is the potential benefits that rechargeable lithium-ion polymer batteries can achieve in both increased energy density and ergonomic packaging. The objective of this study is to evaluate and demonstrate novel approaches to package a rechargeable battery in a wearable configuration for integration into specific military applications. A silicon-based micro-machined thermoelectric generator chip for energy harvesting from body heat was proposed by Weber et al. (2002b).

Mobile and wireless electronic devices are increasingly popular and rapidly developing. Entertainment, voice and data communication, medical and emergency needs foster the further diversification of applications, devices and services. Whereas continuously evolving wireless protocols enable all sorts of services and communications, a sustainable wireless power supply is obviously rather difficult to accomplish. Clothing-integrated photovoltaics, however, can solve this problem by providing sufficiently large areas in the range of several 100 cm^2 up to about 2000 cm^2, directly on garments or accessories. The integration of photovoltaics into clothes imposes some novel challenges and restrictions which are uncommon to standard photovoltaics. Weber et al. (2002b) presented the status and recent progress in all components for clothing-integrated photovoltaics. In parallel to the development of flexible solar cells, they have demonstrated first prototypes of integrated photovoltaics garments, equipped with commercial amorphous silicon cells and realized in co-operation with partners from clothing industry. However, as the system performance is increased and the system functionality improves, the total power requirements are higher. The energy necessary to power such systems is stored in batteries. Batteries are a significant source of size, weight and inconvenience to present-day portable, hand-held and wearable systems.

Integration of flexible solar cells into clothing can provide power for portable electronic devices. Photovoltaics is the most advanced way of providing electricity far from any mains supply, although it suffers from the limits of ambient light intensity. But the energy demand of portable devices is now low enough that clothing-integrated solar cells are able to power most mobile electronics. By introducing clothing-integrated photovoltaics, their scope and limitations, the status of flexible solar cells, charge controller and system design, as well as prototype solutions for various applications.

Since 2000, design studies on solar cells integrated into clothing have been regularly presented at fairs and exhibitions on *smart textiles* or 'smart clothes', for example the Avantex fairs in Frankfurt, Germany, the Nixdorf Innovation Forum. Axisa et al. (2005) and Hartmann (2000) presented a vision of high-tech fashion including the use of photovoltaic power. Although consumers and the clothing industry seem to be very interested in clothing-integrated photovoltaics, the advent of real products in the market has been hindered and delayed by the limited availability and performance of flexible solar cells (Leckenwalter 2000).

From a customer point of view, a clothing-integrated photovoltaic system should be easy to use, comfortable and reliable; offer a universal socket for the countless different charging adapters and devices and, of course, deliver plenty of energy at an affordable price. If parts of the system need to be visible, they should be attractive and integrate well with the particular design of the garments. Connecting wires, charge controllers and batteries ought to be invisible, lightweight and maintenance-free. As an additional requirement, clothing with integrated electronics and photovoltaics should be as washable as every other textile. The most decisive and restricting demand, however, is the conformal flexibility of solar cells used for clothing integration. Existing cells on plastic or metal foils with protective laminates can only bend in one direction rather than exhibiting full conformal flexibility like a woven textile.

Since the idea of a ubiquitous power supply from the photovoltaic conversion of ambient light is very appealing, one needs to investigate how much energy is harvested during a cloudy or rainy day, or during indoor use of integrated photovoltaics. Because of a considerable lack of data in the litera-ture, Gleskova et al. (2002) performed a detailed study to link empirical, long-term data on the incident solar radiation that reaches the earth's surface under different weather conditions, as well as various indoor illumination scenarios, with the spectral and intensity dependence of different types of solar cells.

The interest in flexible solar cells is steadily increasing, since high altitude platforms, satellites for telecommunications and deep-space missions would benefit from rollable or foldable solar generators. The integration of photo-voltaics with textiles is not only interesting for powering portable devices, which address here, but also opens a wealth of opportunities for the inte-gration of electronic features with architectural fabrics. The U.S. company United Solar Ovonic manufactures flexible triple junction a-Si-based mod-ules on steel foil for building integration with a total power output above 45 MW/year. Since these modules are designed for long-term outdoor stabil-ity, the final laminates are comparatively rigid and not suitable for large-area clothing integration. Current thin-film solar cells consist of a layer stack that is continuous in two dimensions and very thin in the third. Because of their planar substrates, these cells bend but do not crinkle. Solar cells have been formed on Cu wires for fabrics made of photovoltaic fibres. Without a con-tinuous planar substrate, however, the fibres arbitrarily move against each other, which give rise to many problems like moving interconnect, shadow-ing and cancellation of the electric output of single fibres. Considering all the unsolved problems of manufacture and interconnection of such photovolta-ics fibres, we do not see *woven solar modules* to be technically feasible in the near future. Finding a compromise between the minimum total thickness and, hence, maximum conformal flexibility of integrated photovoltaics mod-ules on the one hand, and washability, mechanical resilience and durability on the other, is an important task that has not yet been solved for most of the flexible cell technologies described here. A lot of hope is focused on organic and dye-sensitized solar cells that could, in principle, be printed on polymer foils. In a major breakthrough in organic solar cell technology, Brabec (2004) demonstrated bulk hetero-junction cells with a conversion efficiency of 5%. The main challenge for all these organic and dye-sensitized cell concepts, however, is in their encapsulation and long-term stability.

1.11 Future Research Scope

Advancements in wireless communications and mobile technology have increased the importance of personalized health care. The process of transmit-ting and analyzing health data has become simple and inexpensive with even the most inexpensive devices of today. However, the sensing key physiologi-cal parameters, from heart rate, nutritional intake to physical activity, remains a challenge, limited by both technology and user behavioural compliance. For wireless health to reach its full potential, improvements in unobtrusive sensor technology are still required. This chapter has focused on addressing the sensory limitation primarily within a cardiac and neural framework. Physiological data from ECGs and EEGs potentially offer insight into a variety of coronary and neurological disorders; especially information can be aggregated over both long time periods and across large populations. Traditional outpatient cardiac monitor devices, however, are typically constrained to just a day or two for high-resolution acquisition and just 2 weeks for basic rhythm detection. Outpatient EEG monitoring, for applications such as sleep or epilepsy diagnosis, is yet to become practical due to the difficulty in applying and maintain gel electrodes. Non-contact biopotential sensing offers a superior user experience by dispensing with the traditional need for adhesives, gels and even skin-contact, making it suitable for mobile, long-term home use. Previous efforts have demonstrated proof-of-concept non-contact sensors using discrete, off-the-shelf components for both ECG and EEG. Benchmarking of the new sensor against clinical adhesive Ag/AgCl electrodes as well as the older discrete designs showed a considerable improvement in the accuracy in signal accuracy. Nevertheless, unresolved problems with noise and motion artefacts remain a stumbling block. The signal from the non-contact sensor, while much more accurate than before, was still much more noisy despite the significant improvements in the noise specifications of the integrated sensor.

Previous attempts at modelling the noise sources in a non-contact sensor in the literature have always considered the insulation gap between body and sensor to be an ideal capacitor, leading to the erroneous conclusion that decreasing the noise floor was simply a matter of better amplifier components. Closer examination of the all the noise contributors, in this work, showed that the non-contact interface between the body and input sensor was actually the largest contributor of noise. Although disappointing that much of the noise with non-contact sensors is intrinsic to the interface and irreducible, this revelation nevertheless is a important new piece of knowledge in this field. The techniques and knowledge introduced represent the state-of-the-art in noncontact sensor design. High-resolution, near-clinical-grade, ECG sensing, through clothing, have been demonstrated. Additional exemplary applications also include EOG-based eye-tracking system as well as a non-contact, through-hair, brain–computer interface. With the electronic design considerations now well described and understood, it is expected that future improvements in this field will be within the mechanical construction and signal-processing domain. Combining the circuit innovations in research with new industrial design and artefact rejection algorithms will finally enable non-contact sensing as a practical tool for mobile health, fitness and brain–computer interfaces.

Present technological advances make possible development of intelligent wireless sensors that could be used for medical applications, such as heart rate monitors or gastrointestinal tract inspection camera using small camera and wireless link. Concepts of hierarchically organized network of wireless sensors are studied. Current prototype includes sensors capable of collecting ECG, breathing, and movement signals. EEG and GSR need to be incorporated into the next generation of WISE sensors and that is the need of hour. Main research issues include dynamic resource allocation, bandwidth utilization and secure protocol. Novel solutions are required to improve application performance and to reduce power consumption. System coupled with cellular phone link or wireless PDA and global positioning system (GPS) devices are future development platform for telemedical applications.

1.11.1 Next Stage in Man–Machine Interaction

1.11.1.1 Thermo-Generator for Harvesting Electric Power from Body Heat

Numerous wearable devices such as small remote wireless sensor units for medical applications dissipate only a small amount of power. The human body produces several 10 W of heat energy. Miniaturized thermoelectric generators can harness part of this energy and convert it into electrical power. These generators are built of a large number of thermocouples that are electrically connected in series and arranged in meanders to make best use of a given area. They consist of bars of different materials joined at one end. Most available thermo-generators are realized using compound semiconductors such as bismuth telluride. However, those are expensive, difficult to produce, not compatible with standard silicon chip fabrication processes and non-disposable. They will thus not be the optimal choice for low cost applications such as wearables. In line with these arguments, silicon appears to be a better choice. Figure 1.11 shows a technological cross-section revealing the micromachining technology. Recently, the potential of a chip with 16,000 thermocouples on a silicon chip measuring 7.0 mm^2 has been studied. Output power was measured as a function of the temperature difference between both sides of the chip. A quadratic dependence of the output power versus the temperature difference ΔT occurs. In order to gain a large difference between body and ambient temperature, a generator is integrated directly into the fabric of the clothes with good thermal contact to the skin. It has been found that an effective ΔT of up to 5 K can be achieved. For this value, an output power of 1.6 μW/cm^2 is obtained which is sufficient to power devices like a wristwatch.

1.11.1.2 Interwoven Antenna for a Transponder System in Textile Fabric

Today, radio-frequency identification tags (RFID tags) are among the smallest and least expensive electronic systems. They consist of a tiny (typically 1 mm^2) silicon chip embedded into an inlay with a planar antenna structure. The two antenna contacts are connected to the respective contacts on the chip. Total cost is below €1. Neither external leads nor a battery are necessary. Both power supply and I/O are performed by the antenna (Figure 1.12a). These functions are performed by a reader device that emits electromagnetic

FIGURE 1.11
Speech-controlled MP3 player demonstrator systems designed into a sports jacket.

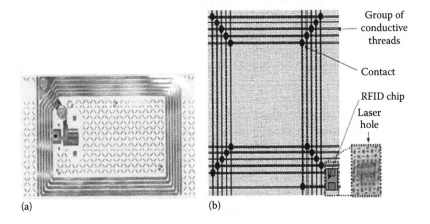

(a) (b)

FIGURE 1.12
(a) Photograph of a modern RFID product with a printed antenna on plastic substrate connected to a small silicon chip; (b) concept of a woven antenna structure for a textile transponder system.

waves at a specific frequency received and modulated by the tag. RFID tags are self-contained systems allowing a hermetically sealed package. This fact facilitates their integration into fabrics subject to harsh cleaning procedures. For integration of RFID tags into textile fabrics, the concept of a transponder system with a woven antenna coil structure is proposed. Applications include the item management in laundries or in logistics supply chains, the protection of branded goods and security applications such as access controls. Due to the self-contained nature of RFID tags, they have excellent properties in withstanding elevated temperatures, pressures, chemicals and mechanic stress. Existing RFID antenna structures are not suited for the rough environment in textile applications. A conducting spiral can be realized by connecting orthogonally oriented conductive warp and weft threads in a fabric, as shown ins Figure 1.12. By this means, the antenna structure is fully embedded into the fabric in an unobtrusive and robust way.

1.11.1.3 Fault-Tolerant Integration of Microelectronics in Smart Textiles

In the following, the concept of technical textiles containing active microsystems spatially distributed over the fabric is presented. The variety of applications for such a system is huge: pressure sensors and display units in floor cloths for surveillance or guidance systems in public buildings, intelligent sheets for monitoring vital signals of patients in hospitals, defect detection for condition monitoring in textile concrete constructions and many more. Each of these applications poses high technical requirements regarding functionality, reliability and ease and cost-efficiency of manufacturing.

The idea was realized by microelectronics integration in a coarsely meshed fabric. This fabric could be a basic or intermediate layer of floor or wall covering or any kind of technical textile. Figure 1.13 shows a schematic of an embedded network of processing elements in a textile fabric. In Figure 1.14, a photograph of the concept study of a smart carpet is depicted. A small microelectronics module is connected at the crossover points of the wires. In this way, each module is connected at each of its four sides to the supply and data lines. After encapsulation, a smart fabric is realized featuring a regular grid of integrated microelectronics modules. For ease of use, a self-organizing technique based on simple locally processed algorithms is used. Small microprocessors on the integrated modules control the data flow. After an initial learning phase, each module in the fabric network knows its exact physical position within the grid. Moreover, the network automatically configures data flow paths through the grid, routing sensor or display information even around defective regions. Figure 1.15 shows the final status: the channels for data flow within the grid are established, automatically routing around defective regions. The unique feature of the proposed integration concept is its fault-tolerance: by means of a self-organization technique, the fabric network automatically recognizes defective regions. It therefore remains operational even if a micro-device fails, or

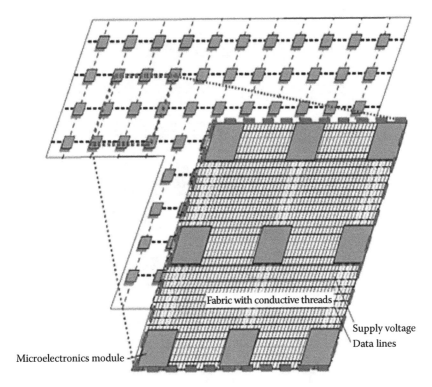

Fabric with conductive threads

Supply voltage
Data lines

Microelectronics module

FIGURE 1.13
Schematic of an embedded network of processing elements in a textile fabric.

FIGURE 1.14
Concept demonstrator of a network of electronic units connected via conducting leads in the fabric.

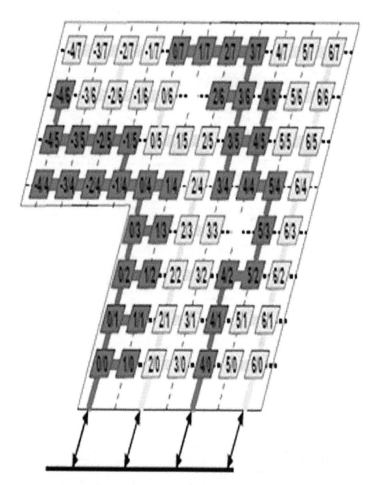

FIGURE 1.15
Final status of the self-organization: the channels for data flow are established, automatically routing around defective regions.

conductive threads are damaged. The fabric may even be cut to arbitrary shapes to fit into a given room while maintaining its function. Moreover, due to the self-organization, booting is automatic. There is no need for a manual installation of the grid microelectronics. The realization of 'textile electronics' is a result of the convergence of microelectronics with textiles surrounding us in our daily life, be it clothing, home textiles or technical textiles. This process requires the development of enabling key technologies. Various technology demonstrators were proposed which consistently aim for improving the interaction between the human individual and information technology. Open the way to promising scenarios like ambient intelligence that may lead to a completely new market of microelectronic technologies in just a few years time.

1.12 Summary

Health monitoring is among the most attractive application fields for wearable electronics and has been studied by many research groups. A variety of wearable devices for monitoring physiological parameters are commercially available today with many others in research and development stage. However, the majority of such devices are aimed at the entertainment market and are not suitable for medical monitoring of high-risk patients. Today's consumer is demanding interactivity, connectivity, ease of use and a *natural* interface for information processing. The enabling technologies of electronics, sensors, computing and communications are becoming pervasive. Since textiles is pervasive from clothes for newborns to senior citizens, from fabrics in automobile seats to tents in battlefields and presents a *universal* interface, it has the potential to meet this emerging need of today's dynamic individual. Moreover, an individual is likely to be forgetful and leave a PDA behind, but is unlikely to walk out of the home without clothes. Therefore, there is a critical need to integrate the enabling technologies into textiles so that the traditionally passive, yet pervasive, textiles can be transformed into an interactive, intelligent information infrastructure for the demanding end user to facilitate pervasive information processing. Such a system can facilitate personalized mobile information processing and give new meaning to the term *human–machine symbiosis* in the context of pervasive/invisible computing.

Bibliography

Adanur, S., *Handbook of Weaving*, Technomic, Lancaster, PA (2001).

Alberto, G. and A. Bonomi, Sensing emotions, physical activity recognition using a wearable accelerometer, *Phill. Res. Book Ser.*, 12, 41–50 (2011).

Amon, A wearable multiparameter medical monitoring and alert system, *IEEE Trans. Inf. Technol. Biomed.*, 8(4), 415–427 (2001).

Anastopoulou, S., Investigating multimodal interactions for the design of learning environments, A case study in science learning, PhD thesis, School of Electronics, Electrical and Computing Engineering, The University of Birmingham, Birminham, England (2004).

Asada, H.H., P. Shaltis, A. Reisner and S. Rhee, Mobile monitoring with wearable photoplethysmo graphic biosensors, *Eng. Med. Biol. Mag.*, 22(3), 28–40 (2003).

Ashokkumar, L. and A. Venkatachalam, Electrotextiles: Concepts and challenges, *Nat. Conf. Funct. Text. Appar.*, 2(3), 75–84 (2007).

Axisa, F., P.M. Schmitt, C. Gehin, G. Delhomme, E. McAdams and A. Dittmar, Flexible technologies and smart clothes for citizen medicine, home healthcare, and disease prevention, *IEEE Trans. Inf. Technol. Biomed.*, 9(3), 325–336 (2005).

Aziz, O., L. Atallah, B. Lo, M. ElHelw, L. Wang, G.Z. Yang and A. Darzi, A pervasive body sensor network for measuring postoperative recovery at home, *Surg. Innov.*, 14(2), 83–91 (2007).

Bahl, P. and V.N. Padmanabhan, RADAR: An in-building RF-based user location and tracking system, In *INFOCOM 2000, 19th Annual Joint Conference of the IEEE Computer and Communications Societies*, Israel, Vol. 2, pp. 775–784 (2000).

Bakay, R.E., P.R. Kennedy and B. Siuru, A brain/computer interface, *Electronics Now*, March, 55–56 (1999).

Bassil, M., J. Davenas and M. EL Tahchi, Electrochemical properties and actuation mechanisms of polyacrylamide hydrogel for artificial muscle application, *Sens. Actuat. B*, 134, 496–501 (2008).

Beith, M., Wearable wires, *Newsweek International*, 30(7), 35–41 (2003).

Belardinelii, G., R. Palagi, A. Bedini, V. Ripoli, A. Macellari and D. Franchi, Advanced technology for personal biomedical signal logging and monitoring, In *Proceedings of the 20th Annual International Conference of the IEEE Engineering in Medicine and Biology Society*, Hong Kong SAR, China, Vol. 3, pp. 1295–1298 (1998).

Berry, P., G. Butch and M. Dwane, Highly conductive printable inks for flexible and low-cost circuitry, In *Proceedings of the 37th International Symposium on Microelectronics*, California, IMAPS (2004).

Bianchi, F., S.J. Redmond, M.R. Narayanan and S. Cerutti, Barometric pressure and triaxial accelerometry-based falls event detection, *IEEE Trans. Neural Syst. Rehabil. Eng.*, 18(6), 619–627 (2010).

Bonderover, E. and S. Wagner, A woven inverter circuit for e-Textile applications, *IEEE Electr. Device Lett.*, 25(5), 145–152 (2004).

Bonderover, E., S. Wagner and Z. Suo, Amorphous silicon thin film transistors on kapton fibers, *Proc. Mater. Res. Soc. Symp.*, 736, 109–114 (2003).

Bourke, A.K., O. Donovan, J. Karol, J. Nelson, O. Laighin and M. Gearoid, Fall-detection through vertical velocity thresholding using a tri-axial accelerometer characterized using an optical motion-capture system, *Eng. Med. Biol. Soc.*, 28(2), 2832–2835 (2008).

Bouten, C.V.C., K.T.M. Koekkoek, M. Verduin, R. Kodde and J.D. Janssen, A triaxial accelerometer and portable data processing unit for the assessment of daily physical activity, *IEEE Trans. Biomed. Eng.*, 44(3), 136–147 (1997).

Brabec, C.J. Organic photovoltaics: Technology and market, *Sol. Energy Mater. Sol. Cells*, 83(2), 273–292 (2004).

Buechley, L., and M. Eisenberg, Fabric PCBs, electronic sequins, and socket buttons techniques for e-textile craft, *Pers. Ubiquit. Comput.*, Springer-Verlag London Limited, 13, 133–150 (2007).

Calhoun, B.H., J. Lach, J. Stankovic, D.D. Wentzloff, K. Whitehouse, A.T. Barth, J.K. Brown, L. Qiang, O. Seunghyun, N.E. Roberts and Y. Zhang, Body sensor networks: A holistic approach from silicon to users, *Proc. IEEE Spect.*, 100(1), 91–106 (2012).

Campbell, N.W., W.P. Mackeown, B.T. Thomas and T. Troscianko, Automatic interpretation of outdoor scenes, In *British Machine Vision Conference*, Birmingham, England (1995).

Carmo, J.P., P.M. Mendes, C. Couto and J.H. Correia, 2.4 GHz wireless sensor network for smart electronic shirts, In *Smart Sensors, Actuators, and MEMS II, Proceedings of SPIE*, Bellingham, WA, Vol. 5836, pp. 579–586 (2005).

Carroll, D. and M. Duffy, Demonstration of wearable power generator, *Power Elect. Appl.*, 10, 101–107 (2005).

Carter, G.S., M.A. Coyle and W.B. Mendelson, Validity of a portable cardio-respiratory system to collect data in the home environment in patients with obstructive sleep apnea, *Sleep Hypnosis*, 6(2), 85–92 (2004).

Chi Ed.H., A. Kittur, B. Suh, B.A. Pendleton, Social network activity, In *Proceedings of the SIGCHI Conference on Human Factors in Computing Systems*, San Jose, CA, 3(1), (2010).

Choi, S. and Z. Jiang, A novel wearable sensor device with conductive fabric and PVDF film for monitoring cardio-respiratory signals, *Sens. Actuat. App. Phys.*, 128, 317–326 (2006).

Choon, S.N., G. Dagang, M.D.B. Hapipi, N.M. Naing, W.J. Shen and A. Ongkodjojo, Overview of MEMS Wear II-incorporating MEMS technology into smart shirt for geriatric care, *J. Phys. Conf. Ser.*, 34, 1079–1085 (2006).

Clark, J.W., The origin of biopotentials, In Webster, J.G. (ed.), *Medical Instrumentation – Application and Design*, John Wiley & Sons, New York, pp. 121–301 (1998).

Corbishley, P. and E. Rodriguez-Villegas, Breathing detection: Towards a miniaturized, wearable, battery-operated monitoring system, *Biomed. Eng.*, 55(1), 196–204 (2008).

CSEM – Centre Swisse d'Electronique et Microtechnique, SA, 2008, www.csem.ch/site/card.asp, accessed 10 May, 2010.

Crowe, J., B. Hayes-Gill and M. Sumner, Modular sensor architecture for unobtrusive routine clinical diagnosis, In *International Workshop on Smart Appliances and Wearable Computing*, Tokyo, Japan, pp. 33–35 (2004).

Curone, D., E.L. Secco, A. Tognetti, G. Loriga, G. Dudnik, M. Risatti, R. Whyte, A. Bonfiglio and G. Magenes, Smart garments for emergency operators: The ProeTEX project, *Inf. Technol. Biomed.*, 14(3), 694–701 (2010).

Dario, P., R. Bardelli, D. de Rossi, L.R. Wang and P.C. Pinotti, Touch-sensitive polymer skin uses piezoelectric properties to recognise orientation of objects, *Sens. Rev.*, 2(4), 194–198 (1985).

De Rossi, D., F. Carpi, F. Lorussi, A. Mazzoldi, R. Paradiso, E.P. Scilingo and A. Tognetti, Electroactive fabrics and wearable biomonitoring devices, *AUTEX Res. J.*, 3(4), 125–132 (2003).

De Vaul, R.W., M. Sung, J. Gips and S. Pentland, MIThril 2003: Applications and architecture, In *Proceedings of the Seventh IEEE International Symposium on Wearable Computers*, White Plains, NY, pp. 4–11 (2003).

De Vries, J., J.H. Strubbe, W.C. Wildering, J.A. Gorter and A.J.A. Prins, Patterns of body temperature during feeding in rats under varying ambient temperatures, *Physiol. Behav.*, 53, 229–235 (1993).

Dhawan, A., A.M. Seyam, T.K. Ghosh and J.F. Muth, Woven fabric-based electrical circuits – Part I, evaluating interconnect methods, *Textile Res. J.*, 74, 913–919 (2004).

Dishman, E., Inventing wellness systems for aging in place, *IEEE Comp.*, 37(5), 34–41 (2004).

Dornhege, G., J.R. Millan, T. Hinthrbeger, J.D. McFarland and K.R. Muller, *Toward Brain-Computer Interfacing*, MIT Press, Cambridge, MA, p. 200, (2007).

Doyle, E.A., S.A. Weinzimer, A.T. Steffen, J.A.H. Ahern, M. Vincent and W.V. Tamborlane, A randomized, prospective trial comparing the efficacy of continuous subcutaneous insulin infusion with multiple daily injections using insulin glargine, *Diabetic Care*, 27(7), 1554–1558 (2004).

Duchamp, D., K.F. Steven and Q.M. Gerald, Software technology for wireless mobile computing, *IEEE Netw. Mag.*, 12(18), 218–230 (1991).

Dudde, R., T. Vering, G. Piechotta, R. Hintsche, Computer-aided continuous drug infusion: Setup and test of a mobile closed-loop system for the continuous automated infusion of insulin, *Inf. Technol. Biomed.*, 10(2), 395–402 (2006).

Dunne, L.E., S. Brady, B. Smyth and D. Diamond, Initial development and testing of a novel foam-based pressure sensor for wearable sensing, *J. NeuroEng. Rehabil.*, 2, 4 (2005).

Dunne, L.E., A. Toney, S.P. Ashdown and B. Thomas, Subtle integration of technology: A case study of the business suit, In *Proceedings of the First International Forum on Applied Wearable Computing (IFAWC)*, Bremen, Germany, pp. 35–44 (2004).

Edmison, J., M. Jones, Z. Nakad and T. Martin, Using piezoelectric materials for wearable electronic textiles, In *Proceedings of the Sixth International Symposium on Wearable Computers*, Washington DC, pp. 41–48 (2002).

Einsmann, C., M. Quirk, B. Muzal, B. Venkatramani, T. Martin and M. Jones, Modeling a wearable full-body motion capture system, In *Proceedings of the Ninth IEEE International Symposium on Wearable Computers*, Osaka, Japan, pp. 144–151 (2005).

Elton, S.F., Ten new developments for high-tech fabrics & garments invented or adapted by the research & technical group of the defence clothing and textile agency, In *Proceedings of the Avantex, International Symposium for High-Tech Apparel Textiles and Fashion Engineering with Innovation-Forum*, Frankfurt-am-Main, Germany, pp. 27–29 (2000).

Farwell, L.A. and E. Donchin, Talking off the top of you head: Toward a mental prosthesis utilizing even-related brain potentials, *Electroenceph. Clin. Neurophysiol.*, 70, 510–523 (1988).

Fensli, R., E. Gunnarson and T. Gundersen, A wearable ECG-recording system for continuous arrhythmia monitoring in a wireless tele-home-care situation, In *Proceedings of the IEEE International Symposium on Computer-Based Medical Systems (CBMS)*, Dublin, Ireland, pp. 407–412 (2005).

Fiedler, P., S. Griebel, P. Pedrosa, C. Fonseca, F. Vaz, L. Zentner, F. Zanow and J. Haueisen, Multichannel EEG with novel Ti/TiN dry electrodes, *Sens. Actuat. A: Phys.*, 221(1), 139–147 (2015).

Firoozbakhsh, B., N. Jayant, S. Park and S. Jayaraman, Wireless communication of vital signs using the Georgia Tech Wearable Motherboard, In *IEEE International Conference on Multimedia and Expo*, New York, Vol. 3, pp. 1253–1256 (2000).

Gandhi, N., C. Khe, D. Chung, Y.M. Chi and G. Cauwenberghs, Properties of dry and non-contact electrodes for wearable physiological sensors, In *International Conference on Body Sensor Networks*, Dallas, TX, pp. 107–112 (2011).

Gevins, A., The future of electroencephalography in assessing neurocognitive functioning, *Electromyogr. Clin. Neurophysiol.*, 106, 165–172 (1998).

Gevins, A.S., S. Bressler, B. Cutillo, J. Illes, J. Miller, J. Stern, and H. Jex, Effects of prolonged mental work on functional brain topography, *Electroenceph. Clin. Neurophysiol.*, 76, 339–350 (1990).

Giansanti D., V. Macellari and G. Maccioni, Telemonitoring and telerehabilitation of patients with Parkinson's disease: Health technology assessment of a novel wearable step counter, *Telemedicine*, 14(1), 76–83 (2008).

Gibbs, P.T. and H.H. Asada, Wearable conductive fiber sensors for multi-axis human joint angle measurements, *J. Neuro. Eng. Rehabil.*, 2(1), 197–203 (2005).

Gleskova, H. and S. Wagner, Amorphous silicon thin-film transistors on compliant polyimide foil substrates, *IEEE Electron Device Lett.*, 20, 473–475 (1999).

Gleskova, H., S. Wagner, W. Soboyejo and Z. Suo, Electrical response of amorphous silicon thin-film transistors under mechanical strain, *J. Appl. Phys.*, 92(10), 6224–6229 (2003).

Gleskova, H., S. Wagner and Z. Suo, Failure resistance of amorphous silicon transistors under extreme in-plane strain, *Appl. Phys. Lett.*, 75 (19), 3011–3103 (2002).

Golden D.P., A spectral analysis of the normal resting electrocardiogram, *IEEE Trans. Biomed. Eng.*, 20(5), 366–372 (1973).

Gorlick, M.M., Electric suspenders: A fabric power bus and data network for wearable digital devices, In *Proceedings of the Third International Symposium on Wearable Computers*, IEEE, San Francisco, CA, Vol. 18, pp. 114–121 (1999).

Gould, P., Textiles gain intelligence, *Materials Today*, 6(10), 38–43 (2003).

Gruetzmann, A., S. Hansen and J. Müller, Novel dry electrodes for ECG monitoring, *Physiol. Meas.*, 28, 1375 (2007).

Guler, M. and S. Ertugrul, Measuring and transmitting vital body signs using MEMS sensors, In *IEEE International Symposium of Wearable Computers*, New York, pp. 182–194 (2003).

Harter, A. and A. Hopper, A distributed location system for the active office, *IEEE Netw.*, 8(1), 62–70 (1994).

Hartmann, W.D., *High-Tech Fashion*, Heimdall Verlag, Witten, Germany (2000).

Healey, J. and B. Logan, Wearable wellness monitoring using ECG and accelerometer, In *Proceedings of the Ninth IEEE International Symposium on Wearable Computers ISWC*, Osaka, Japan, pp. 220–221 (2005).

Hoyt, R.W., J. Reifman, T.S. Coster and M.J. Buller, Combat medical informatics: Present and future, In *Proceedings of the AMIA Annual Symposium*, San Antonio, TX, pp. 335–339 (2002).

Hwang, K., *Advanced Computer Architecture: Parallelism, Scalability, Programmability*, McGraw-Hill, New York (1993).

Istepanian, R.S.H., E. Jovanov and Y.T. Zhang, Guest editorial introduction to the special section on M-health: Beyond seamless mobility and global wireless health-care connectivity, *IEEE Trans. Inf. Technol. Biomed.*, 8(4), 405–414 (2004).

Jayaraman, S., Designing a textile curriculum for the '90s: A rewarding challenge, *J. Textile Inst.*, 81(2), 185–194 (1990).

Jones, M., T. Martin and Z. Nakad, A service backplane for e-textile applications, In *Workshop on Modeling, Analysis, and Middleware Support for Electronic Textiles*, MAMSET, Atlanta, GA, pp. 15–22 (2002).

Jovanov, E., A. Milenkovic, C. Otto and P.C. de Groen, A wireless body area network of intelligent motion sensors for computer assisted physical rehabilitation, *J. NeuroEng. Rehabil.*, 2(6), (2005).

Jovanov, T., E. Martin and D. Raskovic, Issues in wearable computing for medical monitoring applications: A case study of a wearable ECG monitoring device, In *Fourth International Symposium on Wearable Computers*. IEEE Computer Society, Atlanta, GA, pp. 43–49 (2000).

Jung, S., Ch. Lauterbach, M. Strasser and W. Weber, Enabling technologies for disappearing electronics in smart textiles, In *IEEE International Solid-State Circuits Conference*, San Francisco, CA, Vol. 1, pp. 386–387 (2003a).

Jung, S., T. Sturm, C. Lauterbach, G. Stromberg and W. Weber, Applications of microelectronics and sensors in intelligent textile fabrics, In *Proceedings of the Tech Textiles*, Frankfurt, Germany, pp. 87–92 (2003b).

Kallmayer, C., New assembly technologies for textile transponder systems, In *Electronic Components and Technology Conference*, New Orleans, LA, pp. 1123–1126 (2003).

Kaminska, B. and W. New, Wearable biomonitors with wireless network communication, *IEEE Eng. Med. Biol.*, 12(1), 1542–1548 (2005).

Kandel, E.R. and J.H. Schwartz, *Principles of Neural Science*, McGraw-Hill, New York, (2000).

Kendall, K., M. Palin, A small solid oxide fuel cell demonstrator for microelectronic applications, *J. Power Source*, 71(2), 268–274 (1998).

Kim, G.B.C., G.G. Wallace and R. John, Electroformation of conductive polymers in a hydrogel support matrix, *Polymer*, 41, 1783–1790 (2000).

Knight, F., A. Schwirtz, F. Psomadelis, C. Baber, W. Bristowand and N. Arvanitis, The design of the SensVest, *Pers. Ubiquit. Comput.*, 9(1), 6–19 (2005a).

Knight, J.F., H.W. Bristow, S. Anastopoulou, C. Baber, A. Schwirtz and T.N. Arvanitis, Uses of accelerometer data collected from a wearable system, *Pers. Ubiquit. Comput.*, 11, 117–132 (2005b).

Lanata, A., E.P. Scilingo, E. Nardini, G. Loriga, R. Paradiso and D. De-Rossi, Comparative evaluation of susceptibility to motion artifact in different wearable systems for monitoring respiratory rate, *Inf. Technol. Biomed.*, 14(2), 378–386 (2010).

Lanz, O. and S. Messelodi, A Sampling algorithm for occlusion robust multi target detection, *Adv. Video Sig. Based Surv.*, 2(4), 346–351 (2009).

Lauterbach, C., M. Strasser, S. Jung and W. Weber, Smart clothes self-powered by body heat, In *Proceedings of the Second International Avantex Symposium*, Frankfurt, Germany, pp. 10–13 (2002).

Leckenwalter, R., Innovations in new dimensions of functional clothing, In *Proceedings of the Avantex International Symposium for High-Tech Apparel Textiles and Fashion Engineering with Innovation-Forum*, Frankfurt am Main, Germany, pp. 3–16 (2000).

Lee, J.B. and V. Subramanian, Organic transistors on fiber: A first step toward electronic textiles, In *IEDM '03 Technical Digest. IEEE International*, Washington, DC, pp. 199–202 (2003).

Li, H. and J. Tan, An ultra-low-power medium access control protocol for body sensor network, *Eng. Med. Biol. Soc.*, 3, 2451–2454 (2005).

Lind, E.J., R. Eisler, G. Burghart, S. Jayaram, S. Park, R. Rajamanickam and McKee, A sensate liner for personnel monitoring applications, In *Proceedings of the First International Symposium on Wearable Computers*, Cambridge, MA, pp. 98–105 (1997).

Lo, B., S. Thiemjarus, R. King and G.Z. Yang, Body sensor network – A wireless sensor platform for pervasive healthcare monitoring, In *Adjunct Proceedings of the Third International Conference on Pervasive Computing*, Munich, Germany, pp. 77–80 (2005).

Locher, I. and G. Troster, Screen-printed textile transmission lines, *Text. Res. J.*, 77, 837 (2007).

Locher, I., T. Kirstein, and G. Troster, Routing methods adapted to e-textiles, In *Proceeding of. 37th Int. Symp. Microelectron.* IMAPS (2004).

Loomis, J.M., Digital map and navigation system for the visually impaired, Unpublished paper, Department of Psychology, University of California, Santa Barbara, CA (1985).

Lorincz, K., D. Malan, T.R.F. Fulford-Jones, A. Nawoj, A. Clavel and V. Shnayder, Sensor networks for emergency response: Challenges and opportunities, *IEEE Pervasive Comput. (special issue on Pervasive Computing for First Response)*, 3(4), 16–23 (2004).

Lorussi, F., W. Rocchia, E.P. Scilingo, A. Tognetti and D. De Rossi, Wearable, redundant fabric-based sensor arrays for reconstruction of body segment posture, *IEEE Sens. J.*, 4(6), 1325–1332 (2004a).

Lorussi, F., W. Rocchia, E.P. Scilingo, A. Tognetti and D. De Rossi, Wearable, redundant fabric-based sensor arrays for reconstruction of body segment posture, *IEEE Sens. J.*, 4(6), 807–818 (2004b).

Lorussi, F., E.P. Scilingo, M. Tesconi, A. Tognetti and D. De Rossi, Strain sensing fabric for hand posture and gesture monitoring, *IEEE Trans. Inf. Technol. Biomed.*, 9(3), 372–381 (2005).

Lumelsky, V.J., M.S. Shur and S. Wagner, Sensitive skin, *IEEE Sens. J.*, 1(1), 41–51 (2001).

Maglaveras, N., T. Stamkopoulos, C. Pappas, M.G. Strintzis, An adaptive backpropagation neural network for realtime ischemia episodes detection: Development and performance analysis using the European ST-T database, *IEEE Trans. Biomed. Eng.*, 45 (7), 805–813 (1998).

Mann, S., Mediated reality, Technical Report 260, MIT Media Lab, Perceptual Computing Group, Cambridge, MA (1994).

Mann, S., Smart clothing: The shift to wearable computing, *Commun. ACM*, 39(80), 23–24 (1996).

Mann, S., An historical account of the 'WearComp' and 'WearCam' inventions developed for applications in 'Personal Imaging', In *First International Symposium on Wearable Computers*, Cambridge, MA, pp. 66–73 (1997a).

Mann, S., Wearable computing: A first step toward personal imaging, *Computer*, 30(2), 25–32 (1997b).

Marculescu, D., Electronic textiles: A platform for pervasive computing, *Proc. IEEE*, 91(12), 1995–2018, (2003).

Marculescu, D., R. Marculescu and P.K. Khosla, Challenges and opportunities in electronic textiles modeling and optimization, In *Proceedings of the 39th Design Automation Conference*, New Orleans, LA, pp. 175–180 (2002).

Martin, T., E. Jovanov and D. Raskovic, Issues in wearable computing for medical monitoring applications: A case study of a wearable ECG monitoring device, In *Proceedings of the Fourth IEEE International Symposium on Wearable Computers*, Atlanta, GA, pp. 43–49 (2000).

Mathie M J., B.G. Celler, N.H. Lovell and A.C.F. Coster, Classification of basic daily movements using a triaxial accelerometer, *Med. Biol. Eng. Comput.*, 42(5), 679–687 (2004).

Matthews, R., N.J. McDonald, P. Hervieux, P.J. Turner and M.A. Steindorf, A wearable physiological sensor suite for unobtrusive monitoring of physiological and cognitive state, *Eng. Med. Biol. Soc.*, 2007, 5276–5281 (2007).

Mazzoldi, A., D. De Rossi, F. Lorussi, E.P. Scilingo and R. Paradiso, Smart textiles for wearable motion capture systems, *AUTEX Res. J.*, 2(4), 199–203 (2002).

Mei, T., Y. Ge, Y. Chen, L. Ni, W. Liao, Y. Xu and W.J. Li, Design and fabrication of an integrated three-dimensional tactile sensor for space robotic applications, In *Twelfth IEEE International Conference on Micro Electro Mechanical Systems, MEMS, '99*, Orlando, FL, pp. 112–117 (1999).

Minbang L. and K. Heping, A thin-film strain gauge tactile sensor array for robots, In *ICAR Fifth International Conference on Advances Robotics Robots in Unstructured Environments*, Pisa, Italy, Vol. 2, pp. 1734–1737 (1991).

Mitcheson, P.D., Architectures for vibration driven micropower generators, *J. Microelectromechanical Systems*, 13(3), 429–440 (2004).

Mitcheson, P.D., T.C. Green, E.M. Yeatman, and A.S. Holmes, Microeletromechal Systems, *IEEE/ASME J. Sys.*, 13(3), 429–440 (2004).

Mizell, D., Message from the Chair, In *Proceedings of the Third International Symposium on Wearable Computers*, IEEE, San Francisco, CA, pp. 9–10 (1999).

Moore, S.K., Plastics, organic semiconductors—flexible displays a reality, *IEEE Spectrum.*, 14(3), 55–59 (2002).

Mühlsteff, J., Olaf such and Philips research Aachen, dry electrodes for monitoring of vital signs in functional textiles, In *Proceedings of the 26th Annual International Conference of the IEEE EMBS*, San Francisco, CA, Vol. 1, pp. 2212–2215 (2004).

Mundt, C. W., K.N. Montgomery, U.E. Udoh, V.N. Barker, G.C. Thonier, A.M. Tellier, R.D. Ricks et al., A multi-parameter wearable physiologic monitoring system for space and terrestrial applications, *IEEE Trans. Inf. Technol. Biomed.*, 9(3), 382–391 (2005).

Ockerman, J.J. and A.R. Pritchett, Task guidance in aircraft inspection, In *Proceedings of the Second International Symposium on Wearable Computers*, Pittsburgh, PA, pp. 33–40 (1998).

Orr, R.J. and G.D. Abowd, The smart floor: A mechanism for natural user identification and tracking, In *Proceedings of the Conference on Human Factors in Computing Systems*, Atlanta, GA, pp. 275–276 (2000).

Osada, Y. and J.P. Gong, Soft and wet materials: Polymer gels, *Adv. Mater.*, 10(11), 827–837, (1998).

Otto, C., J.P. Gober, R.W. McMurtrey, A. Milenković and E. Jovanov, An implementation of hierarchical signal processing on wireless sensor in TinyOS environment, In *Proceedings of the 43rd ACMSE*, Kennesaw, GA, pp. 49–53 (2005).

Otto, C., A. Milenkovic, C. Sanders and E. Jovanov, Final results from a pilot study with an implantable loop recorder to determine the etiology of syncope in patients with negative noninvasive and invasive testing, *Am. J. Cardiol.*, 82(1), 117–119 (1998).

Otto, C., A. Milenkovic, C. Sanders and E. Jovanov, System architecture of a wireless body area sensor network for ubiquitous health monitoring, *J. Mobile Multimedia*, 1(4), 307–326 (2006).

Pacelli, M., G. Loriga, N. Taccini and R. Paradiso, Sensing fabrics for monitoring physiological and biomechanical variables: E-textile solutions, In *Proceedings of the Third IEEE-EMBS*, Santorini Island, Greece, pp. 1–4 (2006).

Pan, J. and W.J. Tompkins, A real time QRS detection algorithm, *IEEE Trans. Biomed. Eng.*, 32(3), 230–236 (1985).

Pandian, P.S., K. Mohanavelu, K.P. Safeer, T.M. Kotresh, D.T. Shakunthala, P. Gopal and V.C. Padaki, Smart vest: Wearable multi-parameter remote physiological monitoring system, *Med. Eng. Phys.*, 30(4), 466–477 (2008).

Paradiso, J., New ways to play: Electronic music interfaces, *IEEE Spect.*, 34(12), 18–30 (1997).

Paradiso, J. and M. Feldmeier, A compact, wireless, self-powered pushbutton controller, In *Ubicomp 2001: Ubiquitous Computing, International Conference*, LNCS 2201, Springer-Verlag, New York, pp. 299–304 (2005).

Paradiso, R., G. Loriga and N. Taccini, Wearable health care system for vital signs monitoring, In *Mediterranean Conference on Medical and Biological Engineering*, Ischia, Italy, pp. 430–434 (2004).

Park, S. and S. Jayaraman, Adaptive and responsive textile structures, In Tao, X. (ed.), *Smart Fibers, Fabrics and Clothing: Fundamentals and Applications*, Woodhead, Cambridge, U.K., pp. 226–245 (2001a).

Park, S. and S. Jayaraman, Textiles and computing: Background and opportunities for convergence, In *Proceeding ACN CASAS'01*, Atlanta, GA, pp. 186–187 (2001b).

Park, S. and S. Jayaraman, Smart textiles: Wearable electronic systems, *MRS Bull.*, 28(8), 585–591 (2003).

Patel, S., R. Hughes, T. Hester, J. Stein, M. Akay, J.G. Dy and P. Bonato, A novel approach to monitor rehabilitation outcomes in stroke survivors using wearable technology, *Proc. IEEE Sens. Actuat.*, 98, 450–461 (2010).

Patel, S., H. Park, P. Bonato, L. Chan and M. Rodgers, A review of wearable sensors and systems with application in rehabilitation, *J. NeuroEng. Rehabil.*, 9, 21 (2012).

Patterson, J.A.C., D.G. McIlwraith and G.-Z. Yang, A flexible, low noise reflective PPG sensor platform for ear-worn heart rate monitoring, *Wear. Implant. Body Sens. Netw.*, 3(5), 286–291 (2009).

Pentland, S., Healthwear: Medical technology becomes wearable, *IEEE Comp.*, 37(5), 34–41 (2004).

Perng, J.K., B. Fisher, S. Hollar and K.S.J. Pister, Acceleration sensing glove, In *Proceedings of the Third International Symposium on Wearable Computers*, IEEE, San Francisco, CA, 18–19 October, pp. 178–180 (1999).

Pfurtscheller, G., D. Flotzinger and J. Kalcher, Brain-computer interface – A new communication device for handicapped persons, *J. Microcomput. Appl.*, 16, 293–299 (1993).

Piezo Film Sensors, *Technical Manual*, Measurement Specialties, Inc., Norristown, PA (2003).

Post, E.R. and M. Orth, Smart fabric, wearable clothing, In *Proceedings of the First International Symposium on Wearable Computers*, IEEE, Cambridge, MA, pp. 167–168 (1997).

Post, E.R. and M. Orth, E-broidery: Design and fabrication of textile-based computing, *IBM Syst. J.*, 39(3–4), 840–860 (2000).

Prajapati, S.K., W.H. Gage, D. Brooks, S.E. Black and W.E. McIlroy, A novel approach to ambulatory monitoring: Investigation into the quantity and control of everyday walking in patients with subacute stroke, *Neurorehabil. Neural Rep.*, 25, 6–14 (2011).

Prance, R.J., S. Beardsmore-Rust, A. Aydin, C.J. Harland and H. Prance, Biological and medical applications of a new electric field sensor, *Electrostatics*, 12(5), 1–14 (2000a).

Prance, R.J., S.T. Beardsmore-Rust, P. Watson, C.J. Harland and H. Prance, Remote detection of human electrophysiological signals using electric potential sensors, *Appl. Phys. Lett.*, 93(3), 39–46 (1998).

Prance, R.J., A. Debray, T.D. Clark, H. Prance, M. Nock, C.J. Harland and A.J. Clippingdale, An ultra-low-noise electrical-potential probe for human-body scanning, *Meas. Sci. Techn.*, 11(3), 291–298 (2000b).

Priniotakis, G., P. Westbroek, L. Van Langenhove and C. Hertleer, Electrochemical impedance spectroscopy as an objective method for characterization of textile electrodes, *Trans. Inst. Meas. Control*, 29(3/4), 271–281 (2007).

Priya, S., Advances in energy harvesting using low profile piezoelectric transducers, *J. Electroceram.*, 19(1), 167–184 (2007).

Priyantha, B., A. Chakraborty and H. Balakrishnan, The cricket location-support system, In *6th ACM International Conference on Mobile Computing and Networking*, Boston, MA, pp. 32–43 (2000).

Proulx, J., R. Clifford, S. Sorensen, D.-J. Lee and J. Archibald, Development and evaluation of a bluetooth EKG monitoring sensor, In *Proceedings of the IEEE International Symposium on Computer-Based Medical Systems (CBMS)*, Salt Lake City, UT, pp. 507–511 (2006).

Rantanen, J. and H.N. Ninen, Data transfer for smart clothing: Requirements and potential technologies, In Tao, X. (ed.), *Wearable Electronics and Photonics*, Woodhead Publishing in Textiles, Houston, TX, p. 198 (2005).

Richardson, S.W., M.C. Gilbert and P.M. Bell, Experimental determination of kyanite-andalusite and andalusite sillimanite equilibria; the aluminum silicate triple point, *Am. J. Sci.*, 267(3), 259–272, (1969).

Riistama, J., J. Väisänen and S. Heinisuo, Wireless and inductively powered implant for measuring electrocardiogram, *Med. Biol. Eng. Comput.*, 45(12), 1163–1174 (2007).

Riley, R., S. Thakkar, J. Czarnaski and B. Schott, Power-aware acoustic beamforming, In *Proceedings of the Military Sensing Symposium Specialty Group on Battlefield Acoustic and Seismic Sensing, Magnetic and Electric Field Sensors*, Laurel, MD, pp. 17–21 (2002).

Russell, D.A., S.F. Elton, J. Squire, R. Staples and N. Wilson, First experience with shape memory material in functional clothing, In *Proceedings of the Avantex, International Symposium for High-Tech Apparel Textiles and Fashion Engineering with Innovation-Forum*, Frankfurt-am-Main, Germany, pp. 8–14 (2000).

Salarian, A., F.B. Horak, C. Zampieri, P. Carlson-Kuhta, J.G. Nutt and K. Aminian, Sensitive and reliable measure of mobility, *IEEE Trans. Neural Syst. Rehabil. Eng.*, 18, 303–310 (2010).

Salarian, A., H. Russmann, F.J. Vingerhoets, C. Dehollain, Y. Blanc, P.R. Burkhard and K. Aminian, Gait assessment in Parkinson's disease: Toward an ambulatory system for long-term monitoring, *IEEE Trans. Biomed. Eng.*, 51, 1434–1443 (2004).

Satava, R.M., Virtual reality and telepresence for military medicine, *ANNALS Acad. Med. (Singapore)*, 26(1), 118–120 (1997).

Savell, C.T., M. Borsotto, J. Reifman and R.W Hoyt. Life sign decision support algorithms, In *MedInfo*, IOS Press, Amsterdam, the Netherlands, pp. 1453–1457 (2004).

Sazonov, E., L. Haodong, D. Curry and P. Pillay, Self-powered sensors for monitoring of highway bridges, *Sens. J. IEEE*, 9(11), 1422–1429 (2009).

Schalk, G., D.J. McFarland, T. Hinterberger and N. Birbaumer, BCI2000: A general-purpose brain-computer interface (BCI) system, *IEEE Trans. Biomed. Eng.*, 51, 1034–1043 (2004).

Schmidt, A., H.W. Gellersen and M. Beigl, A wearable context-awareness component, In *Proceedings of the Third International Symposium on Wearable Computers*, IEEE, San Francisco, CA, 18–19 October, pp. 176–177 (1999).

Scilingo, E.P., A. Gemignani, R. Paradiso, N. Taccini, B. Ghelarducci and D. De Rossi, Performance evaluation of sensing fabrics for monitoring physiological and biomechanical variables, *IEEE Trans. Inf. Technol. Biomed.*, 9(3), 345–352 (2005).

Sensatex, Inc. Smart Shirt physiological monitoring device. http://www.sensatex. com, accessed 13 April, 2002.

Shaltis, P.A., A. Reisner and H.H. Asada, Wearable, cuff-less PPG-based blood pressure monitor with novel height sensor, *Eng. Med. Biol. Soc.*, 30(3), 908–911 (2006).

Shen, C.-L., T. Kao, C.-T. Huang and J.-H. Lee, Wearable band using a fabric-based sensor for exercise ECG monitoring, In *10th International Symposium on Wearable Computers*, IEEE, Montreux, Geneva, pp. 143–144 (2006).

Shih, E., V. Bychkovsky, D. Curtis and J. Guttag, Continuous medical monitoring using wireless micro sensors, In *Proceedings of the Second International Conference on Embedded Networked Sensor Systems*, ACM, New York, p. 310 (2004).

Shnayder, V., B.-R. Chen, K. Lorincz, T.R.F. Fulford-Jones and M. Welsh, Sensor networks for medical care, Technical report, Harvard University, Cambridge, MA (2005).

Silva, A.F., F. Gonçalves, L.A. Ferreira, F.M. Araújo, P.M. Mendes and J.H. Correia, Optical fiber sensors integrated in polymeric foils, unpublished.

Siuru, B., Applying acoustic monitoring to medical diagnostics, *Sensors Magazine*, pp. 51–52 (1997).

Smailagic, A. and D.P. Siewiorek, The CMU mobile computers: A new generation of computer systems, In *Proceedings of the IEEE COMPCON 94 International Conference*, IEEE Computer Society Press, Los Alamitos, CA, pp. 467–473 (1994).

Snyder, D., Piezoelectric reed power supply for use in abnormal tire condition warning systems, US patent 4,510,484, to Imperial Clevite, Inc., Patent and Trademark Office, (1985).

Snyder, W.E. and J. Clair, Conductive elastomers as sensor for industrial parts handling equipment, *IEEE Trans. Instrument. Meas.*, 27(1), 94–99 (1978).

Starner, T and J.A. Paradiso, Human-generated power for mobile electronics, In Piguet, C. (ed.), *Low-Power Electronics Design*, CRC Press, Boca Raton, FL, Chapter 45, pp. 1–35 (2004).

Stelarc, From psycho to cyber strategies: Prosthetics, robotics and remote existence, *Cult. Value.*, 1(2), 241–249 (1997).

Strasser, M., R. Aigner, M. Franosch and G. Wachutka, Miniaturized thermoelectric generators based on Poly-Si and Poly-SiGe surface micromachining, *Sens. Actuat. A*, 97C–98C, 528–535 (2002).

Taheri, B.A., R.T. Knight and R.L. Smith, Dry electrode for EEG recording, *Electroencephal. Clin. Neurophysiol.*, 90(5), 376–381 (1994).

Teplan, M., Fundamentals of EEG measurement, *Meas. Sci. Rev.*, 2(2), 1–11 (2002).

Thielgen, T., F. Foerster, G. Fuchs, A. Hornig and J. Fahrenberg, Tremor in Parkinson's disease: 24-hr monitoring with calibrated accelerometry, *Electromyogr. Clin. Neurophysiol.*, 44, 137–146 (2004).

Thorp, E.O., The invention of the first wearable computer, In *Second International Symposium on Wearable Computers*, Pittsburgh, PA, pp. 4–8 (1998).

Todd, T.D., Token grid: Multidimensional media access for local and metropolitan networks, In *INFOCOM '92, Eleventh Annual Joint Conference of the IEEE Computer and Communications Societies*, IEEE, Florence, Italy, Vol. 3, pp. 2415–2424 (1992).

Troster, G., T. Kirstein and P. Lukowicz, Wearable computing: Packaging in textiles and clothes wearable computing, In *14th European Microelectronics and Packaging Conference and Exhibition*, Friedrichshafen, Germany, pp. 23–25 (2003).

Uswatte, G., W.L. Foo, H. Olmstead, K. Lopez, A. Holand and L.B. Simms, Ambulatory monitoring of arm movement using accelerometry: An objective measure of upper-extremity rehabilitation in persons with chronic stroke, *Arch. Phys. Med. Rehabil.*, 86, 1498–1501 (2005).

Valchinov, E., and N. Pallikarakis, An active electrode for biopotential recording from small localized bio-sources, *BioMed. Eng. On-line*, 3(1), 25–30 (2004).

Valchinov, E.S. and N.E. Pallikarakis, A wearable wireless ECG sensor: A design with a minimal number of parts, *Med. Biol. Eng. Comput.*, 29, 288–291 (2010).

Wagner, S., E. Bonderover, W.B. Jordan and J.C. Sturm, Electro textiles: Concepts and challenges, *Int. J. High Speed Electron. Syst.*, 12, 1–9 (2002).

Want R., A. Hopper, V. Falcao and J. Gibbons, The active badge location system, *ACM Trans. Inform. Syst.*, 10(1), 91–102 (1992).

Warren, S., J. Lebak, J. Yao, J. Creekmore, A. Milenkovic and E. Jovanov, Interoperability and security in wireless body area network infrastructures, In *Proceedings of the 27th Annual International Conference of the IEEE Engineering in Medicine and Biology Society*, Shanghai, China pp. 3837–3840 (2005).

Weber, J., K. Potje Kamloth, F. Haase, P. Detemple, F. Völklein and T. Doll, Coin-size coiled-up polymer foil thermoelectric power generator for wearable electronics, *Sens. Actuat.*, 132(1), 325–330 (2002a).

Weber, W., R. Glaser, S. Jung, C. Lauterbach, G. Stromberg and T. Sturm, Infineon technologies AG, corporate research, laboratory for emerging technologies, In *Electronics in Textiles the Next Stage in Man Machine Interaction*, Frankfurt, Germany, pp. 435–439 (2002b).

Weiss, A., T. Herman, M. Plotnik, M. Brozgol, I. Maidan, N. Giladi, T. Gurevich and J.M. Hausdorff, Can an accelerometer enhance the utility of the Timed Up & Go Test when evaluating patients with Parkinson's disease? *Med. Eng. Phys.*, 32, 119–125 (2010).

Wheelock, B., Autonomous real-time detection of silent ischemia. MS thesis, University of Alabama, Huntsville, AL (1999).

Wilhelm, F.H., M.C. Pfaltz and P. Grossman, Continuous electronic data capture of physiology, behavior and experience in real life: Towards ecological momentary assessment of emotion, *Inter. Comput.*, 18, 171–186 (2006).

Wolpaw, J.R., N. Birbaumer, D.J. McFarland, G. Pfurtscheller and T.M. Vaughan, Brain-computer interfaces for communication and control, *Clin. Neurophysiol.*, 113, 767–791 (2002).

Yao, J., R. Schmitz and S. Warren, A wearable point-of-care system for home use that incorporates plug-and-play and wireless standards, *IEEE Trans. Inf. Technol. Biomed.*, 9(3), 363–371 (2005).

Yeh, S.-Y., K.-H. Chang, C.-I. Wu, H.-H. Chu and J.Y.-J. Hsu, GETA sandals: A footstep location tracking system, *Pers. Ubiquit. Comput.*, 11(6), 451–463 (2007).

Zheng, J.W., Z.B. Zhang, T.H. Wu and Y. Zhang, A wearable mobihealth care system supporting real-time diagnosis and alarm, *Med. Bio. Eng. Comput.*, 45(9), 877–885 (2007).

2

Textile Fabric Development for Wearable Electronics

2.1 Introduction: Core Spun Yarn Production Methods

Today, wearable electronics play a vital role in many applications such as in the military, medical application, telecommunications and in health-care garments. Conductive textiles also provide a wide range of applications in civilian and military areas. In this research work, an attempt has been made to design and develop core-sheath conductive yarns with copper filament as their core and cotton as sheath using Dref-3 friction spinning system. Copper core conductive yarns are used to develop conductive fabrics. These fabrics have excellent scope for the development of many applications such as electromagnetic-shielding wearable textiles, mobile phone charging and body temperature-sensing garments. The electro-mechanical characteristics of the copper core conductive yarns and fabrics have been studied. In this chapter, different types of yarn and fabric production techniques, modifications carried out in the conventional textile machines for developing the fabrics using specialty yarns, problems faced during the production processes and the solutions are discussed.

2.1.1 Friction Spinning

Friction (DREF) spinning system is an open-ended and/or core-sheath type of spinning system. Along with the frictional forces in the spinning zone, yarn formation takes place. The DREF spinning system is used to produce yarns with a high delivery rate (about 300 mpm). It is yet to gain its importance along with the growth of technical textiles. Amongst the spinning systems, DREF provides a good platform for production of core spun yarns due to its spinning principle. There is less spinning tension to the core, and the core is positioned exactly at the yarn's centre. The development of DREF core-spun yarns creates a path for new products including those of high-performance textiles, sewing threads and in apparels due to its exceptional strength, outstanding abrasion resistance, consistent performance in sewing operation, adequate elasticity for stretch requirements, excellent resistance to perspiration, ideal wash-and-wear performance and permanent press.

2.1.1.1 Principle of Friction (DREF-3) Spinning Systems

Friction spinning system consists of opening and singling fibres from slivers, reassembling the single fibres, twisting and winding of yarn. The DREF-3 spinning principle is described (Figure 2.1) where the opened fibres are rolled with the aid of a mechanical roller for reassembling and twisting. Due to the separate yarn winding and method of twist insertion, its production rate can be high. The DREF-3 machine is DREF-2's next version created to improve yarn quality that appeared in the market in 1981. The system can spin yarns up to 18s Ne. The Dref-3000 spinning system (Figure 2.2) produces bundled yarn according to the friction-spinning principle. Basically, it is a Dref-2000 process extrapolated to accommodate a drafting arrangement (2) before the spinning drums (4). A draw frame sliver (1) with a linear density of 2.5–3.5 ktex is passed into this three-line double-apron drafting arrangement (2). The strand (3) from the draft of about 100–150 fibers proceeds from the delivery of the drafting arrangement to the convergent region between the two perforated drums (4). A pair of take-off rollers (7) draws this strand through the convergent region of the perforated drums and out of the spinning zone. The coherent fibre strand is nipped at the take-off rollers (7) and the drafting arrangement (2) and is rotated between these points by a pair

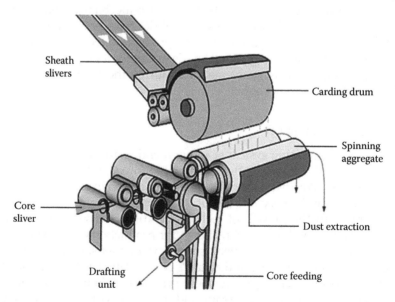

FIGURE 2.1
DREF-3 friction spinning unit.

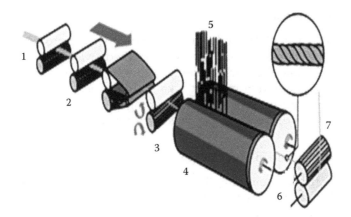

FIGURE 2.2
Principle of DREF-3 friction spinning (core-sheath yarn formation).

of perforated drums (4). It is therefore false-twisted between the nips. This means that turns of twists are present between the drafting arrangement and the drums, but not between the drums and the withdrawn rollers. If this action were to continue, the strand would fall apart. Before this happens, staple fibres are fed in free flight from above (5) into the convergent region between the drums. Because of the rotation of the perforated drums, these incoming fibres wrap themselves around the horizontally moving strand. A bundled yarn is formed. The fibre cloud (5) from above emerges from a second drafting arrangement with two open rollers. This arrangement is fed with four to six draw frame slivers with a linear density of 2.5–3.5 ktex. This is a core-sheath type of spinning arrangement. The sheath fibres are attached to the core fibres by the false twist generated by the rotating action of the drums. Two drafting units are used in this system, one for the core fibres and other for the sheath fibres. This system produces a variety of core-sheath type structures and multi-component yarns through selective combination and placement of different materials in the core and the sheath. Delivery rate is about 300 m/min.

2.1.1.2 Fibres Integration

Through the feed tube, the fibres assemble onto a yarn core/tail within the shear field and are provided with two rotating spinning drums, and the yarn core is in between them. The shear causes the sheath fibres to wrap around the yarn core. The orientation of the fibre is highly dependent on the decelerating fibres arriving at the assembly point through the turbulent flow. The fibres in the friction drum have two probable methods to integrate the incoming fibres to the sheath. In one, the fibres assemble completely on the perforated drum before their transfer to the rotating sheath. In the other method, the fibres are laid directly onto the rotating sheath.

2.1.1.3 Twist Insertion

There has been a lot of research going on in the twisting process in friction spinning. In friction spinning, the fibres applied twist with more or less one at a time without cyclic differentials in tension in the twisting zone. Therefore, fibre migration may not take place in friction spun yarns. The twist insertion for core-type friction spinning and open-end friction spinning mechanisms differ. These two mechanisms are described later.

2.1.1.4 Twist Insertion in Core-Type Friction Spinning

In core type friction spinning, the core is made of a filament or a bundle of staple fibres is false twisted by the spinning drum. The sheath fibres are deposited on the false-twisted core surface and are wrapped helically over the core with varying helix angles. It is believed that the false twist in the core ends once the yarn emerges from the spinning drums, which means this yarn has virtually twisted fewer cores. However, it is quite possible that some false twists remain because the sheath entraps it during yarn formation in the spinning zone.

2.1.1.5 Yarn Structure

The yarn tail can be considered as a loosely constructed conical mass of fibres, formed at the nip of the spinning drums. It is a very porous and lofty structure. The fibres rotate at very high speed and located at an appendage protruding from the open-end of the yarn tail and called as the tip of the tail. Viewed through the perforated drums, this tip is found to be very unstable, flickering about like a candle flame. The yarn tail is enlarged and torpedo-shaped, squashed by the nip of the perforated drums, and the fibres on its surface are loosely wrapped. Moving away from the tip, these wrappings tend to become tighter. Further, the surface structure of the tail consists of outstanding fibres, which stand out almost radically.

2.1.1.6 Advantages of Friction Spinning System

The forming yarn rotates at a high speed compared to other rotating elements. It can spin yarn at very high twist insertion rates (i.e. 300,000 twist/min). The yarn tension is practically independent of speed, and hence, very high production rates (up to 300 m/min) can be attained. The yarns are bulkier than rotor yarns. The DREF-2 yarns are used in many applications. DREF fancy yarns are used for interior decoration, wall coverings, draperies and filler yarn. Core spun yarns through this friction spinning are used in shoes, ropes and industrial cable manufacturing. Filler cartridges for liquid filtration are also effectively made with these yarns. Secondary backing for tufted carpets can be produced with waste fibres in this spinning system. Upholstery, table cloths, wall coverings, curtains, hand-made carpets, bed coverings and other decorative fabrics can be produced economically by the DREF spinning system. Heavy flame-retardant fabrics, conveyor belts, clutches and brake linings, friction linings for automobile industry, packets and gaskets are some examples where the DREF yarns can be effectively used. The DREF-3 yarns produce fabrics that have been used in many applications like backing fabrics for printing, belt inserts, electrical insulation, hoses, filter fabrics and felts made from mono-filaments core. Fabrics made from these yarns are used in hot air filtration and wet filtration in food and sugar industries. It is also used in clutch lining and brake lining in the automotive industry. The multicomponent yarns manufactured using DREF-3000 technology are mainly employed for technical textiles of the highest quality. They provide heat and wear protection, excellent dimensional stability, outstanding suitability for dyeing and coating, wearer comfort, long service life as well as a range of other qualitative and economic advantages. These include cost savings due to the use of less expensive materials, special fibres and wires as yarn cores. Apart from their strength, DREF-3000 yarns are also notable for their good abrasion-resistance, uniformity and excellent Ulster values compared to previous systems.

2.1.1.7 Limitations of Friction Spinning System

Low yarn strength and extremely poor fibre orientation make the friction spun yarns very weak. The extent of disorientation and buckling of fibres are predominant with longer and finer fibres. Friction spun yarns have higher snarling tendency. High air consumption of this system leads to high power consumption. The twist variation from surface to core is quite high; this is another reason for the low yarn strength. It is difficult to have steady spinning conditions. Drafting and fibre limit the spinning system.

2.1.2 Wrap Spinning

A wrap yam is a composite structure comprising a core of twisted or twist less fibres bound by a yam or continuous filament. The term wrap yam therefore includes yams produced by the hollow spindle method as well as similar structures such as Selfil and Repco-wrapped yams. Structures such as Dref, Novacore and fascinated yams involve a configuration of wrapper fibres around a staple or filament core and these do not employ a continuous wrapper yarn.

2.1.2.1 Wrap Spun Yarn Production Method

The wrap spinning system, although not new, has become the subject of renewed interest in recent years. The process claims to provide the spinner with a highly efficient method of producing yams of high quality for a wide range of applications. High production speeds is the advantage and the resultant yams possess several desirable qualities. Wrap yams were produced more than a century ago by the woven horsehair interlining. A core of virtually untwisted fibres has been wrapped first with a yam in 'S' direction and then with one in '2' direction or vice versa. This was then wound and used as weft in high-quality interlining fabrics. Wrap spun yams are said to be as strong as and more regular and bulkier than conventional ring spun yams, giving good cover and a full handle when converted into fabric. Wrap yams have been successfully employed in the production of a wide range of products including woven and knitted goods, tufted carpets and velour fabrics.

2.1.2.2 Operating Principle

This wrap spinning system is shown (Figure 2.3). A roving or sliver feedstock (1) is drafted in a three-, four- or five-roller drafting arrangement. The fibre strand delivered runs through a hollow spindle (3) without going through a real twist. In order to impart strength to the strand before it falls apart, a continuous-filament thread (4) is wound around the strand as it emerges from the drafting arrangement. The continuous filament thread comes from a small, rapidly rotating bobbin (5) mounted on the hollow spindle. Take-off rollers lead the resulting wrap yarn to a winding device. Thus the wrap yarn always consists of two components, one twist-free staple-fibre component in the yarn core (a) and a filament (b) wound around the core (Figure 2.4). This process has been offered by several manufacturers, for example Leesona, Mackie, etc. The most common wrap spinning system is ParafiL by Suessen Corporation, and this process will be briefly described.

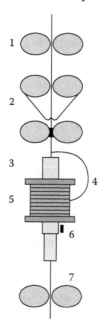

FIGURE 2.3
Wrap spinning principle.

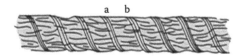

FIGURE 2.4
Wrap-spun yarn structure.

2.2 Development of Cotton-Wrapped Nichrome Yarn

In cotton-wrapped nichrome yarn, the conductive layer and substrate are embedded into each other and the polymer yarn of the substrate forms the carrier frame for the conductive layer. Conductivity of a yarn can be established with different methods. Since coarser count would enable better wrapping with high surface area and also assist in quicker dissipation and to protect thermal shocks, a combed cotton yarn of 34s was used to wrap the nichrome wire. The wrapping of nichrome wire has been carried out using Lohia Star linker braiding machine. Nichrome wire is placed in a bobbin on the centre shaft of a circular braiding machine and a 16 bobbins/2 delivery combination is used to produce nichrome wire-wrapped cotton yarn. The cotton strands from four bobbins intertwine the nichrome wire to form the braided nichrome wire sample. Table 2.1 shows the properties of cotton yarn used to develop nichrome fabric in the braiding technique.

2.2.1 Development of Woven Fabric Embedded with Cotton-Wrapped Nichrome Yarn

The cotton woven fabric was cut as per the sleeve round neck shape T-shirt pattern as shown in Figure 2.5. The cotton-wrapped nichrome wire was laid continuously over the fabric mentioned earlier manually and the laid wires were hand-stitched to retain its laid position on the fabric. Similarly, cotton woven fabric was cut for the sleeveless round neck shape T-shirt pattern. The cotton nichrome yarn was laid on the woven cotton fabric, as shown in Figure 2.6.

TABLE 2.1

Properties of 34s Cotton Yarn

Tenacity (cN/Tex)	20.2
Elongation (%)	4.3
Hairiness index	4.8
U%	11.56
Imperfections per km	244

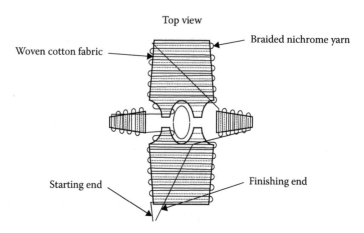

FIGURE 2.5
Woven cotton fabric embedded with nichrome yarn – sleeve pattern.

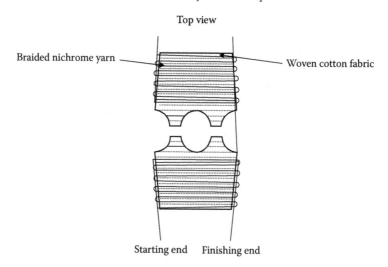

FIGURE 2.6
Woven cotton fabric embedded with nichrome yarn – sleeveless pattern.

2.2.2 Integration of Nichrome Wire-Embedded Cotton Fabric in Knitted Fabric

The nichrome wire embedded cotton fabric is integrated into the cut pattern of the knitted fabric. The total length of the cotton wrapped nichrome wire used in the sleeve pattern is 29.5 m. In a sleeveless pattern pad, the total length of the nichrome wire is 10.4 m. Figure 2.7 shows the nichrome wire-embedded cotton fabric is integrated into the cut pattern of the sleeve knitted fabric, and Figure 2.8 shows that nichrome wire-embedded cotton fabric is integrated into the cut pattern of the sleeveless knitted fabric.

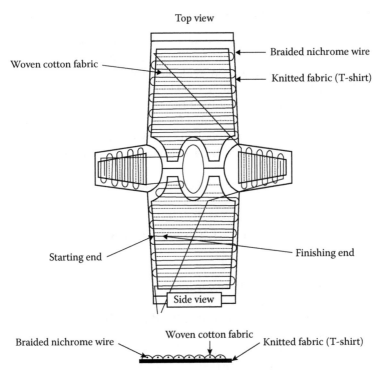

FIGURE 2.7
Integration of nichrome wire-embedded cotton fabric in sleeve knitted fabric.

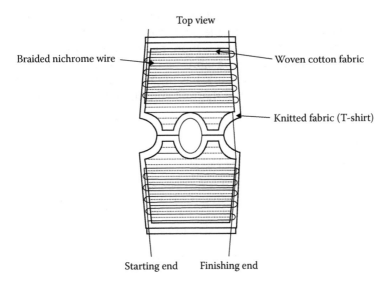

FIGURE 2.8
Integration of nichrome wire-embedded cotton fabric in sleeveless knitted fabric.

2.2.3 Testing of Thermal Conductivity and Nichrome Wire Resistance

Thermal resistance and insulation factor in the thermal conductivity of nichrome wire sleeve fabric pad, double jersey woolen sweater and pad/woven sandwich fabric have been evaluated to assess the thermal characteristics of the developed thermal insulated fabric. As per Lee's disc method, to find out the thermal conductivity of the earlier three different fabrics, the rate of temperature fall (R) is calculated for nichrome wire fabric pad, double jersey woollen sweater and pad/woollen sandwich fabric. Following this, the value of thermal conductivity of the fabric can be determined from the formula given in Equation 2.1.

Thermal conductivity is given by the formula:

$$K = \frac{MSR\,d(2h+r)}{A(T_1 - T_2)(2h+2r)} \ W/mK \tag{2.1}$$

where
K is the thermal conductivity of cardboard
M is the mass of the disc (800×10^{-3} kg)
S is the specific heat of the material of the disc (385)
R is the rate of fall of temperature
h is the thickness of the lower disc (10.125×10^{-3} m)
r is the radius of the lower disc (56.1825×10^{-3} m)
d is the thickness of the fabric
A is the area of cross section of the fabric

From Table 2.2, the insulation is slightly better for nichrome wire fabric pad when compared to double jersey woollen sweater. This is because the double jersey woollen sweater is more porous in the surface area when compared to nichrome wire sleeve fabric pad. It is also interesting to observe that the pad/woven sandwich fabric has higher insulation because the fabric is thicker (3.3 mm) since the thermal conductivity is less for the combined fabric.

TABLE 2.2

Thermal Conductivity of the Fabric

Fabric	Thermal Conductivity (W/m K)	Thermal Resistance (m² K/W)	Insulation of Clothing (Clo) (1 Clo = 0.155 m² K/W)
Nichrome wire sleeve fabric pad	0.02064	0.09687	0.625
Double jersey woollen sweater	0.025575	0.08602	0.555
Pad/woven sandwich fabric	0.02546	0.16492	1.064

TABLE 2.3

Resistances of Cotton-Wrapped Nichrome Wire

Description	Length of Cotton Wrapped Nichrome Wire (m)	Resistance (Ω)
Cotton-wrapped nichrome Wire sleeve pattern	29.5	1500
Cotton-wrapped nichrome Wire sleeveless pattern	10.4	1.5

The insulation of clothes is often measured with the unit 'Clo', where 1 Clo = 0.155 m² K/W. Zero (0) Clo corresponds to a naked person, and One (1) Clo corresponds to a person wearing a typical business suit. The normal Clo value for parkas and sweaters are in the range of 0.35–1, since the earlier results show that if the insulation (Clo) value falls in the range of 0.625–1.064, the fabric is ideally suitable for extremely cold regions.

2.2.4 Testing of Nichrome Wire Resistance

To assess the power supply requirement of the fabric, the measurement of resistance of cotton-wrapped nichrome wire is essential. Table 2.3 shows the resistance of cotton-wrapped nichrome wire sleeve fabric pad and sleeveless fabric pad. Since the resistance value of the pattern with sleeve is high when compared to the sleeveless pattern, power consumption is higher in the sleeve pattern.

2.3 Development of Copper Core Yarn and Fabric

2.3.1 Development of Copper Core Yarn

In this research work, 38 SWG (British Standard Wire Gauge) copper filament of 261 Tex and cotton were used as core material and sheath material, respectively. The Fehrer AG type DREF-3 friction spinning machine was used to produce three different core-sheath ratios of conductive yarns. Four cotton-carded slivers were fed in the first drafting unit and each sliver had 4.22 g/m and sliver irregularity value U% of 3.8%. The copper core filament was fed in the first drafting unit with special guides. The guide device was designed and installed on the first drafting unit to increase the stability of the metal yarn's spinning. The process parameters of the DREF-3 spinning machine such as perforated drum speed and yarn delivery rate were set at 4000 revolutions per min and 70 m/min, respectively, to produce uniform yarn structure and one that has the nominal yarn count of 328 Tex. By varying the draft in the second drafting system, copper/cotton yarns with three different core-sheath ratios of 67/33, 80/20 and 90/10 were produced. Figure 2.9 represents the cross-sectional view of core-sheath DREF-3 friction spun yarn, and Figure 2.10 represents the longitudinal view of the core-sheath DREF-3 friction spun yarn. Three different core-sheath ratios were produced by varying the draft in the second drafting unit; hence, the cotton-wrapping ratio was altered. Also, numerous trials have been undertaken by altering the delivery yarn speed, but it is difficult to attain uniform yarn quality. So the research work was carried out by altering the draft in the first drafting unit. During the spinning process, special care was taken while wrapping copper filament by adjusting the perforated drum speed, yarn delivery rate and draft in the second drafting unit to produce uniform yarn structure. Figure 2.11 shows the photograph of copper core conductive yarns produced by the DREF-3 spinning system wrapped in the double-flanged bobbin.

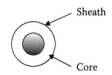

FIGURE 2.9
Cross-sectional view of core-sheath DREF-3 friction spun yarn.

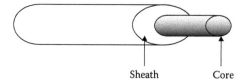

FIGURE 2.10
Longitudinal view of core-sheath DREF-3 friction spun yarn.

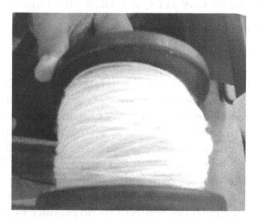

FIGURE 2.11
Illustration of copper core conductive yarn.

Dref-3 is an improvement of Dref-2 to enhance the yarn's quality, productivity and core-sheath yarn production. Core-sheath yarn is created to improve the yarn's quality by aligning part of the laid fibres along the direction of yarn axis in the core. The remaining fibres are wrapped round the core fibres. The sheath fibres are attached to the core fibres by the false twist generated by the rotating action of the drums. Two drafting units are, therefore, used in the system: one for the core fibres and other for the sheath material. After drafting in the first unit, the core filament copper is fed to nip between the spinning drums in a direction parallel to the axis of the drums. The sheath fibres after passage through the second drafting unit are deposited over core filaments by the air stream and wrapped over the core by the rotating action of the drums. The properties of friction, spun yarns, ring-spun yarns and other yarns spun on modern spinning systems, have been compared by several authors and the findings have been at times divergent because of the differences in raw material and processing conditions. The production of core-sheath Dref-3 friction spun yarn made out of cotton and polyester sheath and 820 denier copper filaments as core was reported for development of electromagnetic shielding. Several researchers have pointed out numerous applications in e-textiles for development of sensor wears, electromagnetic shielding and tele-communication using conductive yarns. Development of a complex core spun yarn using stainless steel as core in the ring spinning system was carried out and it was also pointed out that the yarn structure made out of this system was not uniform to get a perfect core-sheath yarn structure. Wireless communications are impacting all aspects of the military, from logistics and training to collaboration and medical support. In military offices, wireless communications keep the personnel mobile who can have access to information anywhere, any time. Laptop computer users attending meetings and conferences now have instant access to data files and the Internet enabling them to obtain instant answers. Outside the office, mobility increases productivity by allowing users to work in previously unproductive situations, such as while travelling or waiting for appointments.

2.3.2 Development of Copper Core Conductive Fabric

The copper core conductive yarn has been developed using DREF-3 friction spinning system with different core-sheath ratios. By using handloom, the copper core cotton sheath conductive yarns were used to produce the following two samples. Figures 2.12 and 2.13 represent the copper core cotton sheath conductive fabric samples and their physical characteristics are given in Table 2.4.

> *Sample I*: Both warp and weft as copper core cotton sheath conductive yarns
>
> *Sample II*: 2/60s Ne cotton as warp and copper core cotton sheath conductive yarns as weft

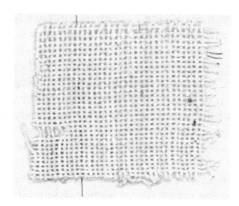

FIGURE 2.12
Copper core cotton sheath conductive fabrics (sample I: both warp and weft as copper core cotton sheath conductive yarns).

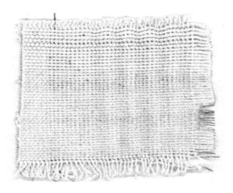

FIGURE 2.13
Copper core cotton sheath conductive fabrics (sample II: 2/60 s Ne cotton as warp and copper core cotton sheath conductive yarns as weft).

TABLE 2.4

Physical Characteristics of Core Conductive Fabrics Made out of Copper Core Yarn

Particulars	Sample I Copper Core Cotton Sheath Conductive Fabric (Warp and Weft)	Sample II Copper Core Cotton Sheath Conductive Fabric (Weft Only)
Ends per inch	16	42
Picks per inch	16	26
Aerial density (GSM)	658.42	705.6
Fabric strength, lbs	137	87
Fabric cover factor	18.59	21.56
Air permeability (cm³/cm² s)	415.66	345.97
Air resistance (kPa s/m)	0.03	0.036
Fabric thickness (mm)	1.38	0.86

2.3.3 Testing of Copper Core Conductive Yarn and Fabric

The physical properties of copper core cotton sheath conductive yarns like copper filament diameter and Tex, fibre length, fineness, tenacity and elongation were tested and shown in Table 2.5.

2.3.3.1 Structural Parameters of Copper Core Cotton Sheath Conductive Yarn

Structural parameters of the parent DREF-3 friction spun yarn and linear density and diameter of core-sheath components are given in Table 2.6. The core and sheath contribution and core-sheath interaction factors to tenacity are derived from the following Equations 2.2 through 2.4

$$\text{Core contribution factor (CC}_T\text{), }\% = \left(\frac{X}{Z}\right) \times 100 \tag{2.2}$$

$$\text{Sheath contribution factor (SC}_T\text{), }\% = \left(\frac{Y}{Z}\right) \times 100 \tag{2.3}$$

$$\text{Core-sheath interaction factor (CSI}_T\text{), }\% = \left[\frac{Z - X - Y}{Z}\right] \times 100 \tag{2.4}$$

where
   X represents the tenacity of the individual core component in cN/Tex
   Y represents the tenacity of cotton sheath in cN/Tex
   Z represents the tenacity of parent yarn in cN/Tex

The fibre strength exploitation is represented as the ratio of tenacity of yarn, core or sheath component to fibre tenacity, expressed as a percentage.

TABLE 2.5

Core and Sheath Material Properties

Property	Values
Copper filament diameter (mm)	0.1524
Copper filament (Tex)	261
Fibre length (mm)	32
Fibre fineness (µg/in.)	3.8
Fibre tenacity (cN/Tex)	3.2–3.6
Fibre elongation %	18.6
Sliver hank (dtex)	2178.5
Sliver U%	3.8
No. of sliver fed	4
Copper core cotton sheath Tex	320

TABLE 2.6

Structural Parameters of Parent DREF-3 Friction Spun Yarn and Core-Sheath Components

Sample Code	Core Component Copper (gpm)	Sheath Component Cotton (gpm)	Parent Yarn (gpm)	Contribution % Core	Contribution % Sheath	Diameter (mm) Copper Filament	Diameter (mm) Parent Yarn	Linear Density (Tex) Parent Yarn	Linear Density (Tex) Copper Filament
Sample A	0.261	0.13	0.391	66.75	33.25	0.1524	0.738	391.05	261
Sample B	0.261	0.058	0.319	81.82	18.18	0.1524	0.667	319.36	261
Sample C	0.261	0.026	0.287	90.94	9.06	0.1524	0.632	286.65	261

0.5 kV   ×40   500 μm   10 37 SEI

FIGURE 2.14
SEM micrographs of conductive yarn (cross-sectional view).

Figure 2.14 shows the cross-sectional view of core-sheath Dref-3 conductive yarn taken by scanning electron microscope (SEM). It shows that the internal wrapping of the core component by cotton fibres is denser than the outside yarn surface. The fibre orientation around the copper filament is found to be random, but the yarn structure of copper core is uniform. During the spinning, the copper filament was fed from the first drafting unit with special guides. The guide device was designed and installed on the first drafting unit to increase the stability of the conductive yarn spinning. To produce a uniform yarn structure, special care was taken while wrapping copper filament by adjusting the perforated drum speed, yarn delivery rate and draft in the second drafting unit.

2.3.3.2 Tenacity

The tensile properties of the parent conductive DREF-3 yarn and their core and sheath components are determined using Uster Tensorapid Model 3 V7.0. The yarn gauge length and the instrument's traverse speed was set at 500

TABLE 2.7

Tensile Properties of Parent DREF-3 Yarns and Core-Sheath Components

Sample Code	Tenacity (cN/Tex)			Breaking Elongation (%)			Contribution Factor		
	Parent Yarn (Z)	Core (X)	Sheath (Y)	Parent Yarn	Core	Sheath	CC_T	SC_T	CSI_T
Sample A	3.27	1.455	1.12	5.87	1.28	2.63	44.52	34.25	21.22
Sample B	2.89	1.455	1.01	4.52	1.28	2.51	50.37	34.98	14.64
Sample C	2.54	1.455	0.85	3.85	1.28	2.27	57.31	33.46	9.21

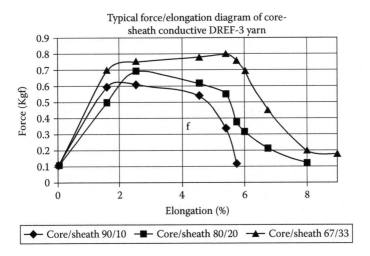

FIGURE 2.15
Force/elongation diagram of core-sheath conductive DREF-3 yarn.

and 5000 mm/min, respectively. The breaking tenacity and elongation % of conductive yarns are given in Table 2.7. Figure 2.15 shows the typical force/elongation behaviour of different core-sheath conductive DREF-3 yarns. It is clear that the major strength of the yarn is being shared by core portion and sheath portion, due to the different structural parameters of their sheath component. It is also evident from Table 2.7, with the increased blend ratio of the sheath component, the core component's contribution in tenacity values shows an upward trend.

2.3.3.3 Breaking Elongation

Table 2.7 shows that the breaking elongation of the individual sheath component for all the samples is almost of a magnitude similar to that of the parent DREF-3 yarn, whereas in the case of the core component, it breaks immediately. It is also observed that the sample A (67/33 core-sheath ratio) showed improved elongation to break when compared to other core-sheath yarns due to improvement in the core-sheath interaction (CSI_T – 21.22%).

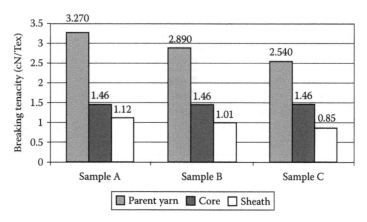

FIGURE 2.16
Breaking tenacity contribution of core-sheath component.

Figure 2.16 shows the typical breaking tenacity of the parent DREF-3 yarn and individual core and sheath components, respectively. In the parent DREF-3 yarn, the sheath fibres create the transverse force to hold the core fibres together. When the yarn extends, the core fibres are initially exposed to most of the stress, so breakage might start from the core section and move quickly through the wrapper section. The breaking tenacity values of the parent yarn, core and sheath components are different for different blend ratios. The core component shows an initial very stiff rise and then a sudden drop in the stress values, followed by a stick-slip movement of the copper wire.

2.3.3.4 Multivariable ANOVA

The breaking tenacity and contribution factor of the core and sheath component (CSI_T) of the DREF-3 friction spun yarns are analysed using multivariable ANOVA analysis. Table 2.8 shows that interaction between copper filament and cotton fibres is minimum because the cohesive nature of copper filament with cotton fibres is very poor due to the frictional characteristics of both the components. It is significant that the core component interaction with the sheath component in all the samples showed different values. Table 2.8 confirms the ANOVA analysis – three

TABLE 2.8

ANOVA Multivariable Data Analysis

Source of Variation	SS	df	MS	F	F_{crit}
Core component	0	2	0	0	6.944
Interaction	1916.661	2	958.330	24.641	6.944
Error	155.560	4	38.890		
Total	2072.222	8			

components such core, sheath and parent yarn breaking tenacity interaction of $F_{2,8}$ values (24.641) greater than F_{crit} (6.944) at $F_{2,8} > 0.05$ level.

2.3.3.5 Electrical Properties

Electrical properties of core-sheath conductive DREF-3 spun yarns such as electrical resistance and current in amps are measured at different applied voltage of 6, 12 and 24 V. The values given in Table 2.9 show that the electrical resistance of conductive yarns vary according to the length of the yarn, and the current in the amps also is directly proportional to the applied voltage. As the yarn length increases, the current values of conductive yarns tend to decrease; it is due to the increasing electrical resistance of the conductive yarns.

The conductive yarn that contained a copper wire as its core was fabricated by using a special guide mechanism on the DREF-3 spinning system. Three different core-sheath ratios of the DREF-3 conductive yarns such as 67/33, 80/20 and 90/10 were compared by varying the draft in the second drafting unit and found to have an average parent yarn linear density of 328 Tex. The 67/33 core/sheath conductive yarn was found to have the highest tenacity of 3.27 cN/Tex and elongation to break at 5.27% when compared to other core-sheath ratios due to its better core-sheath interaction factor CSI_T – 21.22%. It is also observed that the core-sheath interaction factor of conductive yarn sample C (90/10 core-sheath ratio) was found with less core-sheath interaction factor CSI_T – 9.21% and it exhibits lower breaking tenacity and elongation to break, due to the behaviour of the copper filament. The frictional characteristics and the core-sheath components percentage decide the interaction of core-sheath behaviour and mechanical properties of the DREF-3 conductive yarn. The electrical properties of these conductive core spun yarns showed very low resistance (3–28 MΩ) at 6, 12 and 24 V applied voltage.

2.3.3.6 Electromagnetic Shielding Effectiveness

Electromagnetic shielding is the process of limiting the coupling of an electromagnetic field between two locations. The shielding can be achieved using a

TABLE 2.9

Electrical Properties of Core-Sheath Conductive DREF-3 Spun Yarn

Conductive Yarn Length (m)	Electrical Resistance (mΩ)	Applied Voltage		
		6 V	12 V	24 V
		Current (μA)	Current (μA)	Current (μA)
0.5	0.06	100	200	400
1	0.1	60	120	240
2	3	2	4	8
5	20	0.3	0.6	1.2
10	28	0.21	0.42	0.84

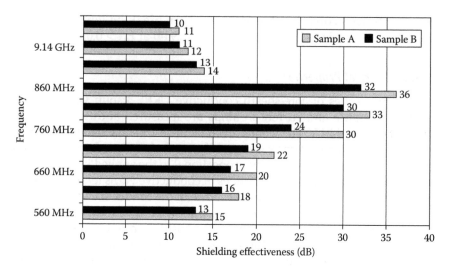

FIGURE 2.17
EMSE versus frequency of conductive fabrics.

material acting as a barrier. The electromagnetic shielding effectiveness (EMSE) of the conductive fabric was measured as per ASTM D4935-99. The method to measure the electromagnetic shielding effectiveness of planar materials was done with the given set-up. The electromagnetic shielding effectiveness of the copper conductive fabrics was made and the effectiveness was measured between the ranges of 560 MHz and 9.22 GHz. Figure 2.17 shows an attenuation of about 32 dB on an average for all frequencies from 760 to 860 MHz for both A (CCY-warp and weft) and B (CCY-weft only) samples. It can also be inferred from the graph that sample A with a core-sheath ratio of 67/33 exhibits the highest shielding effect in the range 760–860 MHz. It is suggested that sample A and B can be used to shield television, computers and similar equipment and also shield gadgets like cellular phones. The variations in the fabrics' shielding effectiveness can be contributed to the varied electrical property of the material depending upon the frequency. So the suggestion is to use the fabric at frequencies where higher attenuation can be obtained. The shielding effectiveness can be increased further by increasing the fabric's cover factor.

2.3.4 Industrial Importance

These copper core conductive yarns with very high value can be used to develop sensor wears and signal-transferring applications such as in defence, medical and technical textiles. Also, this conductive complex of core-spun yarn can be woven and knit into fabrics that can be used in antielectrostatic, electrostatic dissipating, body temperature control jackets, telecommunications and electromagnetic shielding wearable textiles. The process is creative and revolutionary, and has the potential to be commercialised.

2.4 Development of POF Fabric

In this research work, plastic optical fibre (POF) filaments of three different diameters, 0.25, 0.5 and 1 mm, were used to develop POF fabrics of six samples in different combinations to develop communication fabric, illuminated fabric and teleintimation fabric.

2.4.1 Development of Optical Core Conductive Yarn

To develop communication fabric using optical core conductive yarn, two types of POF yarns of 0.5 and 1 mm diameters were selected to produce two different samples. Two types of POFs are used as core materials, namely, ESKA and CK-20, which have diameters of 500 and 1000 μm, respectively, and 2/60s Ne cotton fibres are used as sheath materials for optical core conductive yarns. The Fehrer AG type DREF-3 friction spinning machine was used to manufacture the two different optical core conductive yarns. Four cotton carded slivers were fed in the first drafting unit and each sliver measured 4.22 g/m with sliver irregularity of 3.8% (U%), respectively. The core material (optical fibre) was fed in the first drafting unit with special guides. They were designed and installed in the first drafting unit to increase the conductive yarn during spinning's stability. To produce uniform yarn structure, the process parameters of the DREF-3 spinning machine such as perforated drum speed and yarn delivery rate were set at 4000 revolutions per min and 70 m/min, respectively. The POF filament of ESKA and CK-20 type of 1024 Tex and 281 Tex were used as core materials and cotton used as sheath material in the ratio of 80:20. Figure 2.18 represents the yarn formation

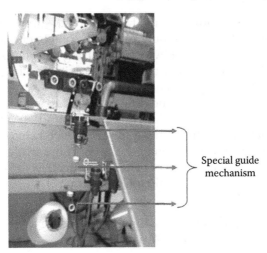

Special guide
mechanism

FIGURE 2.18
Special guide used in DREF-3 spinning machine.

FIGURE 2.19
Wrapping of sheath cotton fibres around POF at perforated drum in DREF-3 spinning machine.

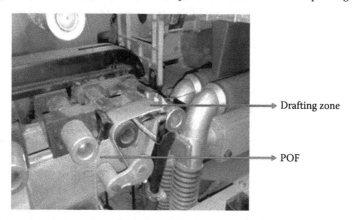

FIGURE 2.20
Drafting zone of DREF-3 friction spinning machine.

and special guide mechanism in the DREF-3 friction spinning system, and Figure 2.19 represents the wrapping of sheath component around the core material to produce core conductive yarns. During the spinning process, the optical fibre is fed from the first drafting unit with special guides. Special care is taken while wrapping the optical fibre by adjusting the perforated drum speed, yarn delivery rate and draft in the second drafting unit to mau-facture uniform yarn structure. The illustration of drafting zone and the spe-cial guide mechanism is shown in Figure 2.20, which helps to spin the yarn and maintain the uniform tension of the optical fibre while unwinding the material from the package. The copper core conductive yarns developed by the DREF-3 spinning system are shown in Figure 2.21.

2.4.2 Development of Communication Fabric Using Optical Core Conductive Yarn

The communication fabrics are developed using handloom with optical core conductive yarns as weft threads and 2/60s Ne cotton threads are used as warp

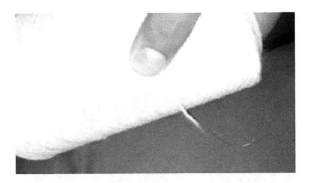

FIGURE 2.21
Copper core conductive yarn.

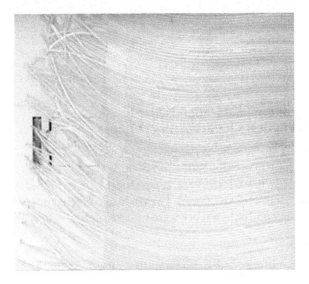

FIGURE 2.22
Sample I: communication fabric with warp as 2/60 s Ne cotton and weft as ESKA POF.

threads. During weaving, special care is taken while inserting the optical core conductive yarn as weft manually and the speed of the loom is also reduced. Figure 2.22 shows the sample developed for communication fabric with warp 2/60s Ne Cotton and weft ESKA POF. Figure 2.23 shows the sample developed for communication fabric with warp 2/60s Ne cotton and weft CK-20 POF.

2.4.3 Testing of Communication Fabric

To assess the physical characteristics and electrical properties of the developed communication optical core conductive fabric samples, various tests such as air permeability, air resistance, fabric thickness, aerial density and signal-transferring capability were conducted.

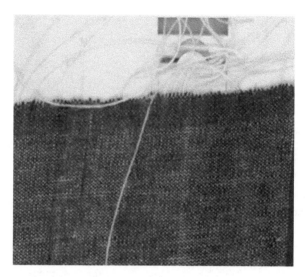

FIGURE 2.23
Sample II: communication fabric with warp as 2/60 s Ne cotton and weft as CK-20 POF.

2.4.4 Testing of Communication Fabric to Assess Physical Characteristics

Table 2.10 shows the results of the physical tests.

The aerial density of the ESKA type optical core conductive fabrics is 52% heavier than the CK-20 type ones due to their higher linear density (1028 Tex) and diameter (1000 μm). For wearable electronic garments, CK-20 type optical core conductive fabrics are more suitable as they have higher air

TABLE 2.10

Physical Characteristics of Communication Fabrics

Particulars	Sample I Communication Fabric (ESKA POF – Core Cotton – Sheath)	Sample II Communication Fabric (CK-20 POF – Core Cotton – Sheath)
Ends per inch		
Warp: 2/60s Ne Cotton	42	42
Picks per inch		
Weft: POF filament core and cotton sheath yarn	13	21
Aerial density (GSM)	564.2	269.4
Fabric cover factor (Kc)	20.07	18.2
Air permeability (cm³/cm² s)	311.375	259.84
Air resistance (kPa s/m)	0.04	0.048
Optical wire weight (g/m)	0.860	0.260
Fabric thickness (mm)	1.24 ± 0.20	0.74 ± 0.13

permeability due to the finer optical fibre (500 μm) and lesser fabric thickness (0.74 ± 0.13 mm) when compared to the thickness of the ESKA type optical core conductive fabrics (1.24 ± 0.20 mm).

2.4.5 Testing of Communication Fabric to Assess Signal-Transferring Capability

POF core conductive fabrics developed using two samples ESKA and CK-20 type were tested by giving different input LEDs for 15 cm length fabric. The light signal-transferring capabilities of ESKA and CK-20 type optical core conductive fabrics are analysed by passing the input LED source at one tip and receive the signal at another tip. If the fibre was damaged during the weaving process, these light signals would be interrupted. The fabric was tested using a test bench with the optical transmitter, which contains light source for transmitting light to the receiver. Power supply for the transmitter circuit is +5 V and it is conveyed to the LED through a limiting resistor. The optical receiver consists of photodiodes to receive signals from the POF. Light source is used at one end of the optical fibre and a photo-diode is used at the other end. When there is a light illumination on the photo-diode, the output of the photo-diode produces 0 V and when there is an illumination, it shows 5 V output in the display unit. Table 2.11 shows test results of the signal-transferring capability of communication fabrics.

It was found that the signal loss in terms of light intensity was higher due to the stress to the optical fibre during the processes of spinning and fabrication. It is evident from Table 2.11 that the signal loss is only 24% in both the communication fabric developed using CK-20 and ESKA POF, when LASER light is used as the input light source. But the percentage of signal loss is higher when red LED is used as the input light source. This is because red LED has the minimum Lux value due to the colour's maximum wavelength. The conclusion derived from the test results is that the communication fabric can be used to manufacture teleintimation garment and in applications where signal transferring is essential.

TABLE 2.11

Test Results of Signal-Transferring Capability of Communication Fabrics

Input Light Source	Input Voltage (V)	CK-20 POF		ESKA POF	
		Output Voltage (V)	Signal Loss (%)	Output Voltage (V)	Signal Loss (%)
Red	5.0	3.1	38	2.9	42
Infrared	5.0	3.1	38	3.2	36
White	5.0	3.6	28	3.5	30
LASER	5.0	3.8	24	3.8	24

2.5 Development of POF Yarn

In this work, the illuminated POF fabric has been developed using poly (methyl methacrylate) (PMMA) filament. The PMMA POF filament of 1040 kgf/cm^2 tensile strength and 970 kgf/cm^2 yield strength has been used in the weft that has a 0.5-mm filament size and 2/80 Ne polyester spun yarn is used in the warp. The end emitting material was converted into side emitting one by providing special surface treatment of 600 grid abrasive papers.

2.5.1 Development of Illuminated Fabric

Since the speed of the power loom does not match to insert PMMA POF filament due to lower elongation of POF filament and slipping nature, handloom has been selected to weave as per the required suitable slow speed. Particulars of illuminated fabric sample's loom type and quality developed using pit loom and PMMA POF filament are given in Table 2.12. The development process of the POF fabric sample is shown in Figure 2.24, and Figure 2.25 shows development of the POF fabric using sateen weave. For better intensity of light in the illuminated fabric, weft should be predominant in PMMA POF filament; this is known as weft float. To achieve this, 5 × 1 sateen and pointed twill weaves were selected to develop the illuminated fabrics.

2.5.2 Characteristics Analysis of PMMA POF Filament

Breaking force (gf), breaking elongation (%) and tenacity (RKM) of 0.5 mm PMMA POF filament have been tested using Premier Tensomaxx 700 V 2.5 instrument to assess the weaving capability of the POF filament. Figures 2.26 through 2.28 show the breaking force, breaking elongation and tenacity values of 0.5 mm POF filament, respectively. It clearly indicates that the average breaking force value is 609 gf, which enables the weaving of the material with special care; this should be considered as this elongation percentage is lesser, and a moderate RKM value floating weave will be better for less weft bending angle in the weave geometry.

TABLE 2.12

Loom Specification and Quality Particulars of Illuminated Fabric

Parameter	Sateen Weave	Twill Weave
Type of loom	Pit loom	Pit loom
Weave design	5 × 1	Pointed
Loom width	24 in.	24 in.
Jacquard hook capacity	—	120
Total ends	—	370
Warp material	2/80 Ne polyester	2/6 Ne cotton

FIGURE 2.24
Fabric sample development of twill using PMMA.

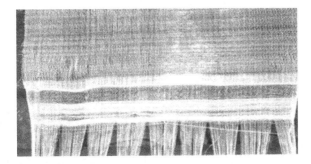

FIGURE 2.25
Fabric sample development of sateen using PMMA.

The material's tenacity is expressed in terms of its breaking force; to weave any textile material, it should have sufficient breaking force and capacity to elongate to withstand the stress and strain in the weaving process.

2.5.3 Influence of Weave Geometry on Illumination

The PMMA POF's illumination was tested using the Lux meter. The light source used for the measurement is carried out using six different LED

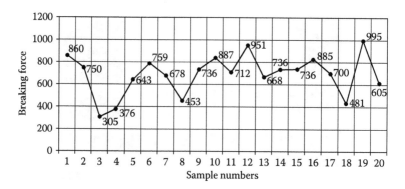

FIGURE 2.26
Breaking force of PMMA.

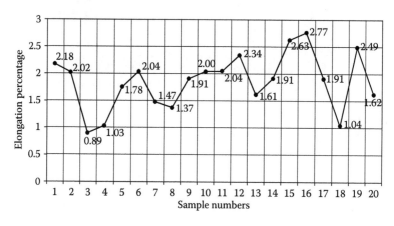

FIGURE 2.27
Breaking elongation of PMMA.

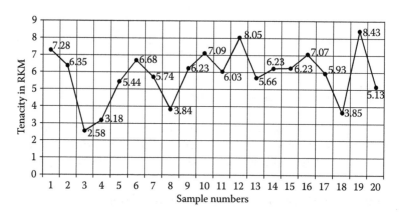

FIGURE 2.28
Tenacity of PMMA.

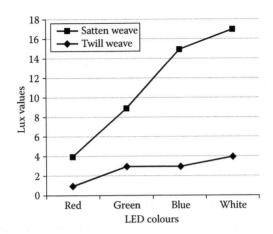

FIGURE 2.29
Lux value of sateen and twill weave using different colour LED.

colours for sateen weave and twill weave as mentioned in Figure 2.29. The pointed twill illuminated fabric sample shows a very low illumination effect because the weft bending angle is more due to the higher warp and weft interlacement. The illumination effect is better in sateen-woven illuminated fabrics compared to pointed twill weave-illuminated fabrics. The different colours of the LED also influence the illumination effect on fabric system. The wavelength of the different colours influences the illumination. Red LED has the minimum Lux value as the colour red has the maximum wavelength.

2.5.4 Influence of LED Colours on Power Consumption in Illuminated Fabric Samples

The PMMA POF-illuminated fabric is connected with 3 V batteries, and to estimate the life of the battery, the LEDs' power requirement was calculated. When compared to red, green and blue LEDs, white required more power because its illumination effect is more. Since the power consumption is in mW, the influence of the LED colour on power consumption in PMMA POF-illuminated fabric samples was noticed to be insignificant. The power consumption by different LEDs is given in Table 2.13.

TABLE 2.13

Power Consumption of LED

LED Colour	Power Consumption (mW)	
	Twill Fabric	Sateen Fabric
Red	5	3
Green	8	5
Blue	14	12
White	24	13

The sateen weave shows better illumination when compared to twill fabrics; the structure of sateen weave has more weft float. Since the weft bending angle is less, it causes more illumination when compared to twill fabrics. The illumination effect on the fabric system is also influenced by the different colours of LEDs; the wavelength of different colours influences the illumination. The colour red LED has the minimum Lux value, which is due to the maximum wavelength of red colour.

2.6 Development of Yarn for Teleintimation Fabric

In this research work, three different POF diametres such as 0.25, 0.5 and 1.0 mm have been chosen to develop teleintimation fabric samples, as shown in Table 2.14. These optical fibres are changed into the fabric form by using three methods, namely, (a) sequential work with POF integrated in the fabric, (b) weaving of the POF using power loom and (c) weaving the POF using handloom. The teleintimation fabric is developed to produce teleintimation garments, which are capable of quickly locating a wounded soldier, triage, monitor vital signs and relevant physiological information and display their status locally or remotely.

2.6.1 Development of Teleintimation Fabric Using the Sequential Technique

It is a method of integrating POF with fabrics by using the embroidery technique. For this, cotton fabric of 2/40s Ne warp and 40s Ne weft woven fabric of 140 GSM were selected. The POF is placed over the fabric in the required matrix format and stitched up to hold the POF on the fabric. The embedding/

TABLE 2.14

Teleintimation Fabric – POF Samples

Name of the POF Fabric	Plastic Optical Fibre (POF) (mm)			Warp		Weft	
	0.25	0.5	1.00	EPI	Material	PPI	Material
POF integrated sequential fabric	√			Integration POF in X–Y direction on the fabric			
Power loom fabric		√	√	24 EPI with 2 EPI of POF	POF and 2/20s Ne cotton	33 PPI with alternate POF	2/20s Ne cotton
Handloom fabric		√	√	24 EPI with 3 EPI of POF	POF and 2/20s Ne cotton	32 PPI with alternate POF	2/20s Ne cotton

integration of POF on the fabric depends on the binding of threads and number of binding points. The major problem faced in embroidery work with POF is withdrawal of POF threads and fibre damage during integration on the fabric and also less serviceability due to hand insertion. Figure 2.30 shows the fabric developed using the embroidery technique with POF arranged in the X–Y direction on the fabric. Figure 2.31 shows the fabric developed using embroidery with POF arranged in the X–Y direction with loops on the fabric.

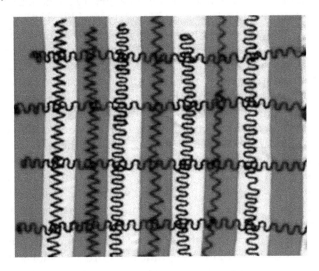

FIGURE 2.30
Integration POF in X–Y direction on the fabric.

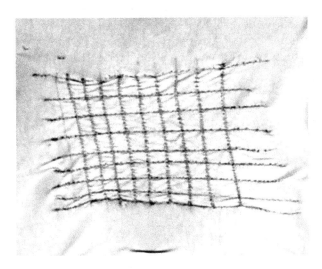

FIGURE 2.31
Integration POF in X–Y direction with loop on the fabric.

2.6.2 Development of Teleintimation Fabric by Power Loom

The teleintimation fabric was created using a power loom; thePOF filament is integrated with the cotton yarn in both warp and weft directions using a rigid rapier loom to produce teleintimation fabrics. The POF fabric developed using rapier looms is in matrix format of the POF pattern. Since the maximum bullet size used in the military is 7.62 mm, the matrix should have the maximum pixel of 5 mm in diameter. The quality particulars of POF fabrics developed using rapier looms are 54 in. fabric width, 32 picks per inch in which every alternative picks are inserted using POF filament with 24 ends per inch, which has 2 ends per inch POF filament and 2/20s Ne cotton. The weave pattern selected for the POF fabric is 1 up and 1 down plain weave and loom running speed of 170 picks per min using Picanal PGW type rapier looms. The major problems faced during weaving of POF fabrics are warping of plastic optical fibre on the beam due to lack of flexibility, mounting the POF beam, drawing and denting of the POF fibre, separation of the warp ends, slough-off of the POF from the spool kept for the weft insertion, slippage of the rapier while picking in higher and as well as lower speed, rubbing of the warp ends while the rapier is in motion, which causes damage to the outer cladding surface of the POF. The warping of POF on the warper's beam is carried out manually by maintaining uniform tension to all POF threads in the warp direction. The beam is properly marked to accommodate the POF with the cotton yarn in the fabric structure, which depends upon the number of wrap ends and POF integration. Figure 2.32 shows the marking of the warp beam and installation of POF in the warp beam, respectively. While placing the POF wounded beam in the normal beam position, loose and slackness of POF threads in warp direction was noticed, so to avoid these problems, the mounting of the beam was done with a little height from the normal beam position which is shown in Figure 2.33a and b respectively.

The tension in the cotton yarn and the one in the POF are entirely different and hence, the drawing of the ends turned out to be a time-consuming process.

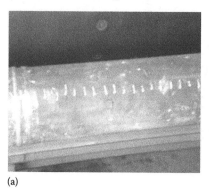 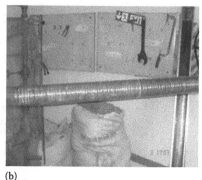

(a)   (b)

FIGURE 2.32
(a) Marking of warp beam and (b) POF wounded on warp beam.

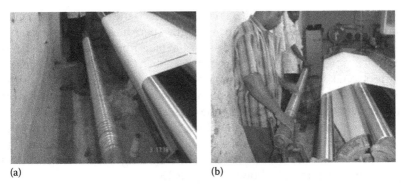

(a)   (b)

FIGURE 2.33
(a) Mounting of the POF beam and (b) mounting of the beam with a height.

There was a chance of the fibre breaking while knotting with the cotton ends. The drawing and denting of POF in rapier looms and control of warp thread tension of both cotton and POF threads are shown in Figure 2.34a through c. One of the major problems with the POF is the fibre sloughing off from the package during weft insertion. Due to this slough-off, the weft insertion is not proper and an arrangement of steel rods is made to avoid this problem as shown in Figure 2.35. Slippage of the POF in the rapier during picking is another major

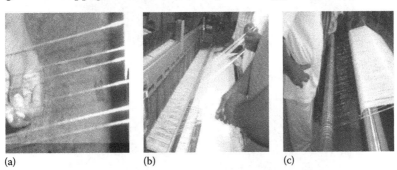

(a)   (b)   (c)

FIGURE 2.34
(a) Drawing of POF, (b) denting of POF and (c) separations of POF warp ends.

FIGURE 2.35
(a) Slough-off of POF in package, (b) steel rod arrangement and (c) unwinding of POF for weft insertion.

problem. While picking, the rapier is unable to catch the POF fibre at the middle of the race board because it slips. To avoid this, there have been numerous trial-and-error processes. Finally, slippage was avoided by making the bend part of the taker rough by means of abrasion as shown in Figure 2.36. One of the major problems is the rubbing of the warp ends resulting from the movement of the rapier. The POF is also present in the warp direction; the movement of the rapier over the warp ends cause the cladding (outer area of the POF) damage, thereby affecting the POF's signalling properties. The POF's signal-transmitting property in weft direction is also affected following the rubbing of the rapier. The teleintimation fabric developed using a rigid rapier loom with POF in specific matrix format is shown in Figure 2.37.

FIGURE 2.36
Abrasion of the rapier using lathe machine.

FIGURE 2.37
POF integrated fabric using power loom.

2.6.3 Development of Teleintimation Fabric Using Handloom

POFs are inserted in the warp and weft direction to produce the desired matrix to detect the signal transmission characteristics using handloom. The quality particulars of POF fabrics developed using handloom are 22 in. fabric width, 32 picks per inch in which every alternative pick is inserted using POF filament with 24 ends per inch, which has 3 ends per inch POF filament and 2/20s Ne cotton. The 1 up and 1 down plain weave pattern is selected for this fabric. A few modifications, namely, warping, drawing and denting of the POF, separation of the warp ends, winding of the POF on the shuttle for pick insertion processes are needed in the preparatory process to weave POF fabrics. Warping of POF on the warper's beam is done by allowing the fibre manually with equal tension. The beam is marked properly to accommodate the POF with the cotton yarn in the fabric depending upon the number of warp ends and the winding of the POF on the warp beam. In order to avoid the tension variation between the POF and cotton warp ends, the drawing and denting of the warp ends are carried out manually, because the width of the loom is small compared to power looms and rapier looms.

Figure 2.38a shows the drawing and denting of the POF along with cotton threads in the handloom. Properly held shaft arrangements are made for the warp ends to accommodate the weaving process along with the cotton yarn so that the tension in the POF remains the same till the weaving ends. The weave pattern is plain weave (1/1) and the POF is inserted for every alternate pick in the weft path. For easy insertion of the pick, the POF is wound in the shuttle with appropriate tension. The winding is done manually and is wound with optimum tension to avoid the POF from sloughing off. Figure 2.38b shows the POF wound in the shuttle. The winding of cotton and POF is done manually with the help of manual winding machines. The weaving of the POF with cotton yarn in both warp and weft directions in the handloom started with the earlier arrangements. The shed opening for

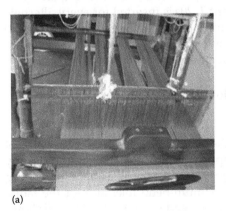
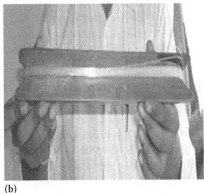

(a) (b)

FIGURE 2.38
(a) Drawing and denting of POF and (b) winding of POF in shuttle.

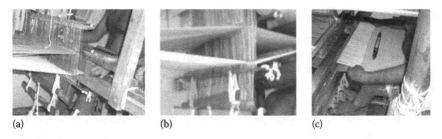

(a)   (b)   (c)

FIGURE 2.39
(a) Weaving of POF fabric (b) shed opening and (c) pick insertion.

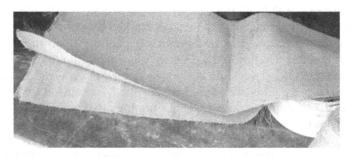

FIGURE 2.40
POF integrated fabric using handloom.

the pick insertion is carried out easily by pressing the paddle at the bottom of the loom as shown in Figure 2.39a. The pick insertion is done with alternate POF and cotton yarn following the count 2/20s Ne. During the shed opening, the first colour cotton yarn will be up and the next colour will be down and vice versa. When the shed is open, the POF is inserted for one pick and the cotton yarn is inserted for the next pick, respectively. The pick insertion is done manually in handlooms with the help of the shuttle wound with cotton yarn and POF. The cotton yarn of count 2/20s is inserted for every alternate pick with POF and fabric width of 22 in. is produced. The shed opening and pick insertion in the handloom are shown in Figure 2.39b and c, respectively. Figure 2.40 represents the teleintimation fabric developed by handloom.

2.6.4 Physical Characteristics of Teleintimation Fabric

2.6.4.1 Bending Rigidity

Bending rigidity measured with KES-FB2 bending tester is a measure of the force required to bend the fabric approximately at 150°. Bending rigidity per unit fabric width value indicates greater stiffness/resistance to bending motions. It is calculated to test the maximum bend tolerance for the fabric with POF to transmit light energy from one end to another. Table 2.15 shows the bending length of the POF measured in POF fabrics in which the arrangement of the POF is different in each fabric.

TABLE 2.15

Testing of POF Bending in Warp and Weft

POF Fabric	Bending Length (cm)
Warp way	7.3
Weft way	5.2

From Table 2.15, it has been observed that the bending tolerances for the POF located in the warp way and weft way are 7.3 and 5.2 cm, respectively. Beyond the bending length, the attenuation of the light signals will increase resulting in loss of signals.

2.6.4.2 Tensile Strength

The tensile test carried out on the KES-FB1 Tensile-Shear Tester measures the stress/strain parameters at the maximum load set for the material being tested. Tensile resilience percent indicates the recovery of deformation from strain, or the ability to recover from stretching when the applied force is removed. Higher values indicate greater recovery from having been stretched. In shear testing, the KES-FB1 Tensile-Shear Tester applies opposing, parallel forces to the fabric till a maximum offset angle of 8° is reached. A pretension load of 10 gf/cm is applied to the specimen. Shearing stiffness is the ease with which the fibres slide against each other, resulting in soft/pliable to stiff/rigid structures. Lower values indicate less resistance to the shearing movement corresponding to a softer material having better drape. It measures the amount of stress applied to a material at its breaking point or the point at which it fails. From Table 2.16, it is noticed that the tensile strength is more in the warp way than in the weft way in both the POF integrated fabric. In case of the khaki fabric (handloom), the arrangement of the POF is in such a way that for every half centimetre, one POF is integrated in both warp and the weft way. Since the arrangement of the cotton yarn and the POF is compact, the strength is more; on the other hand, the blue fabric (rapier loom) is integrated with POF for every one centimetre of the cotton yarn both in the warp and weft way. The cotton yarn of count 2/20s is chosen for both the fabrics to avoid the slippage of the POF during and after weaving of the fabric.

TABLE 2.16

Testing POF Tensile Strength

POF Fabric	POF Fabric	Tensile Strength (lb)
Handloom fabric	Warp way	180
	Weft way	70
Power loom fabric	Warp way	140
	Weft way	45

2.6.4.3 Air Permeability

Air permeability is tested with KES F8 AP1 and can be used to provide an indication of the breathability of weather-resistant and rainproof fabrics, or of coated fabrics in general, and to detect changes during the manufacturing process. The air permeability of a textile fabric is determined by the rate of airflow through a material under a differential pressure between the two fabric surfaces. The prescribed pressure differential is 10 mm of water. The air permeability of a fabric is influenced by several factors: the type of fabric structure, the weave design, the number of warp and weft yarns per centimetre (or inch), the amount of twists in yarns, the size of the yarns and the type of yarn structure. Usually in case of POF fabrics, the air resistance value is measured in terms of kPa cm^2 for both the khaki and blue POF fabrics and the values are given in Table 2.17.

The signal-transferring test results obtained using the signal-transferring unit, which is designed for optical fibres to measure the light-transferring efficiency at various light sources (LED) using various integrated POF fabrics made out of sequential, handloom and power loom have been reported. Four different techniques such as sequential work, handloom, power loom and core conductive fabrics have produced the selected optical fibres to develop the signal-transferring fabrics. The signal-transferring loss of these fabrics has been analysed using a microprocessor with an input voltage of +5 V. The light signal-transferring efficiency of these integrated POF fabrics has been reported. Test results show that the handloom POF integrated woven fabrics have 77.6% signal-transmission efficiency, which is higher than the sequential (60.16%) and power loom fabrics (62.56%) due to lower stress and mechanical fatigue during the fabrication processes. The study shows that the loss in signal transferring of handloom fabrics is less than the other fabrics. The optical core conductive fabrics were produced using the optical core conductive yarns. The physical characteristics of these conductive fabrics are studied.

Through this research work, the light signal-transferring unit for optical fibres has been developed with 02N5777 optical receiver and microcontroller AT89C52 and their performance is analysed using POF with various lengths and diameters. The developed light signal-transferring unit test results are comparable with an optical time domain reflectometer, which is used to obtain the particular characteristics of optical fibres, and also develop a test kit in which microcontroller-based circuit installation and testing charges are lower than the optical time domain reflectometer.

TABLE 2.17

Testing Air Resistance of POF Fabric

POF Fabric	Air Resistance (kPa/cm^2)
Handloom fabric	44
Power loom fabric	18

2.6.5 Testing of POF Filament for Teleintimation Fabric

For teleintimation garments, POFs of three different diameters 0.25, 0.5 and 1 mm were selected. This fibre was tested for signal-transferring efficiency and checked for possible integration of the fibres with the conventional fabric-manufacturing methods. The light transfer distance in the POF was tested with different types of input light sources like white, red, infrared and laser light LED, and they were selected with different levels of light intensity. The hardware setup was designed to transfer the light at one end and receive the light at other end of the POF. Figure 2.41 shows the block diagram of the POF signal transmission test circuit and Figure 2.42 shows the circuit diagram. The lumen level of the light is varied by using adjustable input voltage. The received light signal from the phototransistor (PT) and the variation of the PT signal is amplified. The amplified signals are converted into the digital signal using the analog to digital converter. The digital signal is given to the microcontroller, which has been programmed to display the result in terms of voltage. The signal-transferring test results of the teleintimation fabric of 0.5 and 1.0 mm diameter measure 2 and 10 ft in length at input voltage 4.5 and 6 V, respectively, as shown (Figure 2.43). Figure 2.44 represents the photographs of the POF test carried out at various light sources and photodiodes. POF measuring 0.25 mm was easy to weave, but the light transfer capacity was very low and fixing of sensors/transducers in this fibre was not easy. POF measuring 1 mm was not suitable for weaving as the POF's thickness was very large resulting in less comfort and the fabric less flexible; in its electrical properties, the signal transmission rate and the feedback were good compared to POFs having 0.25 and 0.5 mm diameters. POF measuring 0.5 mm is appropriate for weaving compared to the 1 mm POF, and the

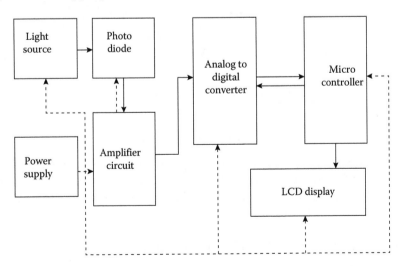

FIGURE 2.41
Block diagram for signal transmission circuit.

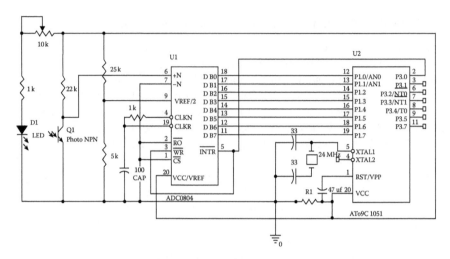

FIGURE 2.42
Circuit diagram of signal transmission testing circuit.

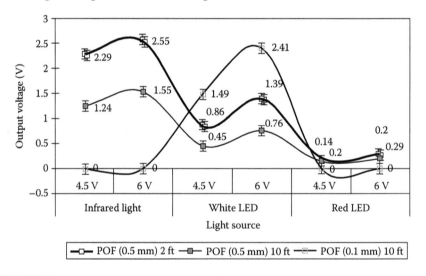

FIGURE 2.43
Signal-transferring test results of POF at various light sources.

signal transmission and light transfer capacity are suitable when compared to 0.25 mm POF. Also 0.5 mm POF has better flexibility and minimum signal loss at bending conditions and excellent signal transmission property, which is evident from the following test results.

2.6.6 Electrical Testing of Teleintimation Fabric

Sequential fabrics, handloom fabrics and woven fabrics were tested with the optical transmitter, which contains a light source to transmit light to a receiver.

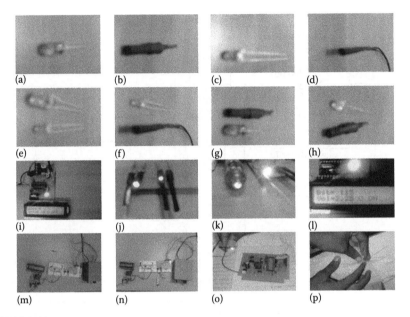

FIGURE 2.44

Various accessories used in testing of teleintimation fabrics: (a) red LED; (b) red LED covered to fix POF; (c) Photodiode; (d) photodiode with rubber covered; (e) white light LED; (f) photodiode to fix POF; (g) red LED with and without cover; (h) photodiode with and without cover; (i) POF light-testing circuit; (j) light source and receiver; (k and l) testing with different light sources and their corresponding O/P; (m and n) POF testing circuits set-up and (o and p) fabric testing with optical transmitter and receiver.

Different light sources like red LED and white LED are used for testing here. The optical transmitter contains a light source to transmit light to a receiver. The very bright white LEDs are chosen in this case as it consumes low power and is of optimised size. Power supply for the transmitter circuit is +5 V and it is given to the LED through a limiting resistor.

2.6.6.1 Testing of POF with IRF14F Optical Receiver

The IRF14F optical receiver consists of photodiodes to receive signals from the POF. A light source is used at one end of the optical fibre and a photo-diode is used at its other end. When there is a light illumination on the photo-diode, the output of the photo-diode produces 0 V and when there is an illumination, it shows 5 V output in the display unit. The circuit diagram to test the optical receiver IRF14F is shown in Figure 2.45. From the previous circuit diagram, it is clear that when there is light illumination on the photo-diode, the output of the photo-diode produces 0 V output. Also when there is no light illumination on the photo-diode, the output of the photo-diode produces 5 V output. Also it is highly sensitive to small variations in the light illumination. The components used in the circuit are: transmitter – rED LED (1 mm); decoder – 74LS151; receiver – photodiode (3 mm high sensitive); microcontroller – AT 89C52; transmission – RF; light source – high

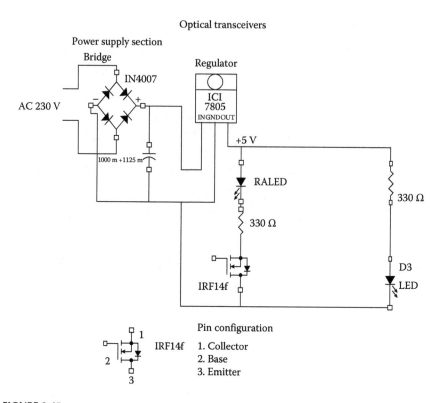

FIGURE 2.45
Testing circuit diagram of optical receiver IRF14F.

bright white LEDs; light receiver – IRF14F – optical. The circuit diagram shown in Figure 2.46 uses an optical receiver of 2N5777. The optical receiver is tested for the light reception characteristics with the POF. The receiver's response to the light illumination is found to be minimum when compared with the light response of the optical receiver IRF14F. The optical receiver IRF14F is highly sensitive to the light illumination. Hence, the final decision has been made to use IRF14F as an optical receiver. AT 89C52 microcontroller is used to test the fabric and 74LS151 decoder is used to decode the signals. The block diagram of the test bench is as shown in Figure 2.47.

2.6.7 Signal-Transferring Efficiency of Teleintimation Fabrics

The signal-transferring efficiency of the handloom, power loom and sequential integrated woven POF fabrics are analysed using the microprocessor ATMEGA 89C51 designed with IRF14F optical receiver as shown in Figure 2.44m and l. Figure 2.48 represents the signal received in terms of output voltage for the various fabrics with various optical fibre lengths such as 15, 20 and 30 cm using the red LED, white light LED and laser light LED with input voltage of +5 V. From the signal values measured in various places

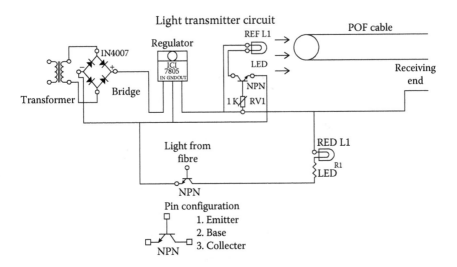

FIGURE 2.46
Optical testing with 2N5777 optical receiver.

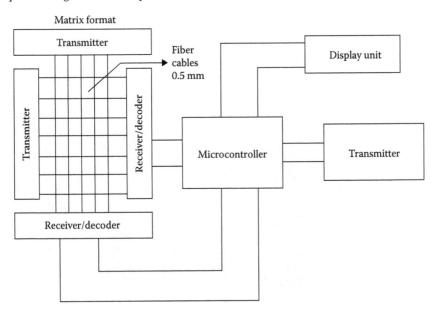

FIGURE 2.47
Block diagram for testing the POF fabric.

of the fabrics at 15, 20 and 30 cm optical lengths are made into cumulative average and their signal output voltages are 3.13, 3.88 and 3.01 V for power loom, handloom and sequential integrated POF fabrics. Figure 2.49 represents the signal-transferring loss percentage of 0.5 mm POF at 2 and 10 ft. fibre length with infrared light, red light and white light for various input voltages

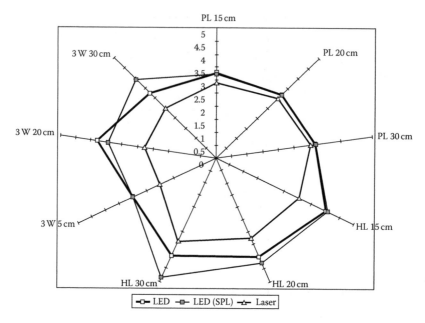

FIGURE 2.48
Signal-transferring behaviour of sequential work fabric (SW), handloom fabric (HL) and power loom fabric (PL) integrated with POF at various light sources.

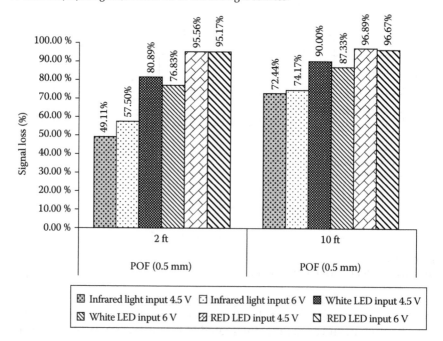

FIGURE 2.49
Signal loss (%) of POF measured at various light sources for 2 and 10 ft length.

of 4.5 and 6 V, respectively. From this signal-transferring analysis, the signal loss percentages for power loom, handloom and sequential integrated POF fabrics are 37.44%, 22.40% and 39.84%, respectively. It is noticed that the signal-transferring loss percentage of handloom fabrics is less when compared to other fabrics.

2.6.8 SEM Analysis of Teleintimation Fabric

The reason for loss of signal-transferring efficiency of these POF woven fabrics are analysed using scanning electron microscopy (SEM). The study shows fracture mode and surface damage of optical fibres have been noticed in the signal-transferring fabrics, which are shown in Figure 2.50. These optical fibres are subjected to mechanical stress during weaving and fabrication processes such as warp preparation and weft preparation under stress conditions (bending). These damages to the fibres are the main reasons for the loss of signal transferring in the fabrics. In the case of handloom POF fabrics, the signal loss is noticed less when compared to power loom and sequential integration of optical fibres because weft optical fibres are inserted manually and mechanical stress is at a minimum.

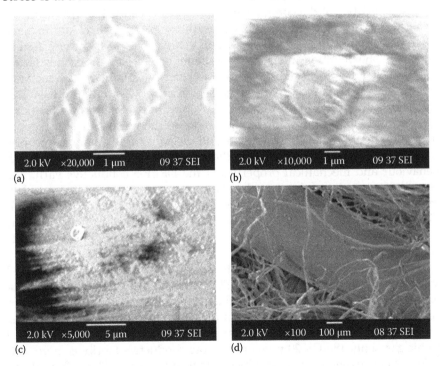

FIGURE 2.50
SEM photographs of POF (a) before weaving, (b) after weaving, (c) fibre at bend condition (maximum stress) and (d) POF with fabric (interlacement).

2.7 Summary

In this chapter, different production techniques of yarn and fabric have been discussed. Cotton-wrapped nichrome yarn is produced using a braiding machine to develop nichrome wire-embedded cotton fabric in knitted fabric for heating garments. Two patterns were made for developing heating garment, out of these, sleeve round neck shape T-shirt pattern is taking high resistance when compared to sleeveless round neck shape T-shirt pattern, because length of the nichrome yarn used in sleeve pattern is higher. As a result, power consumption is higher to produce patterns with sleeves compared to sleeveless patterns. The fabric's comfort has been tested and the insulation (Clo) values ranged 0.625–1.064, which is ideally suitable for extremely cold regions. Following the test results, the sleeveless round neck T-shirt pattern was selected to develop heating garments. The conductive yarn that contained copper wire as its core was fabricated by using a special guide mechanism on the DREF-3 spinning system. Three different core-sheath ratios of the DREF-3 conductive yarns such as 67/33, 80/20 and 90/10 were compared by varying the draft in the second drafting unit, and it was found to have an average parent yarn linear density of 328 Tex. The physical characteristics of the conductive yarn were studied using different tests and it was found that the 67/33 core/sheath conductive yarn had the highest tenacity of 3.27 cN/Tex and elongation to break 5.27% when compared to other core-sheath ratios; it was due to its better core-sheath interaction factor, CSI_T – 21.22%. It was also observed that the core-sheath interaction factor of conductive yarn sample C (90/10 core-sheath ratio) was found to have less core-sheath interaction factor CSI_T – 9.21%, and it exhibited lower breaking tenacity and elongation to break; it was due to the behaviour of its copper filament. The frictional characteristics and the core-sheath components' percentage were the deciding factors in the interaction between the core-sheath behaviour and mechanical properties of the DREF-3 conductive yarn. The electrical properties of these conductive core spun yarns exhibited very low resistance (3–28 MΩ) at 6, 12 and 24 V applied voltage. The conductive complex core spun yarn can be woven and knit into fabrics that can be used as communication garment for measuring body temperature and for mobile charging applications.

The optical core conductive yarn has been developed using the DREF-3 spinning system using two different POF yarns, namely, ESKA and CK-20 of 0.5 and 1 mm, respectively. The aerial density of the ESKA type optical core conductive fabrics is 52% heavier than the CK-20 type ones due to its higher linear density (1028 Tex) and diameter, which is 1000 µm. For wearable electronic garments, the CK-20 type optical core conductive fabrics is more suitable as it has higher air permeability due to its finer optical fibre (500 µm) and thinner (0.74 ± 0.13 mm) fabric when compared to the thickness of ESKA type optical core conductive fabrics (1.24 ± 0.20 mm). It was found that the signal loss in terms of light intensity was higher due to the stress to the optical fibre during spinning and the fabrication process. It is evident from Table 2.11 that the signal loss is only 24% in both communication fabrics developed using CK-20 and ESKA POF when LASER light is used as an input light source. But the percentage of signal loss is higher when red LED is used as the input light source. This is because red LED has the minimum Lux value due to the colour's maximum wavelength. From the test results, it has been concluded that communication fabrics can be used for manufacturing teleintimation garments and for applications where signal transferring is essential. The illuminated fabric has been developed with two different types of weaves, namely, twill and sateen weave using handloom with 0.5 mm PMMA POF. The pointed twill illuminated fabric sample shows very low illumination effect because of the more weft bending angle due to the higher warp and weft interlacement. The sateen-woven illuminated fabrics show better illumination than the pointed twill weave illuminated fabrics because the structure of sateen weave has more weft float and the weft bending angle is less that causes more illumination when compared to twill fabric. The illumination effect on the fabric system is also influenced by the different colours of LEDs; the wavelength of different colours influences the illumination. The colour red LED has the minimum Lux value, which is due to the maximum wavelength of red. Following the test results, the sateen weave fabric has been selected to develop illuminated garments.

The teleintimation fabric has been developed using POF of 0.25, 0.5 and 1 mm diameters. The fabric was developed by the sequential method, handloom and power loom. Teleintimation fabric developed using handloom was considered to be suitable for commercial applications. Though it is also possible to develop using power loom, major modifications have to be carried out and it is not possible to integrate the POF yarn using sequential method, since the POF yarn comes out very quickly when the POF-integrated fabric was made into garment. It is easy to weave 0.25 mm POF, but its light transfer capacity is very low and fixing the sensors/transducers in this fibre was hard. The 1 mm POF is not suitable for weaving as the thickness of the POF is very large, having very less comfort, and the fabric is less flexible; meanwhile, its signal transmission rate and feedback are good compared to the POFs with diameters 0.25 and 0.5 mm. The 0.5 mm POF is suitable for weaving compared to the 1 mm POF, and its signal transmission and light transfer capacity are more suitable than the 0.25 mm POF. Following the tests, it was concluded that the 0.5 mm POF has better flexibility and minimum signal loss at bending conditions besides excellent signal transmission property. The teleintimation fabric's physical characteristics have been tested to assess its properties. It has been noticed that the tensile strength is more in the warp way than in the weft way in both handloom and power loom POF-integrated fabrics. The cotton yarn of count 2/20s is chosen for both the fabric to avoid the slippage of the POF during and after weaving of the fabric, bending rigidity measured with KES-FB2 Bending Tester, the bending length of the POF measured in POF fabrics in which the arrangement of the POF is different in each fabric. From the tests, it was observed that the bending tolerance for the POF located in the warp way and weft way was 7.3 and 5.2 cm, respectively. Beyond the bending length, the attenuation of the light signals will increase, resulting in loss of signals. Air permeability was tested with KES F8 AP1 and it can be used to provide an indication of the breathability of weather-resistant and rainproof fabrics, or of coated fabrics in general, and to detect changes during the manufacturing process. The air permeability of khaki fabric is 44 kPa/cm^2 and power loom fabric is 18 kPa/cm^2.

A tensile test was done on the KES-FB1 Tensile-Shear Tester and it was noticed that the tensile strength was higher in the warp way than in the weft way in both the POF-integrated fabrics. In case of the khaki fabric (handloom), the POF was arranged in such a way that for every half centimetre, one POF was integrated in both the warp and weft way. Since the arrangement of the cotton yarn and the POF is compact, the strength is more, while on the other hand, the blue fabric (rapier loom) is integrated with the POF for every one centimetre of the cotton yarn both in the warp and weft way. The cotton yarn of count 2/20s is chosen for both fabrics to avoid the slippage of the POF during and after the fabric's weaving. The signal-transferring efficiency of the handloom, power loom and sequential-integrated woven POF fabrics have been analysed, and the signal received in terms of output voltage for the various fabrics with varied optical fibre lengths such as 15, 20 and 30 cm using red LED, white light LED and laser light LED with input voltage is +5 V. From the signal values measured in various places of fabrics at 15, 20 and 30 cm, optical lengths are made into cumulative average and their signal output voltages are 3.13, 3.88 and 3.01 V for power loom, handloom and sequential-integrated POF fabrics. From the signal-transferring analysis, the signal loss percentages for power loom, handloom and sequential-integrated POF fabrics are 37.44%, 22.40% and 39.84%, respectively. It is noticed that the signal-transferring loss percentage of handloom fabrics is less when compared to the other fabrics.

Bibliography

Amon, A., Wearable multiparameter medical monitoring and alert system, *IEEE Trans. Inform. Technol. Biomed.*, 8(4), 415–427 (2001).

Ashokkumar, L. and A. Venkatachalam, Electrotextiles: Concepts and challenges, In *National Conference on Functional Textiles and Apparels*, Coimbatore, India, Vol. 2(3), pp. 75–84 (2007).

Aydogmus, Y. and H. Behery, Spinning limits of the friction spinning system (Dref III), *Textile Res. J.*, 69, 925–930 (1999).

Balasubramanian, A., Friction spinning – A critical review, *Indian J. Fibre Text. Res.*, 17, 246–251 (1992).

Beith, M., Wearable wires, *Newsweek Int.*, 7(1), 23–29 (2003).

Blanco, A., Wearable, breathable, water-repellent PUR, *Plastic Eng.*, 58(6), 45–62 (2002).

Bonderover, E., S. Wagner and Z. Suo, Amorphous silicon thin film transistors on kapton fibers, *Proc. Mater. Res. Soc. Symp.*, 1, 736, 109–114 (2003).

Buechley, L. and M. Eisenberg, Fabric PCBs, electronic sequins, and socket buttons techniques for e-textile craft, *Pers. Ubiquit. Comput.*, 69(3), 14–19 (2007).

Chattopadadhyay, R. and S. Banerjee, The frictional behaviour of ring, rotor, and friction-Spun Yarn, *J. Text. Inst.*, 88(3), 59–66 (1997).

Chattopadhyay, R., K.R. Salhotra and S. Dhamija, Influence of core-sheath ratio and core type on Dref-III friction spun core yarns, *Indian J. Fibre Text. Res.*, 25, 256–263 (2000).

Cheng, K.B., S. Ramakrishna and T.H. Ueng, Electromagnetic shielding effectiveness of stainless steel/polyester woven fabrics, *Text. Res. J.*, 71(1), 42–49 (2001).

Cheng, K.B., T.H. Ueng and G. Dixon, Electrostatic discharge properties of stainless steel/polyester woven fabrics, *Text. Res. J.*, 71(8), 732–738 (2001).

Chintz, R., C.C.N. Elsassar, T. Mac, O. Maesschke, J. Meyer, H. Paschen, H. Phoa and J. Schenuit, ITMA review: Spinning, *Mellinand Textilber (Eng Edn.)*, 5, 260–266 (1999).

Choi, S. and Z. Jiang, A novel wearable sensor device with conductive fabric and PVDF film for monitoring cardio respiratory signals, *Sens. Actuators Appl. Phys.* 128, 317–326 (2006).

Dario, P., R. Bardelli, D. De Rossi, L.R. Wang and P.C. Pinotti, Touch-sensitive polymer skin uses piezoelectric properties to recognize orientation of objects, *Sensor Rev.*, 194–198 (1982).

Das, A., S.M. Ishtiaque and P. Yadav, Contribution of core and sheath components to the tensile properties of Dref III yarn, *Text. Res. J.*, 74(2), 134–139 (2004).

De Rossi, D., F. Lorussi, A. Mazzoldi, R. Paradiso, E.P. Scilingo and A. Todnetti, Electro active fabrics and wearable biomonitoring devices, *AUTEX Res. J.*, 3(4), 180–183 (2003).

Dennard, J., Burton, apple introduce wearable electronic jacket, *Text. World*, 153, 2–7 (2003).

Dhawan, A., A.M. Seyam, T.K. Ghosh, and J.F. Muth, Woven fabric-based electrical circuits – Part I, Evaluating interconnect methods, *Text. Res. J.*, 74, 913–919 (2004).

Dunne, L.E., S. Brady, B. Smyth and D. Diamond, Initial development and testing of a novel foam-based pressure sensor for wearable sensing, *J. Neuro Eng. Rehab.*, 2(4), 17–24 (2005).

Dunne, L.E., A. Toney, S.P. Ashdown and B. Thomas, Subtle integration of technology: A case study of the business suit, In *Proceedings of the First International Forum on Applied Wearable Computing (IFAWC)*, 24–25 March, Bremen, Germany, 2004.

Elton, S.F., Ten new developments for high-tech fabrics & garments invented or adapted by the research & technical group of the defence clothing and textile agency, In *Proceedings of the Avantex, International Symposium for High-Tech Apparel Textiles and Fashion Engineering with Innovation-Forum*, 13–17 September, Frankfurt-am-Main, Germany, 2000.

Fehrer, E., Friction spinning: The inventor's analysis, *Text. Month*, 12, 31–34 (1986).

Fehrer, E., Friction spinning: The state of the art, *Text. Month*, 9, 115–116 (1987).

Gleskova, H., S. Wagner and Z. Suo, Failure resistance of amorphous silicon transistors under extreme in-plane strain, *Appl. Phys. Lett.*, 75(19), 3011–3103 (2002).

Gleskova, H., S. Wagner, W. Soboyejo and Z. Suo, Electrical response of amorphous silicon thin-film transistors under mechanical strain, *J. Appl. Phys.*, 92(10), 6224–6229 (2003).

Gorlick, M.M., Electric suspenders: A fabric power bus and data network for wearable digital devices, In *Proceedings of the Third International Symposium on Wearable Computers*, 18–19 October, San Francisco, CA, IEEE Spectrum, pp. 114–121 (1999).

Gowda, M., Friction spinning in new spinning system, *NCUTE Publ.*, 2, 42–50 (2006).

Gruoner, S., The Dref spinning system, *Text. Asia*, 8(3), 44–49 (1977).

Gsteu, M., A spinning process makes the grade, *Int. Text. Bull., Spinning*, 65–82 (1982).

Gsteu, M., High tech yarns with friction spinning technology, *Int. Text. Bull. Yarn Forming*, 1, 36–45 (1989).

Ishtiaque, S.M. and D. Agarwal, The internal structure of sheath fibre in Dref-3 yarn, *J. Text. Inst.*, 91, 546–562 (2000).

Jain, R. and S. Agarwal, Recent innovations in textiles, Part-III: Smart textiles, *Asian Text. J.*, 7, 49–52 (2005).

Kallmayer, C., New assembly technologies for textile transponder systems, In *Electronic Components and Technology Conference*, New Orleans, LA (2003).

Krause, H.W., H.A. Soliman and H. Stalder, The yarn formation in friction spinning, *Int. Text. Bull. Yarn. Forming*, 4, 31–37 (1989).

Leckenwalter, R., Innovations in new dimensions of functional clothing, In *Proceedings of the Avantex International Symposium for High-Tech Apparel Textiles and Fashion Engineering with Innovation-Forum*, Frankfurt am Main, Germany (2000).

Lee, J.B. and V. Subramanian, Organic transistors on fiber: A first step toward electronic textiles, In *International Electron Device Meeting Technical Digest*, 8–10 December, Washington, DC, IEEE Georgia Tech, pp. 199–202, (2003).

Locher, I. and G. Troster, Screen-printed textile transmission lines, *Text. Res. J.*, 77, 837–842 (2007).

Lord, P.R. and J.P. Rust, Fibre assembly in friction spinning, *J. Text. Inst.*, 4, 465–478 (1982).

Lord, P.R. and J.P. Rust, The surface of the tail in open-end friction spinning, *J. Text. Inst.*, 81(1), 100–103 (1990).

Lord, P.R. and J.P. Rust, Fibre assembly in friction spinning, *J. Text. Inst.*, 82, 465–470 (1991).

Lord, P.R., C.W. Joo, and T. Ashizaki, The mechanics of friction spinning, *J.Text. Inst.*, 78(4), 234 (1987).

Lorussi, F., W. Rocchia, E.P. Scilingo, A. Tognetti and D. De Rossi, Wearable, redundant fabric-based sensor arrays for reconstruction of body segment posture, *IEEE Sens. J.*, 4(6), 67–78 (2004a).

Lorussi, F., W. Rocchia, E.P. Scilingo, A. Tognetti, and D. De Rossi, Wearable, redundant fabric-based sensor arrays for reconstruction of body segment posture, *IEEE Sensors J.*, 4(6), 807–818 (2004b).

Lorussi, F., E.P. Scilingo, M. Tesconi, A. Tognetti, and D. De Rossi, Strain sensing fabric for hand posture and gesture monitoring, *IEEE Trans. Inf. Technol. Biomed.*, 9(3), 372–381 (2005).

Lou, C-W., Process of complex core spun yarn containing a metal wire, *Text. Res. J.*, 75(6), 466–473 (2005).

Louis, G.L., H.L. Salaun and L.B. Kimmel, Comparison of properties of cotton yarns produced by Dref 3, ring and open end spinning methods, *Text. Res. J.*, 55, 344–348 (1985).

Lunenschloss, J. and K.J. Brockmanns, Mechanism of OE-friction spinning, *Int. Text. Bull., Yarn Forming*, 3, 29–59 (1985).

Mazzoldi, D., D. De Rossi, F. Lorussi, E.P. Scilingo and R. Paradiso, Smart textiles for wearable motion capture systems, *AUTEX Res. J.*, 2(4), 87–93 (2002).

Mei, T., Y. Ge, Y. Chen, L. Ni, W. Liao, Y. Xu and W.J. Li, Design and fabrication of an integrated three-dimensional tactile sensor for space robotic applications, In *IEEE MEMS*, 17–21 January, Orlando, FL, pp. 112–117 (1999).

Pacelli, M., G. Loriga, N. Taccini and R. Paradiso, Sensing fabrics for monitoring physiological and biomechanical variables: E-textile solutions, In *Proceedings of the Third IEEE-EMBS*, 14–16 September, Drexel University, Philadelphia, PA, pp. 1–4 (2006).

Padmanabhan, A. R. and K. Ramakrishnan, Properties of man-made fibre yarns spun on Dref 3 spinning system, *Indian J. Fibre Text. Res.*, 16, 241–247 (1991).

Park, S. and S. Jayaraman, Adaptive and responsive textile structures, In Tao, X. (ed.), *Smart Fibers, Fabrics and Clothing: Fundamentals and Applications*, Woodhead, Cambridge, U.K., pp. 226–245 (2001).

Park, S. and S. Jayaraman, Smart textiles: Wearable electronic systems, *MRS Bull.*, 28(8), 585–591 (2003).

Perumalraj, R., B.S. Dasaradan and V.R. Sampath, Study on copper core yarn conductive fabrics for electromagnetic shielding, In *National Conference on Functional Textiles and Apparels*, Vol. 2(1), pp. 187–203 (2007).

Post, E.R. and M. Orth, Smart fabric, wearable clothing, In *Proceedings of the First International Symposium on Wearable Computers*, Cambridge, MA, IEEE, pp. 167–168 (1997).

Post, E.R. and M. Orth, E-broidery: Design and fabrication of textile-based computing, *IBM Syst. J.*, 39, 840–860 (2000).

Rantanen, J. and H.N. Ninen, Data transfer for smart clothing: Requirements and potential technologies, In Tao, X. (ed.), *Wearable Electronics and Photonics*, Woodhead Publishing in Textiles, Houston, TX, p. 198 (2005).

Rust, J.P. and P.R. Lord, Variations in yarn properties caused by a series of design changes in friction spinning machine, *Text Res. J.*, 61, 645–652 (1991).

Scilingo, E.P., A. Gemignani, R. Paradiso, N. Taccini, B. Ghelarducci and D. De Rossi, Performance evaluation of sensing fabrics for monitoring physiological and biomechanical variables, *IEEE Trans. Inf. Technol. Biomed.*, 9(3), 345–352 (2005).

Shen, C.-L., T. Kao, C-T. Huang and J. Lee, Wearable band using a fabric-based sensor for exercise ECG monitoring, In *IEEE International Symposium on Wearable Computers (ISWC 2006)*, 11–14 October, Montreux, Switzerland, Vol. 3, pp. 143–144 (2006).

Thimany, J., Intelligent outfits, *Mech. Eng.*, 12, 126–134 (2004).

Tyagi, G.K., R.C.D. Kaushik, S. Dhamija, B. Biswas and K.R. Salhotra, Influence of core content and friction ratio on properties of Dref-3 polyester yarns, *Indian J. Fibre Text. Res.*, 24, 183–187 (1999).

Wagner, S., E. Bonderover, W.B. Jordan and J.C. Sturm, Electrotextiles: Concepts and challenges, *Int. J. High Speed Electron. Syst.*, 12, 1–9 (2002). www.electrotextiles.com.

Zheng, J.W., Z.B. Zhang, T.H. Wu and Y. Zhang, A wearable mobihealth care system supporting real-time diagnosis and alarm, *Med. Bio. Eng. Comput.*, 45, 877–885 (2007).

3

Design of Circuits and Integration into Wearable Fabrics

3.1 Introduction

Electronic textiles are distinct from wearable computing, because emphasis is placed on the seamless integration of textiles with electronic elements like microcontrollers, sensors and actuators. The basic materials needed to construct electronic textiles are conductive threads and fabrics and digital and communication systems. Most research and commercial electronic textile products are hybrids where electronic components embedded in the textile garments are connected to classical electronic devices or components. Some examples are touch buttons that are constructed completely in textile forms by using conducting textile weaves. In this work, a description of novel sensors developed to be integrated in sensorized garments for monitoring vital signs and body gesture/posture has been presented. The main advantage ensured by these systems is the possibility of wearing them for a long period of time without discomfort. Several issues deriving from the employment of the new technology which has consent the realization of these unobtrusive devices have been addressed. Moreover, the use of these sensorized garments as a valid alternative to existing instrumentation applicable in several health-care areas has been pointed out. Finally, results on the performances of the sensing systems were briefly reported. The work continues towards potential commercial exploitation. Also, this chapter is concerned with one of the most challenging aspects of creating wearable electronic circuits and integration with wearable electronic fabrics. The design of circuits, selection of controllers for different fabrics, types of printed circuit boards (PCBs), programming methods for controlling the various signals and the testing methodology have been discussed.

3.2 Microelectronics for Smart Textiles

For integration into everyday clothing, electronic components should be designed in a functional, unobtrusive, robust, small and inexpensive way. Therefore, small single-chip microelectronic systems rather than large-scale computer boxes are a promising approach. Today, a number of entertainment, safety and communication applications have become possible candidates for deploying onto clothing using off-the-shelf chip systems. Such examples include audio processing chips for MP31 music, fingerprint sensor chips for person identification, Bluetooth chipsets for short-range wireless data exchange or smart electronic labels for logistics purposes. With the ongoing progress of miniaturization, many complex and large electronic systems will soon be replaced by tiny silicon microchips measuring just a few square millimetres. In this section, the implementation of microelectronic components into clothes and textile structures in a reliable and manufacturable way has been discussed. The example presented is a speech-controlled digital music player system. Attention has been paid to an appropriate textile design for the tailoring of smart clothes. Damage of the components by washing processes and daily use must be avoided. Most solutions known so far require removal of complex electronics before starting the cleaning process. Our aim is to demonstrate technological solutions for the interconnect between textile structures and electronics and for a robust electronic packaging, which does not impair wearing comfort and offers ease of use to the customer. Due to the fact that textiles are used in everyday life, smart textiles have to meet certain demands. On one hand, normal textiles cannot be used for the transmission of power or signals, so other materials have to be integrated. This integration of new materials leads to another problem: the price of the machines. The technique of integrating conductive fibres is still rather new, but a weaving technology for implementing conductive fibres into fabrics for textile shields against electromagnetic radiation has already been developed. A goal of the research is that new materials could be used with the old machines of the textile industry. Most materials used as conductive fibres are already used in normal wires (carbon, polymers, polymers with nickel, copper and copper with silver coating). The best material is chosen according to the intended usage of the textile, but the fabric should be lightweight, durable, flexible and cost-competitive. In the majority of cases, copper is used for signal transmission and power. The copper wires are normally silver-plated for corrosion resistance. In the event these smart textiles are exposed to humidity, the wires are insulated by an additional polyester or polyamide coating. This applies to most use cases. Most smart textiles (especially wearable textiles) have to endure forces like sportive activities, rain, snow, the washing machine and many more. One possibility is removing the complex parts (the MP3 player in a jacket) before washing.

3.2.1 System Overview

An overview of the demonstrator system is shown in Figure 3.1. It is composed of four units: the central audio module, a detachable battery and multimedia card (MMC) module, an earphone and microphone module and a flexible keyboard sensor module. All units are electrically connected via ribbon-like narrow fabrics with conductive threads. The core of the digital music player is the audio module. The module contains an audio signal–processing microchip, which can be programmed to perform various functions, as discussed later. Besides this chip, further components, such as programme memory and a few auxiliary devices, are implemented. In principle, the audio module is a simple miniaturized computer system. Microphones, earphones, storage media, keypads, displays, sensors, actuators and a battery can be directly connected to its interfaces. The functionality of the module is determined by the built-in software. Speech recognition, MP3 decoding, text-to-speech conversion and music synthesis are only a few possible applications. An overview on the technical schematic and the existing software of the core microchip integrated with clothing has been developed. The functionality of this main module is tailored for the typical requirements in wearable electronics and smart clothes, since speech recognition allows for hands-free interaction with the system. The system can be extended to establish a connection to standard networks like the Internet (software modem function).

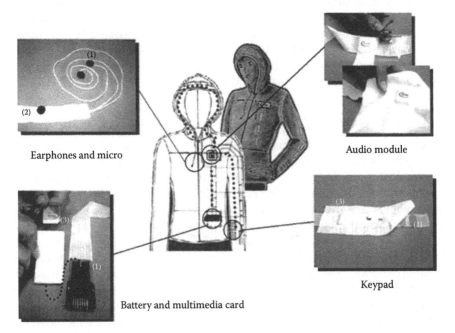

Earphones and micro

Audio module

Battery and multimedia card

Keypad

FIGURE 3.1
Digital music player system integrated with clothing.

The detachable battery and MMC module features a rechargeable lithium-ion polymer battery to supply the necessary electric energy. Its capacity is sufficient for an operating time of several hours. The battery module weighs only 50 g and is fixed to the cloth by means of a simple one-piece connector. A slot for the MMC has been integrated into the housing of the battery module. Both battery and MMC can thus be detached fast and easily. The MMC can store up to 64 Mb of digital music or audio data. When detached from the cloth, the battery and MMC module can be plugged to a PC in order to recharge the battery and download new music to the MMC. The MMC is a standardized memory card, which can also be used with many other digital consumer electronics products, such as digital cameras, PDAs or mobile phones. The flexible keyboard sensor module consists of thin metallic foils brought onto the conductive narrow fabric. The metal foils are fixed using melting adhesives commonly used in garment production. The metal foils are connected to a small sensor chip module that detects whether a finger is close to a specific pad or not. By means of the keys, the user can activate the music player, control the volume or activate voice control. Voice recognition ('stop!', 'start!', 'louder!') is activated after a specific button of the keypad has been touched. This measure avoids unintentional activation of control functions. The audio module recognizes the spoken words, for example the number or title of the music track. The earphone and microphone modules simply consist of a piece of narrow fabric connecting the audio module and conventional earpieces and microphones, as depicted. Special care is necessary for the textile design. All materials are chosen according to maximum wear comfort and environmental compliance. For example, the audio module is fully covered by garment. The wearer still has a comfortable textile touch in case the electronic module gets into direct contact with the skin. The supply voltages of the integrated electronics are as harmless as of a standard Walkman or comparable device.

3.2.2 Interconnect Technology between Textiles and Electronics

Recent advances in microelectronics have enabled the manufacturing of integrated electronic circuits with millions of logic switching elements per square millimetre of silicon. Since the feature sizes of these devices are in the micrometre regime and the typical dimensions in textile and garment technologies are in the order of several millimetres, a novel technology for the electrical interconnects has been developed. The gap of the spatial dimensions can be overcome by two methods that are described in the following. For test purposes, a polyester narrow fabric with several groups of parallel conductive warp threads has been used. In a first approach, the endings of the conductive fabric are prepared by soldering tiny metal contact plates. The module is then connected by electrically isolated bonding wires. In the last step, the module, the wires and the contact plates are moulded for mechanical and chemical protection. A second approach uses a thin flexible circuit board with structured electrodes, which is glued and soldered to the textile structure before being moulded. Copper wires qualified for textile applications coated with silver and polyester are used for the experiments. Figure 3.2 shows microphotographs of the interconnect areas. The electrical isolation of the wires is removed by laser treatment (Figure 3.2a). The resulting holes of the fabric are then filled with tiny metal plates of the same size. These connect the woven wires by soldering. A mechanical connection between metal and the fabric is achieved by melting and resolidifying the synthetic polyester fibres during a short high-temperature soldering step. The contact area can be re-isolated by covering it with a layer of melting adhesive (Figure 3.2b). Finally, Figures 3.2c and d show the textile structure contacted to the electronics according to soldered respectively. Figure 3.2d show module and the contact areas moulded on the textile fabric.

First results, including the implementation of a miniaturized MP3 player system, have been demonstrated, considering manufacturability aspects in a textile environment. The system contains a central microchip module,

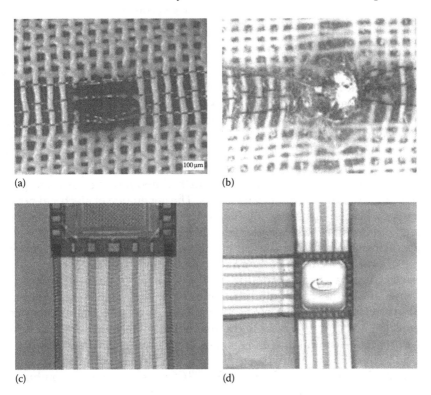

(a)　　　　　　　　　　　(b)

(c)　　　　　　　　　　　(d)

FIGURE 3.2
Microphotographs of interconnect experiments performed on a woven test ribbon. (a) Coating of the wires is removed by laser treatment. (b) Woven wires are soldered to a small metal foil and connected to an electronic circuit by a thin wire. (c) Alternatively, the contact to an electronic module can be established via a flexible circuit board soldered to the ribbon. (d) Module and the contact areas moulded.

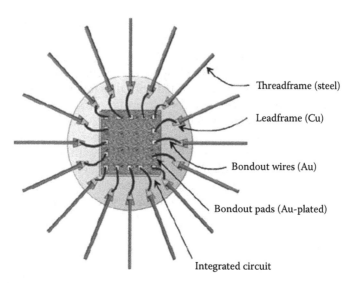

FIGURE 3.3
Internal structure of a PTCC.

headphones, battery and a storage card for music data, all interconnected by conductive textiles. All electronic components are encapsulated in a robust package (Figure 3.3). The principal tasks are performed by a single versatile chip, which is capable of MP3 decoding, speech recognition and text-to-speech conversion. The system has been integrated in specifically designed types of clothing. Specific design aspects have been considered, such as usability, comfort and protection against mechanical stress of the system. In the future, the system is expandable by further modules, such as fingerprint sensor chips or wireless data transceivers. The generation of electrical power from body heat as a possible solution to the power supply problem for low-energy applications has been demonstrated. The manifold of new possible applications shows that smart clothes are a prime example of the convergence of existing technologies with new applications and markets.

Electrical contacts to the conductive fibres woven or embroidered to the surface of the textiles can be made by crimping, soldering or using conductive adhesives. The contact size of the silicon-based microelectronic-integrated circuits (80–100 µm) differs from the size of the contact area of the conductive fibres (160–200 µm); therefore, several techniques have been developed to bridge the gap (normally used for comparatively small microelectronic systems). *The plastic-threaded chip carrier (PTCC)* has long flexible conducting leads and is stitched into an embroidered fabric circuit.

In the *wire bonding technology*, a polyester fabric with electrically isolated copper wires is used (Figure 3.4), the coating is removed by a laser and then the wires are soldered to contact plates and thin bonding wires (Figure 3.5). At last, the whole contact area and the module are encapsulated using polyurethane adhesive for mechanical protection. Similar to the wire bonding

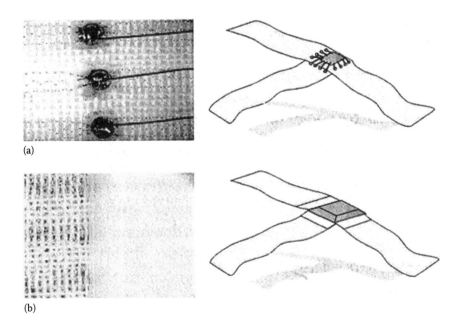

FIGURE 3.4
Wire bonding technology. (a) Polyester fabric with electrically isolated copper wire and (b) wire bonded by soldering.

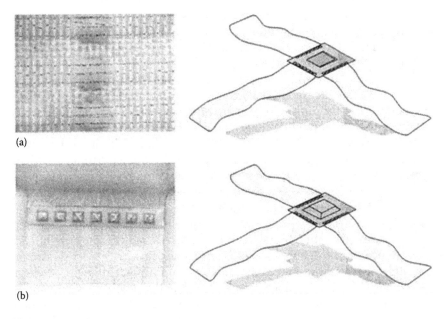

FIGURE 3.5
(a) Contact area prepared by laser treatment; (b) contact area encapsulated by silicone.

technology, the coating of the wires is removed by a laser while using *flexible PCBs*. Then, the flexible board is soldered, and as finishing step, the contact area is encapsulated by silicone.

Another topic of research is *silicon wafers*. They become flexible like foils when thinned below 50 μm, but the handling of these chips is still a challenge. The price of some techniques is still troublesome. Some are very expensive (like laser treatment), and it is therefore not possible to use them for large area textiles. In this field, the adoption of a manufacturing process known from radiofrequency identification technology (RFID) tags is possible (chips are mounted on plastic foils). In this case, chips are connected by the use of conductive adhesive, anisotropic conductive adhesive or non-conductive adhesives. Ambient energy may be used as power supply for microchips. Flexible solar cells could be integrated into textiles to use the energy of the sun. Movement is used for the generation of electricity in power boots. Piezoelectric transmitters generate electricity under mechanical stress and thermo-generators use body heat for power generation.

3.3 Flexible Wearable Electronics

With the technology miniaturization and wireless revolution, electronics have found new applications in the form of flexible wearable for monitoring purposes. Ongoing research on wearable electronic systems has resulted in a number of prototype garments that can monitor and relay various types of data. In the continuous pursuit of innovation, academic and industrial researchers alike have opened up floodgates in the development of wearable electronics. The progress was distinctly stepwise. First, there were large non-portable electronics, computing and communication devices. Then, smaller and lighter portables with the integration of some multiple functions have been focused. Now, in the era of the multi-purpose unobtrusive wearable, basic criteria such as weight, comfort, durability, energy management and wireless communication are the essential building blocks of the product. Tao's definition of wearable electronics is that of 'apparel with unobtrusively built-in electronic functions' (Tao 2005). More specifically, Koncar et al. (2005) sum up various functions as providing '… intelligent assistance that augments memory, intellect, creativity, communication and physical senses'. In the last few years, such types of electronics have been developed for various applications. In the entertainment industry, garments with integrated iPods, speakers, radios, etc., have already been commercialized by the likes of Gap, O'Neill, Rosner, Spyder, Textronics or Scottevest, working in collaboration with Philips, Infineon, Eleksen, SOFTswitch, Konarka Technologies and others. In the health and safety industry, a number of research groups have more challengingly developed monitoring garments, which, when worn, should be able to continuously or discreetly sense physical bodily or positioning data and relay them to a communication centre. Practical issues about the 'wearability', accuracy and reliability of the developed prototypes are being considered and addressed, and early steps for the commercialization of some of these products have been taken. While this is still in its infancy, it is perhaps now timely to bring together the various research results from the international community and review the current progress in flexible wearable monitoring electronics (Lam et al. 2007).

3.3.1 Textiles and Clothing as Flexible Supports

Textiles are ubiquitous in our society and provide the ideal base or support for wearable monitoring electronics. Because textile garments or accessories can be worn close to the skin, integrating electronics within the textiles gives benefits unrivalled by other systems. Measurements of data requiring close contact with the skin are possible, with minimal effects on the comfort of the wearer, and with minimal disruption to his/her day-to-day activities. To date, a number of wearable electronic textiles have been developed. Many are multi-layer systems, consisting of at least an internal layer (in contact with the skin) and an external one, with connected electronics and circuitry. The torso garment (Dunne et al. 2005) is an example of a double-layer prototype. Stylios and Luo (2003) by contrast describe a multiple layer system consisting of an innermost layer for comfort, an electromagnetic mask layer to shield the body from radiation, an electronic layer and an outermost layer, where power management systems such as solar cells can be incorporated. Rantanen and Hannikainen (2005) describe a three-layer system, consisting of a skin layer for physiological measurements, an inner clothing layer as a platform for the electronic components and an outer layer for environmental and positioning sensors and equipment. Structurally, both knitted and woven fabrics have been used for wearable electronic clothing for monitoring applications. The main advantage with textiles is their flexibility, which relates to some extent to wearing comfort. Knitted fabrics have the advantage of being stretchable and deformable to some extent and have been used where the fabrics need to be close to the body, or close fitting, such as for leotards or other sportswear. Comparatively, woven fabrics provide more dimensional stability and are more suitable where large movements of the body are not an essential factor to consider. Exploring the duality of functions, the interlacing of warps and weft in woven fabrics have been explored as networks for electrical circuits, in addition to being the supporting material for the integrated electronics (Dhawan et al. 2004). At this point, the influences of fabric structure and geometrical construction and of fibre type on the monitoring performance and accuracy have not been considered to be of significance. Most of the research has instead focused on constructing systems and ensuring that it works efficiently. Fabrics of reported prototype monitoring garments have been made of various types of fibres, including natural fibres, polyester, acrylic, elastane, nylon and optical fibres (De Rossi et al. 1999; Dunne et al. 2005).

3.3.1.1 Flexible Strain Sensors

A key component for wearable electronics systems for monitoring applications is the sensor, which gathers data from the wearer and relays the information to a processing unit. Recent works in monitoring vests and garments have integrated multiple sensors that provide physiological data such as body temperature, heart rate, skin conductivity, etc., and location data, using satellite facilities. Depending on their size and requirements, a number of traditional sensors made of conducting materials such as metals or carbons can be incorporated in smart clothing. Current research trends, however, are to develop flexible sensors out of textile materials, making use of the new range of polymers capable of changing properties with the environmental conditions, for example change in pressure, moisture, temperature, etc. Novel flexible sensors have thus been developed specifically for use in smart clothing or textiles and will be discussed throughout this paper. The type, position and number of sensors used naturally depend on the application end-use of the smart clothing. Recent examples of integrated sensors include the Life Shirt (tracking physiological measurements), an electronic bra for breast cancer detection and baby pyjamas to assist in preventing cot death.

3.3.1.2 Detection of Posture and Movement

Various prototypes for monitoring the posture and movement have been developed for improving body postures, reducing sports and other injuries and assisting in rehabilitation. Previously, such types of monitoring were done by applying sensors directly on the body, with the disadvantages of being cumbersome, uncomfortable to wear and easy to displace. With the advent of the concept of smart textiles, sensors are now being integrated within textiles, or securely attached to it. Most are based on the principles that the electrical resistance of the flexible sensor changes during stretching. A good account of early attempts at developing strain and stress sensors for fabrics and was one of the firsts to investigate flexible strain sensors from piezo-electric polyvinylidene fluoride (PVDF) polymer films. However, the developed films exhibited limitations such as sensitivity to temperature and electromagnetic interference, tensile stiffness and transient output signals, which precluded their use in wearable garments. A more recent, simple but successful example is a fabric strain sensor attached to a knee sleeve that acts as a biofeedback device, monitoring the wearer's knee movements during physical activity and sending back information via a beeping sound. This type of sensor is, however, limited in its lifespan and has been developed as a disposable device. Prototypes of more complex and durable garments and accessories such as jackets, leotards and gloves with surface-attached sensors have also demonstrated the ability to sense the position of different parts of the body and monitor physical activity (De Rossi et al. 2003; Farringdon et al. 1999). Normally, a number of sensors have to be positioned over specific points of the body, where angles of joints and stretch distances can be measured. From a design point of view, the surface attachment of many strips of sensors on the garment be capitalized as a decorative element, but is also a source of design limitations and restrictions. Other systems have been developed whereby the sensors are embedded between fabric layers (Dunne et al. 2005) or into the fabric itself (Kirstein et al. 2005). Many of the developed flexible strain sensors are based on using coated fabric technology. The discovery of conducting electro-active polymers such as polypyrrole (PPy), polyaniline and polythiophene has opened doors in the development of various types of textile sensors. PPy has been particularly well explored in coated fabrics (De Rossi et al. 1999, 2003) and foams (Dunne et al. 2005). Such materials have been found to have transducing properties as strain gauges, exhibiting a drop in electrical resistance with a physical deformation and topology change. De Rossi et al. (1999) reported gauge factors similar to that of nickel. However, they also reported that the sensor resistance shows strong variations with time and that the material has a high response time. Other elongation sensors include conductive fibre potentiometers, conductive fluids and carbon-filled rubber-coated fabrics, which have been shown to have some properties similar to that of metals in the 1%–13% strain range (De Rossi et al. 2003). Kirstein et al. (2005) summarizes the pros and cons of each system, highlighting the simple and cheap manufacturing process of PPy-coated fabrics and carbon-filled rubber, but their high hysteresis and dependence on prehistory and stretch velocity. Conductive fluids (strain gauges made from liquid metals such as mercury, or electrolytes such as copper sulphate contained in a rubber tube) are claimed to have better transducer performance, but the potential problem of permeability of the rubber tube, its limited lifespan and its lack of textile quality are the main drawbacks.

3.3.1.3 Biometric Measurements

A number of smart wearable electronic garments have incorporated existing, adapted or developed sensors for the measurement of biometric factors such as body temperature, heart rate, respiration rate, skin conductivity, etc. (Catrysse et al. 2004; De Rossi et al. 2003; Dunne et al. 2005; Goulev et al. 2004). Applications for such types of garments or accessories cover the healthcare, clinical, military, rescue or sports sectors, where the monitoring of vital signs is essential. In the medical field, smart monitoring suits can be used for long-term continuous monitoring of patient's conditions and can also be a potential aid in the fight against cot death in babies. In the military field or in a rescue situation, acquiring and relaying the vital signs of a soldier or fire fighter in action can help the medical team prepare for any required treatment. In sports, smart suits can be used as a tool to study conditions for optimum performances, but can also be used as tools in extremes and life-saving situations such as avalanches (Michahelles et al. 2003). Unlike posture and movement sensors, in most cases for biometric sensors, they have to be positioned next to or very close to the skin, which limits their incorporation in large loose garments such as jackets. For the majority of prototypes, the smart garment is close fitting and worn as an undergarment. As such, textile sensors, with better tactile and comfort properties, are being developed in order to improve the 'wearability' of the electronics. Textile electrodes, also called 'Textrodes', have been developed out of knitted stainless steel fibres for the measurement of electrocardiograms (Catrysse et al. 2004). Trials showed that the Textrodes are less irritating to the skin (in contrast to gel electrodes), and that they are able to provide reliable monitoring despite having high skin-electrode impedance. In the same spirit, Catrysse et al. (2004) also developed a textile alternative to conventional methods of measuring respiration rates. Made of stainless steel yarn and knitted in an elastane-containing belt, their 'Respibelt', when worn around the thorax, is able to measure thoracic changes in perimeter and cross-section, through changes in inductance and resistance. In another approach, Dunne et al. (2005) investigated the use of PPy-coated pressure-sensitive polyurethane foams as sensors for breathing. Limitations with regard to ageing of the material, oxidative degradation and hysteresis effects have been observed; however, the coated foam still showed interesting promises for monitoring breathing rates. PPy-coated materials also have good thermal sensitivity for temperature monitoring. De Rossi et al. (1999) found that the temperature sensitivity of PPy-coated Lycra is of the same order as that of ceramic thermistors. Pre-oxidized carbonized woven polymers such as the trademark Gorix can also be used as a temperature sensor (Kirstein et al. 2005).

3.3.1.4 Location Monitoring

The incorporation of unobtrusive tracking devices in garments to monitor the location and whereabouts of the wearer is now being explored commercially. Thanks to rapid development in mobile phone technology and in long-range communication, the technology for remote localization is well established and can be readily transferred to wearable electronics. The Japanese were amongst the first to explore openly the global positioning systems (GPS) technology in clothing. Small GPS tracking devices have already been integrated in school uniforms and prisoners' outfits, as a safety and security measure. There is no doubt that a GPS positioning system can be useful in certain situations (kidnapping, prisoner escape, avalanche or other types of rescue, accidents, etc.); however, given the inconspicuous nature of the devices, some issues can be raised on privacy and human and moral rights.

3.3.1.5 Pressure Sensors

Pressure sensors integrated in apparel can have different applications – detecting pressures during day-to-day, physical or sports activities, or providing an input interface for the wearer, for example in the form of an integrated keyboard or touch pad (Jones 2004; Swallow and Thompson 2001). Significant development in the field of flexible pressure sensors, with improved tactile properties and weight (compared to conventional keyboards and pressure sensors), has been achieved and commercialized, based on various types of technologies. Dunne et al. (2005) investigated the use of PPy-coated polyurethane foam as pressure-sensitive materials and applied them for monitoring the body movement and breathing monitoring. Most other textile pressure sensors work from multi-layer principles, for example two conductive textile layers sandwiching a non-conductive one. The middle layer can be an electro-resistive composite that changes resistance when compressed, for example as per the commercially available Softswitch technology (Jones 2004), or a mesh-type structure, which would allow physical contact between the two conductive layers with pressure (Swallow and Thompson 2001). Another commercial system comprises a five-layer structure with conductive carbon fabric layers and insulating separation mesh layers separating a partially conductive layer that conducts locally when compressed (Gilhespy 2004). The technology is now developed enough to provide detailed information on the pressure positions and forces. Most applications of pressure sensors to date have been used for textile electronics in the entertainment industry, but applications in the medical and health-care fields are also being explored.

3.3.1.6 Ballistic Penetration and Fabric Damage

The military and safety sectors have been key stakeholders in the development of smart monitoring textiles. The functionalities provided by sensors described earlier are attractive for the support of frontline personnel, and much work has been carried out in the pursuit of smart clothing for fire fighters, rescue workers, policemen, soldiers and other special forces. In one smart garment prototype developed with the military in mind, optical fibres form a conductive circuit backbone in a grid of rows and columns. Upon penetration of a bullet, the conductive path of the thread is broken and detected by the microprocessor, which can identify the exact coordinates of the penetration. The information can then be quickly relayed to medical teams. A similar concept can potentially be used for other type of fabric damage, including for chemical and biological protective garments, whereby a warning signal can be emitted when a hole or tear occurs in the fabric.

3.3.1.7 Actuators and Output Signals

Various types of textile-based or other actuators that can transform electrical signals into a physical phenomenon have been incorporated in wearable electronics to generate some form of output, which can be audio, visual or tactile. In the piloted sensing knee sleeve, the output of the information sensed (angle of knee bent) is in the form of an audible sound, warning the wearer when the correct knee angle is achieved during landing. Audio signals are simple but effective warning signals, which can also be varied by changing pitch, tone and other audio factors, to convey different messages. A vast amount of research is ongoing in the area of flexible visual displays, which are able to provide more information to the wearer. One group of flexible visual displays uses colour-changing mechanisms in dyes and pigments to create colour, and hence visual outputs. In this respect, many sensors can also act as actuators and automatically respond by a physical change, when sensing an environmental change, for example chromatic dyes changing colour with temperature, light, pressure, etc. Other chromatic materials can also react with an electric current and hence change colour as and when required to create a colourful output. Tao (2005) and Philips Research (2002) were provided a brief overview of current technologies in flexible displays, from the academic and industrial points of view. Displays made of polymer light-emitting diodes (polyLED) produced on flexible plastic substrate can lead to extra flexibility but have the disadvantage that the plastic backing permits degradation of the light-emitting properties of the polyLED. Various semi-conducting polymers have been found to have potential as polyLEDs, for example polyanilines, polythiopenes and polypyridines, and much work still continues in this area. Textile-based display prototypes are produced using electro-luminescent materials that are triggered with an electric voltage, as with the commercially developed vision displays. The electro-luminescent material is used to thinly coat wire conductors, which are then woven into a rows-and-columns electrodes network. Emission of light is triggered when a voltage is applied to a specific row and column, hence creating an electric field at their intersection. Another technology is to use encapsulated charged particles within microcapsulses and attract or repel them using an electric field. Yet another system is to use optical fibres in woven textile structures. Koncar et al. (2005) describe an optical fibre flexible display (OFFD) that has been prototyped on a jacket. Using a patented process to generate microperforations in optical fibres such as poly(methylmethacrylate [PMMA]), fibres with multiple point lateral illumination are obtained. These fibres are woven into a number of surface units, or large 'pixels', that can be lit individually to create specific shapes, texts or logos. A final visual and tactile type of actuator are shape memory materials such as alloys and polymers, which change shape upon receiving a stimulus such as temperature or electrical field. Such materials have been explored for use in apparel and interior textiles (Winchester and Stylios 2003), but not yet in a function related to monitoring. As with chromatic materials, shape memory materials can act as both sensors and actuators and may thus have interesting uses in the wearable monitoring field.

3.3.1.8 Data Management and Communication

With basic sensing/actuating materials, such as chromatic dyes and shape memory materials, the wearer may be able to monitor specific parameters without the need for a processor for the data, and communication network to relay the information. As example, consider a photochromatic patch on an outdoor vest, swimwear or sportswear. The patch may be able to monitor the level of ultraviolet rays responsively by sensing and reacting with the incident light. With this type of system, no data management, storage and communication are necessary. In the case of wearable electronics, the complexity of the system is higher, because the electronic components will need to convert electrical signals into output and/or input into electrical signals. Processors such as transistors, diodes and other non-linear devices are needed to amplify signals, process simple arithmetic operations and store data gathered. Attempts have been made to create electroactive textile fibres that can act as transistors. Few researchers have made attempts on demonstrating the patterning of transistors directly on fibres using a novel weaving-based lithography, a technique that lends itself to textile manufacturing. The fibre transistors showed good and stable electrical characteristics, with performance suitable for applications in smart textiles, but were limited by enhanced dielectric leakage and reduction in reliability. Central processing units for smart clothing systems often include small 8–16-bit microcontrollers (Rantanen and Hannikainen 2005). The monitoring system consists of a combination of several such units, positioned at different locations to perform specific functions and communicate with each other.

3.3.1.8.1 Short-Range Communication

One of the challenges of interactive wearable electronic textile systems is the effective transfer of data or power between different modules of the system. For practicality, weight and design purposes, it is often convenient to spread out the sensors, actuators and microcontrollers over different strategic locations on the body. This distribution of information microhubs requires an effective and comfortable network of communication. Earliest attempts included the use of wires. Infrared and personal area networks (PANs), such as Bluetooth technology, were also suggested as possible alternatives. Recently, fabric area network (FAN) and the transfer of information through the human body has been investigated (Hum 2001). The various PAN systems have attracted much attention in the area of wireless interactive wearable electronics, but security and privacy still need to be worked on. The need for good data encryption systems, and for a universal communication and encryption standard to enable different products to work together wirelessly, has been identified as essential area of research (Gould 2003). For inter-device physical connections, a large amount of work has been carried out in the area of conductive fibres, yarns and textile circuitry. The development of flexible circuitry to interconnect various elements of wearable electronics has explored various approaches across the world. Conductive yarns and coatings played a significant role in enabling various textile circuitry technologies. Anderson and Seyam (2002) have provided an account of the various methods of imparting conductivity and semi-conductivity to textiles.

Conductive textile circuits are now in a position to compare favourably to conventional flexible circuits. As mentioned by Post et al. (2000), there are various methods for making flexible circuits, most of which rely on the metallization of a flexible high-temperature-resistant polymer substrate such as polyimides, which can withstand soldering temperatures. The flexibility of such circuitry is, however, generally limited to the substrate – soldered joints are the weak points during mechanical stress.

The challenges of creating circuitry that can be mechanically stressed in all directions and positions led to the development of textile-based circuitry. The earliest attempts suggested the use of printing and coating technologies to create a circuit pattern on a textile substrate using conductive polymers such as PPy, polyaniline and polythiophthene. Post et al. (2000) outlined various alternatives to create textile-based circuitry. In one line of research, the use of metallic silk organza was explored, and electrical components were soldered directly on the fine fabric. A particular characteristic of organza that makes it ideal for conducting electrical signal is that its silk warp threads are wrapped in a metal foil helix, which has a high conductivity. The labour-intensive piecework produced, however, highlighted several limitations, including the fragility of the metal wrap and its propensity for corrosion. Further, soldering components still led to poor mechanical properties at joints (considering the stresses expected in apparel) and raised issues of toxicity when worn next to the skin. Embroidery proved to be a good solution to textile-based circuitry. Whilst stitching on the metallic organza was not an option because of the fragility of the wrapping foil, using conductive yarns on other types of fabrics gave good results. A number of conductive yarns were shown to have varying degrees of conductivity, durability and performance: yarns made of continuous or short stainless steel fibres, blends of stainless steel and polyester, nylon-stainless steel core-wrapped structure and metal-clad Aramid fibres. However, in the case of embroidered circuitry, the yarns of continuous metal fibres were not suitable for machine stitching because of the low yarn elasticity and build-up of tension during stitching. Post et al. (2000) remarked that with embroidery, multi-layer textile circuitry becomes easier to develop, compared to conventional multi-layer circuitry. In addition to embroidery, weaving has also been shown to be a potential solution for flexible circuitry, including multi-layer ones. The interlacing of conductive and non-conductive yarns can be used in complex networks, with the right interconnecting and disconnecting points for conductive yarns to route the electrical signals (Dhawan et al. 2004). Interconnection of unconnected yarns for specific circuit designs can be done by resistance welding using microprobes and air splicing. Disconnection can be done with microcutters as well as during resistance welding, to cut and connect at the same time. One of the limitations of textile (and other flexible) circuitry is a lack of signal integrity under certain conditions. This happens, for example when there is a rapid voltage change on a line, leading to cross-talk or electromagnetic interference to an adjacent line. Dhawan et al. (2004) investigated the use of different types of yarn structures to reduce cross-talk and found that interference is reduced when coaxial and twisted pair copper threads are used instead of bare ones. Seyam (2003) reviewed the potential of the latest developments in woven fabric technology for electro-textiles, highlighting the possible auto-mation, high-speed productions, versatility and quality. For disconnecting and interconnecting conductive yarns into a designed woven circuit, the incorporation of small robotic devices and optical sensors enables automa-tion of the process. One example is the Wearable Motherboard™, which is woven in one piece, bypassing the cut and sew operation. The wearable computer incorporates and integrates plastic optical fibres (POFs), creating a whole garment circuit.

3.3.1.8.2 Long-Range Communication

For external data transfer, for example via the Internet or a network proto-col, the technologies have been well developed in the recent two decades. A range of communication systems is already available: the Global System for Mobile Telecommunication (GSM) is widely used for small-sized data transfer (e.g. voice transmission); the third-generation (3G) wireless sys-tem is now being used to transfer larger files such as photographs and videos. On the downside, the energy requirements for the more powerful data transfer systems are higher, and as such, efficient power management systems are required. The main energy requirements for a wearable elec-tronic textile system include low power consumption, light, easy to carry around, long-lasting and discreet power sources. In the case of monitoring systems that will encounter harsh environmental conditions or movement (e.g. for use in sports activities, fire-fighting exercises, military conditions or extreme environmental conditions), the robustness of the system is also an essential criterion. Batteries such as the lithium–polymer ones are at pres-ent the most common types of power sources, but because of the high con-sumption of energy for electronic components in wearable electronics, they may not be the most suitable source for long-term conditions, for example mountaineering or rescue expeditions. Alternative solutions include using longer lifespan power sources such as microfuel cells, or harvesting continu-ous and semi-continuous power sources such as the sun (photovoltaics), the human body (kinetic or heat energy) or microwaves (Tao 2005). The smart textiles sector has taken a keen interest on the use of solar energy for power-ing the electrical components in outdoor clothing. Recently, commercially available flexible photovoltaic materials made from copper indium galium diselenide (CGIS) have been used in the development of a solar-powered jacket. Silicone-based thin photovoltaic films are also under investigation for use in woven military uniforms and other applications. On the research front, various methods of producing textile-based photovoltaics are being investigated, including depositing nanocrystalline silicon on woven or non-woven substrates, weaving and knitting photovoltaic fibres developed with photoreactive dyes and titanium dioxide, and adapting multi-layer thin film technology for cylindrical materials such as fibres. Organic and polymer photovoltaics have also raised interest for applications in textiles (Krebs et al. 2005). Although still in its early stages of development, polymeric photovoltaics have been applied as coating on textiles, with encouraging performances, but with issues to be addressed regarding stability and durability.

3.4 Wearable Motherboard

Virtual reality (VR) has been making inroads into medicine in a broad spectrum of applications, including medical education, surgical training, telemedicine, surgery and the treatment of phobias and eating disorders. The extensive and innovative applications of VR in medicine, made possible by the rapid advancements in information technology, have been driven by the need to reduce the cost of health care while enhancing the quality of life for human beings. In this chapter, we discuss the design, development and realization of an innovative technology known as the Georgia Tech Wearable Motherboard (GTWM), or the 'Smart Shirt'. The principal advantage of GTWM is that it provides, for the first time, a very systematic way of monitoring the vital signs of humans in an unobtrusive manner. The flexible data *bus* integrated into the structure transmits the information to monitoring devices such as an EKG machine, a temperature recorder, a voice recorder, etc. GTWM is lightweight and can be worn easily by anyone, from infants to senior citizens. We present the universal characteristics of the interface pioneered by the Georgia Tech Wearable Motherboard and explore the potential applications of the technology in areas ranging from combat to geriatric care. The GTWM is the realization of a personal information–processing system that gives new meaning to the term *ubiquitous* computing. Just as the spreadsheet pioneered the field of information processing that brought 'computing to the masses', it is anticipated that the Georgia Tech Wearable Motherboard will bring personalized and affordable health-care monitoring to the population at large. VR has paved the way for a new methodology that enhances the use of information in a broad spectrum of medical applications, including medical education, surgical training, telemedicine, surgery and the treatment of phobias and eating disorders. The extensive and innovative applications of VR in medicine have been made possible by the rapid advancements in information technology and human–computer interaction. For example, the abdominal simulator is an instructional tool that enables the surgeon to practise endoscopic surgical techniques. It is also an important tool for teaching the true anatomic relationships of intra-abdominal structures to students. Virtual reality therapy (VRT) has been used to overcome some of the difficulties inherent in the traditional treatment of phobias such as the fear of flying, fear of heights and fear of public speaking. VRT can provide stimuli for patients who have difficulty in imagining scenes and/or are too phobic to experience real situations. VR has also been used for the treatment of eating disorders. Thus, these examples of VR in medicine illustrate the importance of utilizing advancements in innovative technology for the benefit of humankind. This approach of developing and harnessing technology to save lives and/or enhance the quality of life is especially critical in today's context of increasing health-care costs. Innovative technology holds the key to addressing the twin challenges of health-care and battlefield management. According to Andy Grove, the Chairman of Intel Corp., the health-care industry is facing an Internet-driven 'strategic inflection point' or a time in which extreme change forever alters the competitive landscape of an industry, creating new opportunities and challenges.

3.4.1 Need for a Technology Solution

The health-care industry must meet the challenge of balancing cost containment with maintenance of desired patient outcomes. Consequently, health-care professionals are trying to provide patient care more efficiently and, whenever possible, in the least expensive setting, be that an intensive care unit (ICU), a hospital general care unit, a skilled nursing facility or an outpatient clinic. Even a patient's home is becoming the site for many types of care or monitoring that once were provided only by hospitals. This has created a demand for portable, versatile medical devices that can be moved easily from the ICU all the way to a homecare setting.

3.4.2 Design and Development of the Wearable Motherboard

In this section, we discuss the methodology for the design and development of the GTWM.

3.4.2.1 Performance Requirements

The goals of the research project undertaken at Georgia Tech have been to conceptualize a system that would meet the two broad performance requirements, design the system applying the principles of concurrent engineering, produce the garment and demonstrate the realization of the performance requirements. The first step in this process was to identify clearly the various characteristics required by the customer (the U.S. Navy) in the product being designed. Therefore, using this information on the two key performance requirements, an extensive analysis was carried out. A detailed and more specific set of performance requirements was defined, and these requirements are functionality, usability in combat, wearability, durability, manufacturability, maintainability, connect ability and affordability. The next step was to examine these requirements in depth and to identify the key factors associated with each of them.

For example, functionality implies that the GTWM must be able to detect the penetration of a projectile and should also monitor body vital signs – these are the two requirements identified in the broad agency announcement from the navy. The factors deemed critical in battlefield conditions are under usability in combat. These include providing physiological thermal protection, resistance to petroleum products and electromagnetic interference (EMI), minimizing signature detectability (thermal, acoustic, radar and visual), and offering hazard protection while facilitating electrostatic charge decay and being flame and directed energy-retardant. Wearability implies that the GTWM should be lightweight, breathable, comfortable (form-fitting), easy to wear and take off and provide easy access to wounds. These are critical requirements in combat conditions so that the soldier's performance is not hampered by the protective garment. The durability of the GTWM is another important performance requirement. It should have a wear life of 120 combat days and should withstand repeated flexure and abrasion, both of which are characteristic of combat conditions. Ease of manufacture is another key requirement, since eventually the design (garment) should be produced in large quantities over the size range for the soldiers; moreover, it should be compatible with standard issue clothing and equipment. Maintainability of the GTWM is an important requirement for the hygiene of the soldiers in combat conditions; it should withstand field laundering, dry easily and be easily repairable (for minor damages). The developed GTWM should be easily connectable to sensors and the personal status monitor (PSM) on the soldier. Finally, affordability of the proposed GTWM is another major requirement so that the garment can be made widely available to all combat soldiers to help ensure their personal survival, thereby directly contributing to the military mission as force enhancers.

3.4.2.2 GTWM Design and Development Framework

GTWM design and development framework and it encapsulate the modified QFD-type (Quality Function Deployment) methodology was developed for achieving the project goals. The requirements are then translated into the appropriate properties of GTWM: sets of sensing and comfort properties. The properties lead to the specific design of the GTWM with a dual structure meeting the twin requirements of 'sensing' and 'comfort'. These properties of the proposed design are achieved through the appropriate choice of materials and fabrication technologies by applying the corresponding design parameters. These major facets in the proposed framework are linked together as shown by the arrows between the dotted boxes. This generic framework can be easily modified to suit the specific end-use requirements associated with the garment. For instance, when creating GTWM for medical monitoring only, the requirement of 'usability in combat' would not apply nor would the functionality of penetration detection. This structured and analytical process eventually led to the design of the structure. The Wearable Motherboard consists of the following 'building blocks' or modules that are totally integrated to create a garment (with intelligence) that feels and wears like any typical undershirt. The modules are as follows:

The elegance of this design lies in the fact that these building blocks (like LEGO™ blocks) can be put together in *any* desired combination to produce structures to meet specific end-use requirements. For example, in creating a GTWM for health-care applications (e.g. patient monitoring), the penetration-sensing component will not be included. The actual integration of the desired building blocks will occur during the production process through the inclusion of the appropriate fibres and yarns that provide the specific functionality associated with the building block. In and of itself, this design and development framework represents a significant contribution to systematizing the process of designing structures and systems for a multitude of applications.

3.4.2.3 Production of GTWM

The schematic of one variation of the Wearable Motherboard is shown in Figure 3.6. This design was woven into a single-piece garment (an under-shirt) on a weaving machine to fit a 38–40 in. chest. The various 'building blocks' integrated into the garment along with their relative positions in the garment are also shown in Figure 3.6. Based on the in-depth analysis of the properties of the different fibres and materials, and their ability to meet the performance requirements, the following materials were chosen for the building blocks in the initial version of the Smart Shirt:

- Meraklon (polypropylene fibre) for the comfort component
- POFs for the penetration-sensing component
- Copper core with polyethylene sheath and doped nylon fibres with inorganic particles for the electrical conducting component
- Spandex for the form-fitting component
- Nega-Stat™ for the static dissipating component

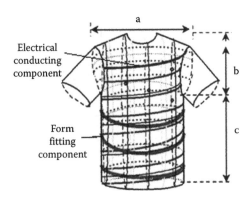

FIGURE 3.6
Schematic of the Woven Wearable Motherboard. (From Gopalsamy, C. et al., *J. Virtual Reality*, 4, 152, 1999.)

The POF is spirally integrated into the structure during the fabric production process without any discontinuities at the armhole or the seams using a novel modification in the weaving process. With this innovative design, there is no need for the 'cut and sew' operations to produce a garment from a two-dimensional fabric. This pioneering contribution represents a significant breakthrough in textile engineering because, for the first time, a fully fashioned garment has been woven on a weaving machine.

An interconnection technology was developed to transmit information to and from sensors mounted at any location on the body, thus creating a flexible 'bus' structure. T-connectors – similar to 'button clips' used in clothing – are attached to the yarns that serve as the databus to carry the information from the sensors (e.g. EKG sensors) on the body. The sensors plug into these connectors, and at the other end, similar T-connectors are used to transmit the information to monitoring equipment or DARPA's personal status monitor. By making the sensors detachable from the garment, the versatility of GTWM has been significantly enhanced. Since human anthropometry is variable, sensors can be positioned on the right locations for all users without any constraints being imposed by GTWM. In essence, GTWM can be truly 'customized'. Moreover, it can be laundered without any damage to the sensors themselves. In addition to the fibre-optic and specialty fibres that serve as sensors, and the databus carrying sensory information from the wearer to the monitoring devices, sensors for monitoring the respiration rate (e.g. RespiTrace™ sensors) have been integrated into the structure. Three generations of the Woven Wearable Motherboard have been produced and a knitted version of the Wearable Motherboard has also been created. Figure 3.7 shows the third-generation Woven Wearable Motherboard. The lighted optical fibres illustrate that GTWM is *armed* and ready to detect projectile penetration. The interconnection technology has been used to integrate sensors for monitoring the

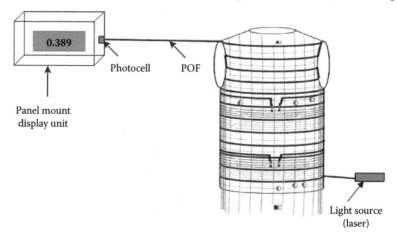

FIGURE 3.7
Third-generation woven GTWM.

following vital signs: temperature, heart rate and respiration rate. In addition, a microphone has been attached to transmit the wearer's voice data to recording devices. Other sensors can be easily integrated into the structure. For instance, a sensor to detect oxygen levels or hazardous gases can be integrated into a variation of GTWM to be used by firefighters. This information, along with the vital signs, can be transmitted to the fire station where personnel can monitor the firefighter's condition continuously and provide appropriate instructions including ordering the individual to evacuate the scene, if necessary. Thus, this research has led to a truly and fully customizable 'Wearable Motherboard' or intelligent garment.

3.4.2.4 Testing of GTWM

The penetration-sensing and vital signs–monitoring capabilities of GTWM were tested. *Penetration sensing*: The bench-top set-up for testing the penetration-sensing capability is shown in Figure 3.8. A low-power laser was used at one end of the POF to send pulses that 'lit up' the structure, indicating that GTWM was armed and ready to detect any interruptions in the light flow that might be caused by a bullet or shrapnel penetrating the garment. At the other end of the POF, a photodiode connected to a power-measuring device

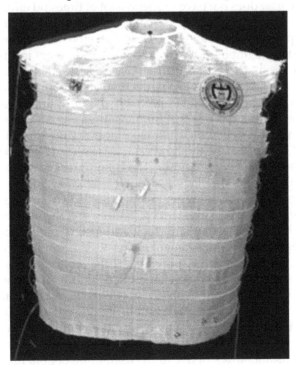

FIGURE 3.8
Bench-top set-ups for projectile penetration.

measured the power output from the POF. The penetration of GTWM resulting in the breakage of POF was simulated by cutting the POF with a pair of scissors; when this happens, the power output at the other end on the measuring device falls to zero. The location of the actual penetration in the POF can be determined by an optical time domain reflectometer, an instrument used by telephone companies to pinpoint breaks in fibre-optic cables. This research on the design and development of GTWM has opened up new frontiers in fields such as personalized information processing, health care and telemedicine. Applications of this technology could also be useful to monitor astronauts during space exploration. Until now, it has not been possible to create a personal information processor that was customizable, wearable and comfortable; neither has there been a garment that could be used for unobtrusive monitoring of humans on earth or in space for vital signs such as temperature, heart rate, etc. In this section, we discuss one of the significant contributions of the research and explore the key applications of GTWM in health care.

3.4.2.5 Wearable Motherboard as a Universal Interface

Humans are used to wearing garments; they enjoy clothing and also know what to do with it; therefore, clothing as an interface is truly universal. Moreover, the user does not have to spend a lot of time and effort in learning how to use it. This universal interface of clothing can indeed be 'tailored' to fit the user's physical needs and desires; at the same time, it must accommodate the constraints imposed by the ambient environment in which the user interacts with the interface, that is different climates, activities and occasions. The interface can also be fitted to suit the financial resources available to the user. In other words, a garment is probably the most universal of human–computer interfaces and is one that humans need, use and are very familiar with, that can be enjoyed and customized for every occasion. The Wearable Motherboard could become this universal interface and give new meaning to the term man–machine symbioses. The benefits from a technology can be harnessed only if the technology is properly deployed. Proper deployment requires education and training in how to utilize the technology; moreover, the easier the interface, the greater the chances that the technology will be utilized. For instance the standard 'look and feel' of programmes that run under the MS-Windows environment makes it easier for users to adopt and utilize the computing capability provided by the programmes and the hardware. Just as the spreadsheet pioneered the field of information processing that brought 'computing to the masses', it is possible that the Smart Shirt or GTWM with its universal interface will bring personalized and affordable health-care monitoring to large segments of the population. In his research on 'wearable computers', Mann defined the following attributes for wearable computers: constant, unrestrictive to the user, unmonopolizing of the user's attention, observable by the user, controllable by the user, attentive to the environment and personal. Mann's criteria for wearable computers include it being eudemonic, existential and in constant operation and interaction. The research on GTWM (carried out in parallel) focused on creating a personal wearable information interface that would be as comfortable as any other garment, rather than just making a computer wearable. However, a comparison with the attributes and criteria defined for wearable computers shows that GTWM meets these attributes and criteria and, indeed, may go beyond wearable computing.

3.4.2.5.1 Range of Potential Applications

The broad range of applications of GTWM in a variety of segments is summarized in Table 3.1. The table also shows the application type and the target population that can utilize the technology. A brief overview of the various applications follows:

3.4.2.5.2 Combat Casualty Care

GTWM can serve as a monitoring system for soldiers that is capable of alerting the medical triage unit (stationed near the battlefield) when a soldier is shot, along with information on the soldier's condition characterizing the extent of injury and the soldier's vital signs. This was the original intent behind the research that led to the development of GTWM.

3.4.2.5.3 Health Care and Telemedicine

The health-care applications of GTWM are enormous and it greatly facilitates the practice of telemedicine, thus enhancing access to health care to patients

TABLE 3.1

The Wearable Motherboard/Smart Shirt: Potential Applications

Segment	Application Type	Target Customer Base
Military	Combat casualty care	Soldiers and support personnel in battlefield
Civilian	Medical monitoring	Patients: surgical recovery, psychiatric care
		Senior citizens: geriatric care, nursing homes
		Infants: SIDS prevention
		Teaching hospitals and medical research institutions
	Sports/performance monitoring	Athletes, individuals scuba diving, mountaineering, hiking
Space	Space experiments	Astronauts
Specialized	Mission critical/hazardous applications	Mining, mass transportation
Public safety	Fire fighting/law enforcement	Firefighters/police
Universal	Wearable mobile information infrastructure	All information-processing applications

in a variety of situations. These include patients recovering from surgery at home (e.g. after heart surgery), geriatric patients (especially those in remote areas where the doctor/patient ratio is very small), potential applications for patients with psychiatric conditions (depression/anxiety), infants susceptible to sudden infant death syndrome (SIDS) and individuals prone to allergic reactions (e.g. anaphylaxis reaction from bee stings).

3.4.2.5.4 *Sports and Athletics*

GTWM can be used for the continuous monitoring of the vital signs of athletes to help them track and enhance their performance. In team sports, the coach can track the vital signs and the performance of the player on the field and make desired changes in the players on the field depending on the condition of the player.

3.4.2.5.5 *Space Experiments*

GTWM can be used for monitoring astronauts in space in an unobtrusive manner. The knowledge to be gained from medical experiments in space will lead to new discoveries and the advancement of the understanding of space.

3.4.2.5.6 *Mission Critical/Hazardous Applications*

Monitoring the vital signs of those engaged in mission critical or hazardous activities, such as pilots, miners, sailors and nuclear engineers is one such application. Special-purpose sensors that can detect the presence of hazardous materials can be integrated into GTWM and enhance the occupational safety of the individuals.

GTWM is an effective and mobile information infrastructure that can be tailored to the individual's requirements to take advantage of the advancements in telemedicine and information processing. Just as special purpose chips and processors can be plugged into a computer motherboard to obtain the required information-processing capability, GTWM is the intelligent garment or information infrastructure into which the wearer can 'plug in' the desired sensors and devices. In short, the 'Wearable Motherboard' fulfils the twin roles of being: (1) a flexible information infrastructure that will facilitate the paradigm of ubiquitous computing; and (2) a system for monitoring the vital signs of individuals in an efficient and cost-effective manner with a 'universal' interface of clothing.

Moreover, GTWM has the potential to revolutionize a wide range of human endeavours and significantly enhance the quality of life. The widespread utilization of the technology can not only enhance the quality and accessibility of health care but also decrease its cost. At the same time, the potential for the new paradigm of wearable, flexible information-processing systems is waiting to be explored and GTWM represents a significant contribution in that direction.

3.5 Vital Signs Monitoring

Three EKG sensors, similar to the ones used in hospitals for EKG testing, were attached to the human subject, two on the left and one on the right side of the chest. The subject put on the GTWM just like any undershirt. The other side of the sensors on the body was 'plugged' into the T-connectors on the GTWM worn by the subject. The three leads from the EKG monitor were connected to the T-connectors at the bottom of GTWM. Thus, the heart rate information collected by the three sensors on the body passed through the T-connectors at the top, through the GTWM and through the leads at the bottom of the GTWM and into the EKG monitor. Initial testing was done at Crawford Long Hospital followed by another more extensive set of tests in the Department of Physiology at Emory University. The EKG trace from one of the tests at Emory University is shown in Figure 3.9. As seen from the figure, there is no difference between the control trace (using regular EKG sensors/setup) at the top and the trace.

In a combat situation, the vital signs and the penetration alert information will be transmitted from the GTWM into DARPA's personal status monitor worn by the soldier for onward trans-mission to the medical triage unit. The soldier's vital signs can be continuously monitored by the triage unit and, if a penetration alert is received (when the soldier is shot), medical assistance can be provided immediately to the individual needing it the most, thus optimizing the use of scarce medical resources in combat.

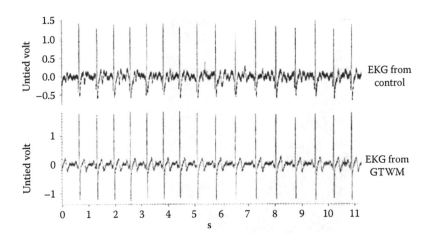

FIGURE 3.9
EKG trace from the GTWM.

3.6 Voice Monitoring

The voice-monitoring capabilities of GTWM are being tested and reported that a microphone is plugged into the T-connectors on the GTWM worn by the subject. When the subject speaks, the voice is transmitted through the GTWM and is tapped at the T-connectors at the bottom of the garment and recorded in a computer. When the voice is played back, it is clear, thus demonstrating the voice monitoring capabilities of GTWM. *Comfort testing:* A participant wearing GTWM continuously for long periods of time evaluated the garment's comfort. The participant's behaviour was observed to detect any discomfort and none was detected. The garment was also found to be easy to wear and take off, since it is similar to a typical undershirt in terms of the weight, the form and fit and the types of fibres in it. For monitoring acutely ill patients who may not be able to wear the Smart Shirt over the head (like a typical undershirt), Velcro™ and zipper fasteners are used to attach the front and back of the shirt, creating a garment with full monitoring functionality (Figures 3.10 and 3.11). Thus, a fully functional and comfortable Wearable Motherboard or Smart Shirt has been designed, developed and successfully tested for monitoring vital signs.

FIGURE 3.10
Velcro attachment on knitted Wearable Motherboard.

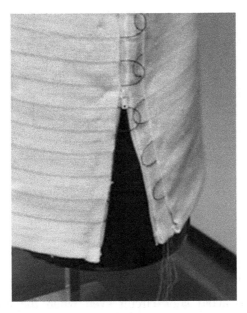

FIGURE 3.11
Zipper attachments on knitted Wearable Motherboard.

3.7 Design of Heat-Generating Circuit for Nichrome Fabric

The heat-generating circuit consists of nichrome wire, temperature sensor, microcontroller, battery and temperature controller. A flexible nichrome wire of resistance 7 Ω has been used as a heating element and is capable of heat up to 1850°F. In this, two LM35 temperature sensors are used for measuring the body temperature and nichrome wire temperature. The special feature of the sensor is: any change in the output voltage directly influences the temperature in linear scale. And also it has wide temperature range of –55°C to +150°C. PIC 12F675 microcontroller has been used to control the temperature. It consists of four channels analogue-to-digital converter (ADC) to convert the analogue-to-digital value. This part is used to measure the body temperature using thermistor. Microcontroller power supply battery of 4 × 1.2 V with 2.3 A h and heating coil power supply battery of 6 V with 10 A have been used and are capable of generating heat up to a maximum of 3 h. The ADC of the microcontroller converts the analogue value from the LM35 temperature sensor into digital value. For the conversion, voltage reference (V_{ref}) to microcontroller should be stable and has been set using a 3.3 V zener diode. LM35 temperature sensor is capable of generating 10 mV/°C, that is it generates 280 mV for a room temperature of 28°C and is used for the calibration of temperature

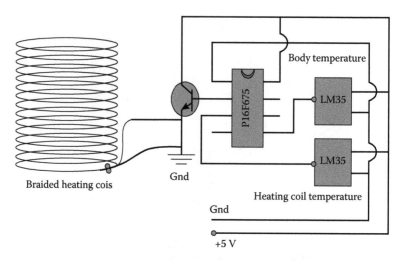

FIGURE 3.12
Temperature control circuit for heating garment.

measurements. To avoid short-circuit even when the microcontroller is switched *off*, the heating power supply is connected with normal open (NO) pin in the 6 V relay. After soldering the components in PCB, it has been coated using resins to coat over the board for waterproof. Figure 3.12 shows the diagram of temperature control circuit used for nichrome fabric.

3.7.1 Operation of Heat-Generating Circuit

When the power supply is switched *on*, green LED glows and it turns to orange within few seconds, indicating the microcontroller is in operation. The LM35-1 sensor, which is fixed to measure body temperature, converts the measured value into digital and compares with preset temperature value in switch-off position of the heating coil. If the value is equal to or above the preset temperature, then there will be no action and the microcontroller will remain idle. During this cycle, LED will glow orange and the operation continues after every second. Figure 3.13 shows the schematic diagram and connection details of the temperature control circuit.

Vdd – Power supply of 5 V from 4 × 1.2 V battery

Vss – Ground

GPIO.0 – Used for serial transmission

GPIO.1 – Used for serial reception

GPIO.2 – Connected with LM35-1 to measure the body temperature

GPIO.3 – Not used

GPIO.4 – Connected with LM35-2 to measure the heating coil temperature

GPIO.5 – Connected with BC547 to switch the power supply

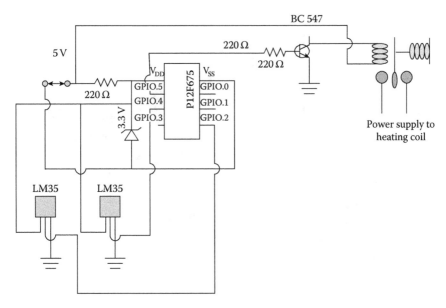

FIGURE 3.13
Schematic diagram of temperature control circuit.

In the aforementioned process, if the measured body temperature drops below preset temperature, microcontroller will switch *on* the heating power supply by switching the 6 V relay from normally open port (NO) to normally closed port (NC). LED status will indicate green during this process and the temperature of the heating coil is also monitored by means of LM 35-2 sensor by the microcontroller. The heating coil generates heat as long as its temperature is within preset temperature. Beyond this temperature, the microcontroller waits for 2 s and switches *off* the power supply to heating coil. The process restarts once the temperature decreases below 60°C. The heat-generating circuit can be integrated with the nichrome fabric for developing the heating garment.

3.8 Design of Communication Circuit for Copper Core Conductive Fabric

The copper core conductive fabric is attached with communication circuit for charging the mobile phone, and also, the fabric is integrated with temperature measurement circuit to measure the body temperature. The circuit diagram of mobile phone charger is as shown in Figure 3.14. The 220–240 V AC mains supply is down-converted to 9 V AC by transformer T1. The transformer output is rectified by BR1 and the positive DC supply

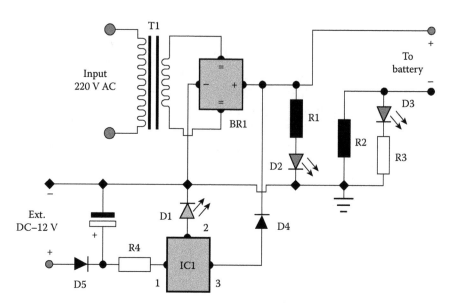

FIGURE 3.14
Mobile phone–charging circuit.

is directly connected to the charger's output contact, while the negative terminal is connected through current-limiting resistor R2. D2 works as a power indicator, with R1 serving as the current limiter and D3 indicates the charging status. During the charging period, about 3 V drop occurs across R2, which turns on D3 through R3. An external DC supply source (e.g. from a vehicle battery) can also be used to energize the charger, where R4, after polarity protection diode D5, limits the input current to a safe value. The three-terminal positive voltage regulator LM7806 (IC1) provides a constant voltage output of 7.8 V DC, since D1 connected between the common terminal (pin 2) and ground rail of IC1 raises the output voltage to 7.8 V DC. D1 also serves as a power indicator for the external DC supply. After constructing the circuit on a PCB, enclose it in a suitable cabinet.

Charging the mobile phone battery is a big problem while travelling as power supply source is generally not accessible. If the mobile phone is switched on continuously, the battery will drain within 5–6 h, making the mobile phone useless. A fully charged battery becomes necessary especially when the travelling distance is long. The circuit developed in this research work replenishes the mobile phone battery within 2–3 h. The mobile phone charger circuit is a current-limited voltage source. Generally, mobile phone battery requires 3.6–6 V DC and 180–200 mA current for charging. Current of 100 mA is sufficient for charging the mobile phone battery at the slow rate. A 12 V battery containing eight pen cells gives sufficient current, that is 1.8 A to charge the battery connected across the output terminals. The circuit also monitors the voltage level of the battery. It automatically cuts off the charging process when its output terminal voltage increases above the predetermined voltage level.

3.8.1 Temperature Measurement Circuit for Copper Core Conductive Fabric

The temperature measurement circuit is integrated with the copper conductive fabric and the fabric is tested for its functionality. Figure 3.15 shows the circuit to measure the body temperature. Here, negative temperature coefficient sensor platinum thermistor – 100 (PT-100) is used in which the resistance value is decreased when the temperature is increased. The thermistor is connected with resister bridge network, and the bridge terminals are connected to inverting and non-inverting input terminals of comparator. The comparator is constructed by TLO74C operational amplifier. Initially, the reference voltage is set to room temperature level, so the output of the comparator is zero. When the temperature is increased above the room temperature level, the thermistor resistance is decreased so that variable voltage is given to comparator. Then, the error voltage is given to next stage of preamplifier. Here, the input error voltage is amplified, and then, the amplified voltage is given to next stage of gain amplifier. The output voltage is given to final stage of DC voltage follower through this the output voltage is given to ADC and then to LCD unit. The mobile phone–charging circuit and the temperature-measuring circuit is integrated with the copper-conductive fabric to develop communication garment.

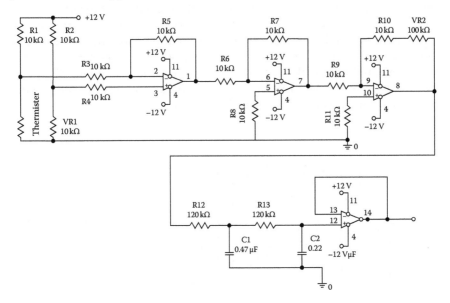

FIGURE 3.15
Circuit diagram for temperature measurement.

3.9 Design of Signal-Transferring Circuit for Optical Core Conductive Fabric

The optical core conductive fabric is integrated with the signal-transferring circuit for detecting the number and place of the bullet wounds. This optical core conductive fabric circuits can be used for developing the teleintimation garment. The block diagram for detecting the number and place of the bullet wound is shown in Figure 3.16. To detect the bullet would and location, it is decided to weave the POF in matrix format. The actual matrix format size for the finished garment will vary, depending upon the size of the garment. The circuit consists of AT 89C52 microcontroller to test the signal loss. Using this circuit, information about the number of bullets and bullet wound location can be derived. The signal collected from the soldier who wears the garment is being transmitted to the remote end server, where the details about the soldiers are kept in a database.

The optical core conductive fabric has been tested with the optical transmitter which contains a light source for transmitting light to receiver. Here, different light sources like red LED and white LED are used for testing purpose. The optical transmitter contains light source for transmitting light to receiver. Power supply for transmitter circuit is +5 V, and it is given to the LED through limiting resistor. The photodiode receives the signals from the optical core conductive fabric. Light source is used at the one end of the optical fibre, and a photodiode is used at another end of the optical fibre. When there is a light illumination on the photodiode, the output of the photodiode produces 0 V and when there is an illumination, it shows 5 V output in the display unit. When there is light illumination on the photodiode, the output of the photodiode produces 0 V output. Also, when there is no light illumination on the photodiode, the output of the photodiode produces 5 V output. The circuit set up to test the signal-transferring capability of optical core conductive fabric is shown in Figure 3.17.

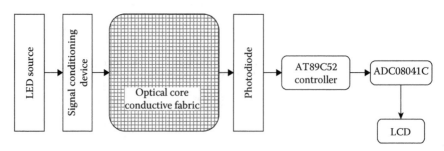

FIGURE 3.16
Block diagram of signal-transferring circuit with microcontroller AT89C52.

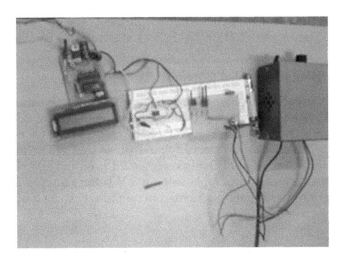

FIGURE 3.17
Signal-transferring circuit set-up.

3.10 Development of Illuminated System for POF Fabric

In the POF fabric, the system has been integrated to illuminate the fabric using different LEDs for different designs. The fabric consists of several bundles of fibres depending upon the illuminated portion of the garment. In this work, three different designs have been carried out, namely POLICE design and duck design. Each design consists of 4–10 bundles of POF fibres. An electro-optic transducer is used at each end of the fibre bundle to convert the electric signal into optical signal. In this system, three different LEDs were used as a source element. The specifications of the LEDs were mentioned in the Table 3.2.

The 3 V battery is connected with the fibre-optic cable which supplies power to the LED. The selected three different designs were positioned at

TABLE 3.2

LED Specifications

Colour	Green	White	Blue
Wavelength (nm)	520	430–700	470
Size (mm)	5	5	5
Directivity (degree)	15	20	15
DC reverse current (A)	50	50	50

the centre portion of the silhouette in each garment. The positioned designs were stitched using class 300 stitches by lock stitch sewing machine.

3.10.1 Design of Illuminated System

The constructional detail of the fibre-optic panel is shown in Figure 3.18. The illuminated system consists of a garment design made out of side emitting POF which is shown as (d) in Figure 3.18. The ends of the POF represented as (c) and (e) are attached to sleeve (b) and (f), so that LEDs can be easily attached to the system. The LEDs are represented as (a) and (g) in Figure 3.18. The LEDs are given power supply using a 3 V battery supply. Depending upon the designs used in the garment, different LEDs can be used to illuminate the fabric.

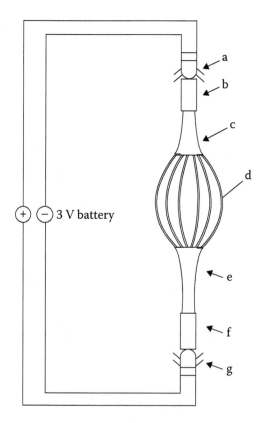

FIGURE 3.18
Fibre-optic panels (a) and (g) LED, (b) and (f) sleeve, (c) and (e) POF and (d) light-emitting portion of the panel.

3.11 Design of Bullet Wound Intimation Circuit for Teleintimation Fabric

The bullet wound intimation circuit was designed and developed to indicate the number and place of bullet wounds. For this purpose, bullet wound intimation circuits were developed using different controllers. These circuits are fabricated using flexible PCB to give comfort and easiness to the soldiers during combat situation. The bullet wound signals are transmitted to the soldier-monitoring station at the remote end. The bullet wound intimation circuits consist of two POF matrix pattern for left and right chest by weaving POF in matrix format: it can detect the location of bullet, where it wounded, and also, it count the number of bullets. Since the maximum bullet size used in military is 7.62 mm, the matrix should have the maximum pixel size of 5 mm of spacing between the fibres. The bullet wound and location detection circuit is developed from various stages, and it was tested with different controllers with different types of circuits to achieve the maximum flexibility and robustness. The circuit concept is divided into five modules as given in Table 3.3. The processed signals from the controller will be sent to the remote station to monitor the status of the soldier.

3.11.1 Circuits Using 8 × 8 Matrix Format

A prototype with 8 × 8 matrix format was developed, since the actual matrix format size for the finished garment varies with the size of the garment. This prototype serves as a basic platform from which required modification could be made. The transmitter unit detects the number of bullets and bullet wound location. This circuit is attached in a fabric, integrated with POF. A light source is used to transmit the signals and the fabric is continuously monitored for the signal transmission. 89C51 microcontroller was used to get the information about the number of bullets and bullet wound location,

TABLE 3.3

Bullet Wound Detection Circuit Types

Circuit Matrix Type	Controller	Transmission System Used
8 × 8	AT89C51	RF
64 × 64	AT89C52	RF
64 × 64 with SMD components	AT89S52	RF
Detachable circuits	PIC 18F877A	RF
80 × 80 flexible PCB	PIC18F4550	GSM

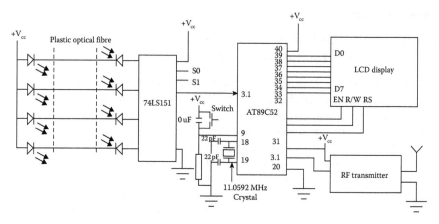

FIGURE 3.19
Circuit diagram for 8 × 8 matrix.

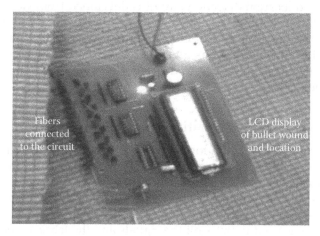

FIGURE 3.20
Transmitter circuit for 8 × 8 matrix.

if there is any signal loss because of the broken POF. Also, the information about the bullet detection is displayed in the LCD, and the same is being transmitted to the remote end receiver. The block diagram and components specifications are as shown in Figures 3.19 and 3.20 that show the circuit integrated with the garment.

3.11.2 Circuit Using 64 × 64 Matrix Format

In this module, the same circuit concept has been brought to 64 × 64 matrix. It has totally IRF14F 128 optical receivers arranged in 64 × 64 formats. The 74LS151 decoder/multiplexer has eight inputs and one output with three control lines; totally 16 decoders were used for this matrix. Each of the eight lines is coupled and given as input to the 74LS151 decoder.

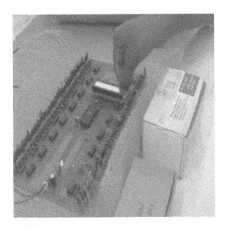

FIGURE 3.21
Column POF removed and its count and location in LCD.

Output from each decoder is given as input to one port of the microcontroller AT89C52. RF communication technique is used to send the signals to the remote location. When a POF is broken, microcontroller will count the number and location of the broken POF and bullet penetration is counted and displayed using LCD as shown in Figure 3.21. This circuit can be remodified with the required input lines according to the exact measurement of the garment.

3.11.3 Circuit Using SMD Components

The circuit size of the 64 × 64 is reduced to give comfort to the wearer by implementing the same 64 × 64 circuit using SMD components and AT89S52 microcontroller was used. The circuit is designed in L shape so that the row and column lines of POF could be directly connected to the circuit. The light source is given at one end of the POF and microcontroller AT89S52 is used with the 20 ports being connected to the 20 decoder units. Each decoder unit will connect to the eight photodiodes which in turn makes 160 photodiodes to be connected to the circuit. The RF module with frequency range of 300–433 MHz is being used at the transmitter end. Figure 3.22 shows the circuit integrated with the POF garment.

3.11.4 Circuits Using PIC Microcontroller with Detachable Circuits

The circuit is developed based on the flexibility and robustness with respect to the wearability issues. The detachable circuits can be easily removed and fixed into the garment. In this module, the bullet wound and location detection circuit using PIC microcontroller was developed and the block diagram is shown in Figure 3.23. The optical transmitter is used at the one end of the POF and another end is connected with the optical receiver. The comparator

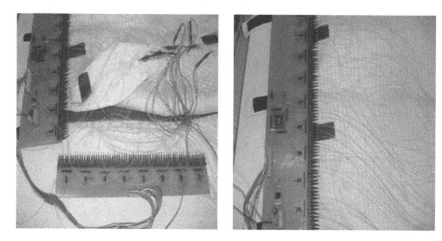

FIGURE 3.22
Teleintimation fabric with circuit.

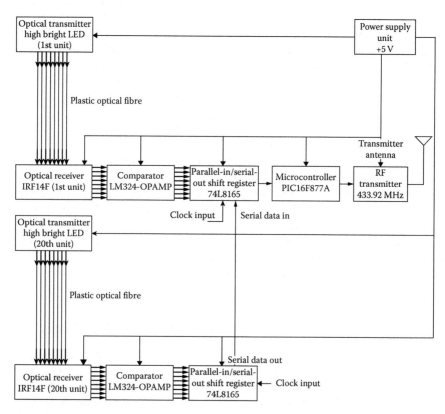

FIGURE 3.23
Block diagram of PIC microcontroller with detachable circuit.

unit compares the reference voltage with the output voltage of the photodiode. Depending on the presence or absence of illumination of light on the photodiode, high or low output is produced from the comparator, respectively. In this module, totally 20 × 8 optical receiver units are being used. The parallel-in–serial-out shift register is used to serialize the data and it is transmitted using 433.92 MHz RF transmitter.

3.11.5 Circuits Using Flexible PCB

In this method, flexible PCB was developed using PIC18F4550 controller to reduce the size of the circuit. All the sensing circuits are connected with microcontroller through 74HC573 latch and 74HC154 decoder are used to detect the wire cut from the connected optical cables. The block diagram is shown in Figure 3.24. It has 22 latches to collect input from photodiodes and the outputs from the latches are fed to controller for processing to determine both the horizontal and vertical latch numbers. GSM module is used to send information to remote station if the controller finds any break in optical cable. GSM module is interfaced to controller through UART interface. The flexible PCB and the circuit integrated garment are as shown in Figure 3.25.

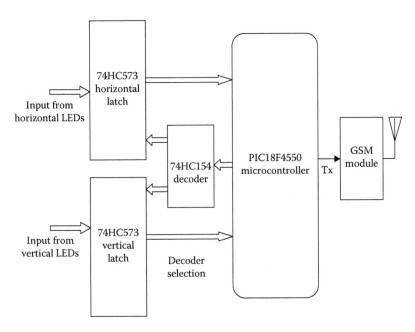

FIGURE 3.24
Block diagram of flexible PCB circuit.

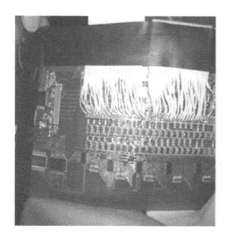

FIGURE 3.25
Flexible PCB integrated in the teleintimation fabric.

3.11.6 Method of Bullet Count

To count the number of bullet wounds in the soldier's body, voltage level at the receiver end of the POF is continuously monitored. If any POF is interrupted between the transmitter and receiver end, logic high signal is given to the microcontroller. A variable with a count increment will be made in the microcontroller if any port gives a logic high signal. This count is taken as number of bullet penetration into the garment, which is the bullet wound in the soldier's body. When there is no breakage, the microcontroller ports will be in logic low. The POF matrix format coordinates could do the bullet detection mechanism. This matrix format is shown in Figure 3.26. Whenever there is a bullet penetration in the garment, the particular co-ordinates will be affected and the location of bullet penetration can be detected.

1,1	1,2	1,3	1,4	1,5	1,6	1,7	1,8
2,1	2,2	2,3	2,4	2,5	2,6	2,7	2,8
3,1	3,2	3,3	3,4	3,5	3,6	3,7	3,8
4,1	4,2	4,3	4,4	4,5	4,6	4,7	4,8
5,1	5,2	5,3	5,4	5,5	5,6	5,7	5,8
6,1	6,2	6,3	6,4	6,5	6,6	6,7	6,8
7,1	7,2	7,3	7,4	7,5	7,6	7,7	7,8
8,1	8,2	8,3	8,4	8,5	8,6	8,7	8,8

FIGURE 3.26
POF matrix coordinates.

3.11.7 Bullet Location and Bullet Count Technique

The number of POF lines required for the efficient number of bullet counts and bullet wound location is 80 POF lines with 0.5 cm distance each in vertical section and 80 POF lines with 0.5 cm in the horizontal section. The data received from the PIC microcontroller are in the form of hexadecimal values. The corresponding binary value for $(07)_{16}$ is $(0000\ 0111)_2$ and it represents three bullets at second, first and zeroth location of the wearer and similarly for the other hexadecimal values ranging from 00 to 0F.

3.11.8 Problems Faced in Detecting the Bullet Location

When a fibre got cut at the location (1,1), the location could be displayed, but at the same time, when a bullet hits at the location (2,2), the new bullet wound location could be displayed. In addition to this, two more locations (1,2) and (2,1) are also displayed even though there is no bullet penetration. This is because already the location (1,1) got wounded; hence, whenever another bullet hits at the different row or column, the previously wounded row and column are taken into account. This gives the false reading of the bullet location. To overcome this issue, the bullet wound location in terms of rows and columns is displayed individually.

3.12 Circuits for Smart Shirt

In the Smart Shirt, to monitor the vital signs like body temperature, pulse rate and respiratory rate circuits were developed to integrate into the garment and the measured signals were sent to the remote station by mobile communication technology using telemonitoring system.

3.12.1 Circuits for Body Temperature Measurement

In the garment, temperature of the wearer can be determined by placing the sensor in the armhole of the soldier. The accurate body temperature measurement was tested using thermistor type – wire wound resistor. This sensor transfers the measured body temperature into analogue voltage, and in turn, it is then converted to digital voltage by means of ADC. The measured temperature signal is displayed in the monitoring device to indicate the body temperature. The conventional system of measuring instrument, thermometer, is replaced by thermistor temperature sensor of model FVT-UI with measurement range from 90°F to 109°F. This sensor is made up of base metal – nickel chromium and tungsten alloy, it is the heat sensor which can

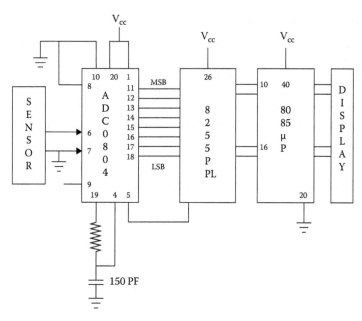

FIGURE 3.27
Temperature measurement circuit diagram.

sense both the maximum and minimum temperature and also transmits these signals in terms of digital values with operating voltage of 1.5 V DC, 300–600 mA. The main advantage of this body heat sensor is that the temperature measured is accurate, and the sensitivity and response time of the sensor is very high compared to other type of sensors. Figure 3.27 shows the circuit diagram to measure the body temperature. The body heat senses measured value in the analogue form and this signal is sent to ADC 0804 converter which converts these analogue signals to digital signals. The output from the ADC is sent to 8085 microprocessor through 8255 PPI. After processing, the microprocessor displays the value in the LCD unit.

3.12.2 Circuits for Pulse Rate Measurement

The pulse rate is measured by IR transmitter and receiver. Figure 3.28 shows the circuit to measure the pulse rate in the blood flow. Infrared transmitter is one type of LED which emits infrared rays generally called IR transmitter. Similarly, IR receiver is used to receive the IR rays transmitted by the IR transmitter. Both the IR transmitter and the receiver should be placed on a straight line to each other. The IR transmitter and the receiver are placed in the pulse rate sensor. To measure the pulse rate, the pulse rate sensor has to be clipped in the finger. The IR receiver is connected to the V_{cc} through the resistor which acts as potential divider. The potential divider output is connected to amplifier section. Figure 3.29 shows the pulse rate sensor

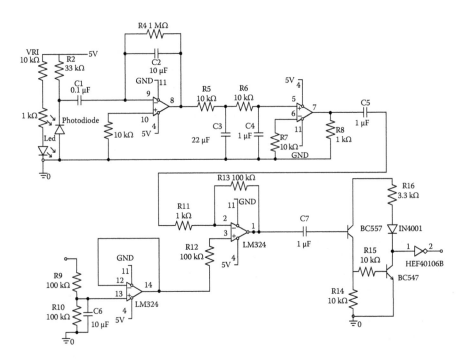

FIGURE 3.28
Circuit diagram of pulse rate sensor unit.

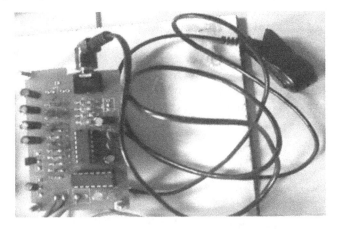

FIGURE 3.29
Pulse rate sensor circuit.

circuit diagram and the testing method. When the supply is *on*, IR transmitter passes the rays to the receiver. Depending on the blood flow, the IR rays are interrupted. Due to that IR receiver, conduction is interrupted and then the final square wave signal is given to microcontroller or other interfacing circuit in order to monitor the pulse rate.

3.12.3 Circuits for Respiration Rate Measurement

The circuit is designed to measure the respiration rate. In this circuit, two thermistors are used for the respiration measurements that are connected in the resistor bridge network. Here, one thermistor is used for the respiration measurement. Another thermistor is used as reference which measures the room temperature. Then, the error voltage is amplified by the next stage of the amplifier, and then the final TTL pulse is given to microcontroller in order to monitor the respiration rate. Figure 3.30 shows the respiration rate sensor circuit diagram.

3.12.4 Circuits for Telemonitoring System

The measured vital signs are transmitted to the remote system using telemonitoring system. Figure 3.31 shows the general block diagram of the telemonitoring systems. The measured vital signs, temperature, respiration rate and pulse rate were sent to microcontroller unit. Microcontroller sends these signals to the LCD display and to modem (mobile) through RS 232 cable using MAX 232 IC, the level logic converter. The remote station can receive these data by using mobile phone and know the status of the wearer. The vital sign–measuring circuitry and the microcontroller unit need a power supply. The microcontroller unit needs 5 V regulated DC source and the measuring circuits require regulated DC source of 12, −12 and 5 V. PIC16F877A microcontroller is used for processing the signals received from the measurement circuits. Figure 3.32 shows the block diagram of microcontroller unit interfaced with LCD to display the measured values.

3.12.4.1 4RS232 Communication Unit

In telecommunication, RS-232 is a standard for serial binary data interconnection between a data terminal equipment (DTE) and a data circuit-terminating

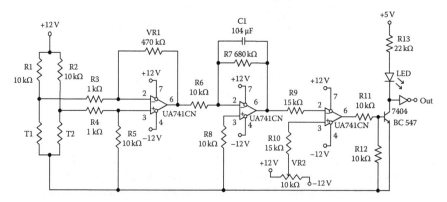

FIGURE 3.30
Respiratory rate measurement circuit.

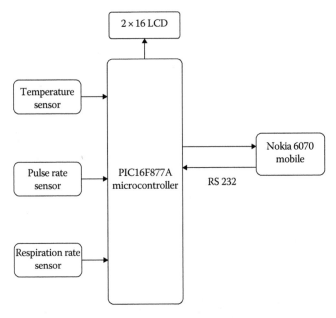

FIGURE 3.31
Block diagram of telemonitoring system.

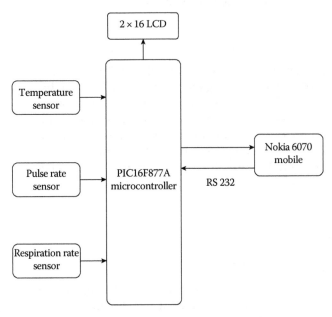

FIGURE 3.32
Microcontroller interfaced with LCD for telemonitoring system.

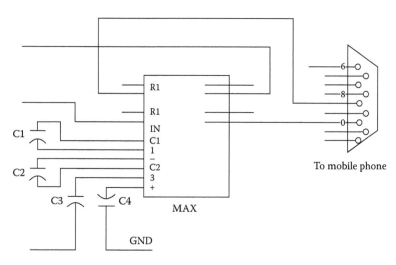

FIGURE 3.33
Circuit diagram of RS232 communication unit.

equipment (DCE). It is commonly used in computer serial ports. In the circuit shown in Figure 3.33, MAX 232 IC is used as level logic converter. The MAX232 is a dual driver/receiver that includes a capacitive voltage generator to supply EIA 232 voltage levels from a single 5 V supply. In this circuit, the microcontroller transmitter pin is connected in the MAX232 T2IN pin which converts input 5 V TTL/CMOS level to RS232 level.

3.12.4.2 Mobile Phone (MODEM)

Mobile phone with GPRS enabled like Nokia 6070 can be used as a modem to send data to other mobile using mobile network. This can be achieved using RS232 cable and AT commands for connection of mobile and microcontroller unit. It is also possible to record and transmit vital signal in combat situations; thus, mobile connection allows continuous monitoring of the soldier. Figure 3.34 shows the complete setup with the mobile phone; 2 × 16 liquid crystal display (LCD) is used for displaying the vital parameters. Figure 3.35 shows the vital signs displayed in the LCD unit.

3.13 Future Research Scope

In the last half of the twentieth century, significant technological advances in various disciplines have led to a plethora of technologies designed to enhance our lifestyles. The 1990s and early 2000s created a multidisciplinary revolution, bringing in together technologies from a wide range of disciplines to

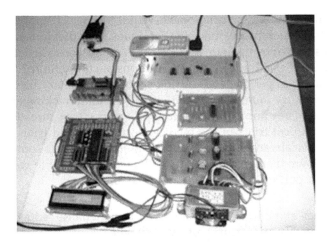

FIGURE 3.34
Complete set-up of telemonitoring system.

FIGURE 3.35
LCD unit displays the vital signs from the telemonitoring system.

create multifunctional products. The smart and interactive clothing is one typical example of the collaboration of a number of different experts and researchers in the electronics, information technology, material science, textile technology and textile and fashion design, among others. Two of the major markets are the American and Japanese ones, both of which having invested (and still investing) large sums on research in this discipline. However, groundbreaking advances are also being reported in Europe, in particular the United Kingdom, Finland, France, Belgium and Italy. China, including Hong Kong, is also expected to bring in a large contribution to the research and industrial community in the area of smart clothing. As it stands, significant progress has already been achieved in the design and development of smart interactive garments for monitoring applications.

Research on the individual components required for such applications (sensors, actuators, microprocessors, etc.) continues with dynamism, fuelled by the interests shown by the industry. Research on the construction of interactive garments is also buoyant for applications such as clinical and general health monitoring, performance during sports and military uses, including life sign and injury monitoring, position and location tracking and communication systems. A number of working prototypes have already been produced, and clinical trials in the medical sector and field trials in extreme conditions have already started. Mass commercialization of some systems is not too far down the line, but work is still needed in order to resolve all the technicalities, particularly with regard to the following:

- Accuracy and reliability of data measurement (noise reduction); reliability, safety and security of data transfer
- Minimization of the number of additional attachments (simplification of the systems by multi-functional textiles)
- Efficiency of power management, including power generation
- Durability of the systems, including washability and long-term accuracy in performance; user-friendliness, including the ability to use/wear the garments without the essential need for assistance
- Mobility, including weight reduction
- Cost versus product lifespan or durability
- Comfort and physical aesthetics, including breathability, absorbency, drape, handle, etc.

The last point raises the importance of aesthetic and product design in what appears to be principally technical and functional textiles. The concept of product development includes the whole package of design, technology, functionality and logistics. While the technology has been established in many areas, creating the end-use functionality is still a subject of major continuing and maturing research. The design component of research for such applications by contrast is still in its early stages, but with the advent of a new generation of multidisciplinary researchers, it is growing rapidly. The logistic and sociological aspects have not been fully explored, primarily because commercialization of such products has not yet occurred on a large scale.

As outlined by Gould (2003), privacy and security of data transfer is an issue that needs to be resolved by the research community. With the current systems for short-range communications, encryption and decryption systems have to be reinforced in order to prevent 'accidental' transfer of personal data between individuals wearing smart interactive devices. The rights to privacy and data protection are issues that are strongly felt by consumers and various human rights groups alike. In this line of thought, there have been some reactions against the inclusion of RFID in high street clothing items due to privacy issues (Ilic 2004). Concerns can also be raised with the illicit or non-consensual use of such devices. There are still many practical, technical and non-technical challenges that have to be met, and it is anticipated that the coming years will witness an even greater level of multidisciplinary collaboration, in particular bringing in now more design, sociological and legal aspects. With the increasing interest of consumers in lifestyle products, it is expected that the applications for interactive monitoring garments will expand out of institutional use (hospitals, military, rescue services), where it is currently focused, into individual consumer use. The growing ageing population could be an influential factor: flexible wearable electronic clothing systems would facilitate the independence of the aged and disabled, by allowing them to get on with their day-to-day activities, while still being monitored. However, in order for this level of commercialization to be reached, convenience, reliability and accuracy have to be improved; costs have to be brought down and the supporting infrastructure for the monitoring and assistance need to be developed. Again, this highlights the necessity for multidisciplinary industrial collaboration.

Given the possibility of embedding both computation and interconnect in e-textiles, the necessity of providing a set of meaningful and reasonable set of 'services' for the application at hand has become important. Two broad categories of e-textile applications are envisioned: wearable and large-scale non-wearable. Many specific applications in the field of wearable computing have been envisioned and realized, though most suffer bulky form factors. In the new field of large-scale non-wearable e-textiles, applications include large-scale acoustic beam–forming arrays, self-steering parafoils and intelligent, inflatable decoys. Both categories of e-textile applications share three common design goals: low-cost, durable and long running.

1. *Low cost* dictates the use of inexpensive, off-the shelf electronic components (COTS) and yarns as well as the design of weaves/architectures that are manufacturable in current or slightly modified textile production systems.

2. *Durable* dictates that the system must tolerate faults, both permanent and transient, that are inherent in the manufacture and use of the device. In addition, there is an expectation that individual components may not be repairable and that system functionality should gracefully decline as components fail.

3. *Long running* dictates that the system must manage power consumption in an application-aware fashion to minimize the need for bulky batteries and/or external power recharge. Power scavenging and distributed power management are essential.

In addition to these constraints, wearable applications should have a comfortable, flexible and unobtrusive form factor if they are to be adopted by a large number of users, while non-wearable applications should be able to operate with hundreds to thousands of components on fabrics that are tens of meters in length. By their nature, wearable e-textiles have a more complex form factor and are in motion more often than their non-wearable counterparts. E-textiles are in a transition from a few isolated computing and sensing elements on the fabric to a network of many computing and sensing elements distributed over the entire textile. Compared to previous work in distributed systems, e-textiles will be physically spread over a relatively smaller space, but will have a greater dependence on physical locality of computation, lower bandwidth for communication and tighter constraints on energy usage. Thus, e-textiles represent an extreme corner of the computing design space. Based on this design space, this section seeks to rationalize a set of services that form a 'backplane' for a wide array of e-textile applications, including wearable and non-wearable applications.

This section describes two e-textile applications and identifies their common software service needs. In the category of wearable computing, a garment designed to provide the user with precise location information within a building is analyzed. A large-scale, non-wearable acoustic beam–forming array is analyzed later in the section. Both textiles are woven as opposed to alternative textile manufacturing techniques such as knitting, embroidering or non-woven technologies. The physical media for communication on these textiles will be pliable, durable bundles of very fine metal fibres. Virtually, all woven e-textiles are expected to use the new fibre batteries and solar cells under development. This will lead to highly distributed, redundant power supplies for e-textiles in which some parts of the textile have more remaining power than others. These two applications were chosen as reasonably representative of the nascent e-textile technology in that each combines a range of sensors, reasonable computing requirements and a small number of actuators. Both applications benefit from the wide distribution and large quantity of sensors that can be distributed across the textile. The mapper garment is currently under design at Virginia Tech with aspects currently in the prototyping phase. Single cluster e-textile beam-forming arrays have been constructed at Virginia Tech and a multi-cluster array is under construction.

1. *Mapper garment*: The mapper garment tracks the motion of the user through a structure by monitoring the user's body position, the user's movement and the distance of the user from surrounding obstacles. Such a garment would allow users to be given directions in a building, maintenance workers to be automatically shown blueprints for their current room or users to automatically map existing structures. The user's body position is measured by a set of piezoelectric strips woven into the clothing; by measuring the deformation along 10s of strips, the physical configuration of the user's body can be detected. The user's activity, such as walking upstairs, climbing a ladder or walking on a flat surface, can be determined. The user's movement rate can be measured by a small set of discrete accelerometers as

well as a digital compass attached to the garment. The distance from obstacles is measured using ultrasonic signals. The primary challenge in this application is interfacing to a large number of sensors and actuators in a reliable fashion. Simply attaching the leads of every sensor/actuator to a single processing unit and power supply would not meet the design goal of a durable e-textile. In the event of a tear in the fabric, single leads running to one collection point could lead to significant rather than graceful degradation in performance. The garment needs multiple points at which analogue data are converted to digital data; these conversion units, likely in the microcontroller or digital signal processor class, would need to communicate within a fault-tolerant network. By carefully managing which sensors and processing units are active, the power requirements of the system can be reduced. For example, the computation of body position can be accomplished with varying numbers of sensors depending on the number and type of positions, among which the garment is trying to distinguish. By selecting sensors from around the garment according to local available power, power use can be balanced across the e-textile. When determining the location of a wall, the garment must activate a transmitter in the direction desired and then sample a receiver for the return signal. This will accurately compute the distance, but gives no information on the *direction* in which it is located. To compute direction, the location of the transmitter on the body, the position of that part of the body relative to the torso and the direction in which the user's torso is pointing must be known. A number of techniques are available for determining the location of one part of the e-textile with respect to other parts of the e-textile; in this garment, the body position is available.

2. *Beam-forming array*: The beam-forming array textile gathers data from a large array of acoustic sensors and analyzes this data to determine the direction of an acoustic emitter (e.g. a moving vehicle or a human voice). Through the use of acoustic beam–forming algorithms, a set of three acoustic sensors can identify the direction of a single emitter if the sensors and the emitter are all in the same plane. Given noise and potential miscalibrations in the acoustic data, the use of redundant acoustic sensors is advisable.

E-textiles may prove to be the next successors of smart cards in allowing people to seamlessly move from one ambient to the other, with the clear advantage of being close to the human body and, thus, being able to provide biometric authentication, inconspicuously and without intrusion. At the same time, e-textiles will be able to provide possibly the best support to impaired individuals by enabling improved sensing, in addition to actuation, both embedded in regular garments. Finally, special segments of the market (such as medical, military, law enforcement, as well as sports training) may also benefit from the already available e-textile-based technology for vital signs monitoring or remote triage.

3.14 Summary

In this chapter, the design of circuits, selection of controllers for different fabrics, types of PCBs, programming methods for controlling the various signals and the testing methodology were discussed. The heat-generating circuit has been developed for heating garment and it consists of nichrome wire, LM35 temperature sensor which is capable of generating 10 mV per degree centigrade, that is it generates 280 mV for a room temperature of 28°C and which is used for the calibration of temperature measurements, and PIC 12F675 microcontroller has been used to control the temperature. The LM35 temperature senor and heating coil power supply battery of 6 V with 10 A is capable of generating heat up to a maximum of 3 h. This heat-generating circuit has been integrated with the nichrome fabric to develop heating garment. The mobile phone–charging circuit and body temperature measurement circuit has been developed to integrate with the copper core conductive fabric for communication garment. The mobile phone–charging circuit consists of three-terminal positive voltage regulator LM7806 (IC1) that provides a constant voltage output of 7.8 V DC. A 12 V battery containing eight pen cells provides 1.8 A to charge the battery connected across the output terminals. The temperature measurement circuit is integrated with the copper conductive fabric, and the fabric is tested for its functionality. The temperature measurement circuit consists of PT-100 temperature sensor to measure the body temperature. Here, negative temperature coefficient is used in which the resistance value is decreased when the temperature is increased. The measured values from the temperature sensor are sent to ADC and then to LCD unit.

The signal-transferring circuit has been developed to integrate with optical core conductive fabric to develop communication garment. It consists of AT 89C52 microcontroller to test the signal loss. Here, different light sources like red LED and white LED are used for testing purpose. Using this circuit, information about the number of bullets and bullet wound location can be derived. The illuminated system has been developed for illuminated garment. It consists of several bundles of fibres and an electro-optic transducer is used at each end of the fibre bundle to convert the electric signal into optical signal. In this system, three different LEDs were used as a source element. A 3 V battery is connected with the fibre-optic cable which supplies power to the LED. The system has been integrated to illuminate the fabric using different LEDs for different designs. In this work, three different designs have been carried out, namely, stop design, police design and duck design. Each design consists of 4–10 bundles of POF fibres. The bullet wound intimation circuit was designed and developed to indicate the number and place of bullet wounds in the teleintimation fabric. For this purpose, bullet wound intimation circuits were developed using different controllers like AT89C51, AT89C52, AT89S52, PIC 18F877A and PIC 18F4550 for different POF matrix arrangements. The bullet wound and location detection circuit is developed from various stages, and it was tested with different controllers with different types of circuits to achieve the maximum flexibility and robustness. The processed signals from the controller have been sent to the remote station using RF and GSM technology for monitoring the status of the soldier. From the test results, it was concluded that the number of POF lines required for the efficient number of bullet counts and bullet wound location is 80 POF lines with 0.5 cm distance each in vertical section and 80 POF lines with 0.5 cm in the horizontal section. The body temperature, pulse rate measurement and respiratory rate measurement circuits have been developed for Smart Shirt. The accurate body temperature measurement was tested using thermistor type – wire wound resistor. The measured temperature signals were processed using 8085 microprocessor to display it in LCD unit. The pulse rate is measured by IR transmitter and receiver, and respiratory rate is measured using the circuit with two thermistors connected in the resistor bridge network. The measured vital signs are transmitted to the remote system using telemonitoring system consisting of PIC16F877A microcontroller that is used for processing the signals received from the measurement circuits. Mobile phone with GPRS enabled like Nokia 6070 can be used as a modem to send data to the remote station.

Bibliography

Anderson, K. and A.M. Seyam, The road to true wearable electronics, In *Proceedings of the Textile Institute's 82nd World Conference*, Cairo, Egypt, pp. 23–27 (2002).

Ashok Kumar, L. and A. Venkatachalam, Electro textiles: Concepts and challenges, In *Proceedings of National Conference on Functional Textiles and Apparels*, PSG Tech, Coimbatore, India, pp. 75–84 (2007).

Ashok Kumar, L. and C. Vigneswaran, Development of signal transferring fabrics using plastic optical fibres (POF), *Non Woven Tech. Textiles*, 1(2), 40–45 (2008).

Belardinelii, G., R. Palagi, A. Bedini, V. Ripoli, A. Macellari and D. Franchi, Advanced technology for personal biomedical signal logging and monitoring, In *Proceedings of the 20th Annual International Conference of the IEEE Engineering in Medicine and Biology Society*, Hong Kong, vol. 3, pp. 1295–1298 (1998).

Berry, P., G. Butch and M. Dwane, Highly conductive printable Inks for flexible and low-cost circuitry, In *Proceedings of the 37th International Symposium on Microelectronics*, IMAPS, Long Beach, CA, pp. 167–181 (2004).

Burchard, B., S. Jung, A. Ullsperger and W.D. Hartmann, Devices, software, their applications and requirements for wearable electronics, In *Proceedings of the ICCE*, Los Angeles, CA, pp. 224–225 (2001).

Calvert, T., B. Corbett and J.D. Lambkin, 80°C continuous wave operation of an AlGaInP based visible VCSEL, *Electron. Lett.*, 83, 222–223 (2002).

Catrysse, M., R. Puers, C. Hertleer, L. Van Langenhove, H. Van Egmond and D. Matthys, Towards the integration of textile sensors in a wireless monitoring suit, *Sens. Actuators*, 114, 302–311 (2004).

Cheng, K.B., S. Ramakrishna and T.H. Ueng, Electromagnetic shielding effectiveness of stainless steel/polyester woven fabrics, *Textile Res. J.*, 71(1), 42–49 (2001a).

Cheng, K.B., T.H. Ueng and G. Dixon, Electrostatic discharge properties of stainless steel/polyester woven fabrics, *Textile Res. J.*, 71(8), 732–738 (2001b).

Choi, S. and Z. Jiang, A novel wearable sensor device with conductive fabric and PVDF film for monitoring cardio respiratory signals, *Sens. Actuators A: Phys.*, 128, 317–326 (2006).

Corbett, B., P. Maaskant, M. Akthter, J.D. Lambkin, B. Beaumont, M.A. Posisson, N. Proust et al., High temperature nitride sources for plastic optical fibre data buses, In *Proceedings of the 10th International Plastic Optical Fibres Conference*, Amsterdam, the Netherlands, pp. 81–87 (2001).

Daum, W., J. Krause, P.E. Zamzow and O. Riemann, *POF Polymer Optical Fibers for Data Communication*, Springer, Berlin, Germany, 2002.

De Neve, H., J. Blondelle, P. Van Daele, P. Demeester, R. Baets and G. Borghs, Recycling of guided mode light emission in planar micro cavity light emitting diodes, *Appl. Phys. Lett.*, 70, 799–801 (1997).

De Rossi, D., A. Della Santa and A. Mazzoldi, Dress ware: Wearable hardware, *Mater. Sci. Eng.*, C7, 31–35 (1999).

De Rossi, D., F. Lorussi, A. Mazzoldi, R. Paradiso, E.P. Scilingo and A. Todnetti, Electro active fabrics and wearable biomonitoring devices, *AUTEX Res. J.*, 3(4), 180–183 (2003).

Delbeke, D., R. Bockstaele, P. Bienstman, R. Baets and H. Benisty, High-efficiency semiconductor resonant-cavity light-emitting diodes: A review, *IEEE J. Sel. Top. Quantum Electron.*, 8, 189–206 (2002).

Dhawan, A., T.K. Ghosh, A. Seyam and J.F. Muth, Woven fabric-based electrical circuits. Part II: Yarn and fabric structures to reduce crosstalk noise in woven fabric-based circuits, *Textile Res. J.*, 74, 955–960 (2004a).

Dhawan, A., A.M. Seyam, T.K. Ghosh and J.F. Muth, Woven fabric-based electrical circuits Part I: Evaluating interconnect methods, *Textile Res. J.*, 74, 913–919 (2004b).

Dobrev, D. and I. Daskalov, Two-electrode telemetric instrument for infant heart rate and apnea monitoring, *Med. Eng. Phys.*, 20(10), 729–734 (1998).

Dumitrescu, M., M. Saarinen, M.D. Guina and M. Pessa, High-speed resonant cavity light-emitting diodes at 650 nm, *IEEE J. Sel. Top. Quantum Electron*, 8, 219–230 (2002).

Dunne, L.E., S. Brady, B. Smyth and D. Diamond, Initial development and testing of a novel foam-based pressure sensor for wearable sensing, *J. NeuroEng. Rehabil.*, 2, 4 (2005). doi:10.1186/1743-0003-2-4.

Edmison, J., M. Jones, Z. Nakad and T. Martin, Using piezoelectric materials for wearable electronic textiles, In *Proceedings of the Sixth International Symposium on Wearable Computers*, Los Alamitos, CA, pp. 41–48 (2002).

Eitan, B. and S. Wagner, A woven inverter circuit for e-Textile applications, *IEEE Electron Device Lett.*, 25(5), 187–198 (2004).

Elton, S.F., Ten new developments for high-tech fabrics and garments invented or adapted by the research & technical group of the defense clothing and textile agency, In *Proceedings of the Avantex, International Symposium for High-Tech Apparel Textiles and Fashion Engineering with Innovation-Forum*, Frankfurt-am-Main, Germany, p. 67 (2000).

Eves, D., J. Green, C. Van Heerden, J. Mama and S. Marzano, *New Nomads – An Exploration of Wearable Electronics*. Uitgeverij, Rotterdam, the Netherlands, p. 10, 2001.

Farringdon, J., A.J. Moore, N. Tilbury, J. Church and P.D. Biemond, Wearable sensor badge and sensor jacket for context awareness, In *The Third International Symposium on Wearable Computers, Digest Papers*, San Francisco, CA, pp. 107–113 (1999).

Federico, L., W. Rocchia, E. Pasquale Scilingo, A. Tognetti and D. De Rossi, Wearable redundant fabric-based sensor arrays for reconstruction of body segment posture, *IEEE Sens. J.*, 4(6), 1241–1256 (2004).

Gemperle, F., C. Kasabach, J. Stivoric, M. Bauer and R. Martin, Design for wearability, In *Proceedings of the Second International Symposium on Wearable Computers*, Pittsburgh, PA, pp. 116–122 (1998).

Gibbs, P.T. and H.H. Asada, Wearable conductive fiber sensors for multi-axis human joint angle measurements, *J. Neuro Eng. Rehabil.*, 2(1), 7–18 (2005).

Gilhespy, N., Enabling soft sensor solutions, In *Proceedings of the Wearable Electronic and Smart Textiles Seminar*, Leeds, IA, pp. 468–471 (2004).

Gnade, B., T. Akinwande, G. Parsons, S. Wagner and R. Shashidhar, Active devices on fiber: The building blocks for electronic textiles, In *International Interactive Textiles for the Warrior Conference*, Cambridge, MA, pp. 17–23 (2002).

Golding, A.R. and N. Lesh, Indoor navigation uses a diverse set of cheap, wearable sensors, In *Proceedings of the Third International Symposium on Wearable Computers*, Washington DC, pp. 29–36 (1999).

Gopalsamy, C., S. Park, R. Rajamanickam and S. Jayaraman, The wearable motherboard: The first generation of adaptive and responsive textile structures (ARTS) for medical applications, *J. Virtual Reality*, 4, 152–168 (1999).

Gorlick, M.M., Electric suspenders: A fabric power bus and data network for wearable digital devices, In *Digital Papers, Proceedings of the Third International Symposium on Wearable Computers*, San Francisco, CA, IEEE, pp. 114–121 (1999).

Gould, P., Textiles gain intelligence, *Mater. Today*, 6, 38–43 (2003).

Goulev, P., L. Stead, E. Mamdani and C. Evans, Computer aided emotional fashion, *Comput. Graph.*, 28, 657–666 (2004).

Haase, J., Neurosurgical tools and techniques – Modern image-guided surgery, *Neurol. Med.*, 38, 303–307 (1998).

Hartmann, W.D., K. Steilmann and A. Ullsperger, *High-Tech Fashion*, Heimdall Verlag, Witten, Germany, 2000.

Hum, A.P.J., Fabric area network – A new wireless communications infrastructure to enable ubiquitous networking and sensing on intelligent clothing, *Comput. Netw.*, 35, 391–399 (2001).

Intanagonwiwat, C., R. Govindan and D. Estrin, Directed diffusion: A scalable and robust communication paradigm for sensor networks, In *Sixth Annual International Conference on Mobile Computing and Networks (MobiCOM 2000)*, Boston, MA, pp. 85–96 (2000).

Jayaraman, S., Designing a textile curriculum for the'90s: A rewarding challenge, *J. Textile Inst.*, 81(2), 185–194 (1990).

Jones, D., Wearable electronics enabled by soft switch technology, In *Proceedings of the Wearable Electronic and Smart Textiles Seminar*, Leeds, U.K., pp. 283–300 (2004).

Jung, S., C. Lauterbach, M. Strasser and W. Weber, Enabling technologies for disappearing electronics in smart textiles, In *Proceedings of IEEE International Solid-State Circuits Conference*, San Francisco, CA, pp. 71–77 (2003).

Jung, S., C. Lauterbach and W. Weber, A digital music player tailored for smart textiles: First results, Presented at the *Avantex Symposium*, Frankfurt, Germany, pp. 51–54 (2002).

Kallmayer, C., New assembly technologies for textile transponder systems, In *Proceedings of Electronic Components and Technology Conference*, New Orleans, LA, pp. 48–67 (2003).

Kim, G.B.C., G.G. Wallace and R. John, Electro formation of conductive polymers in a hydrogel support matrix, *Polymer*, 41, 1783–1790 (2000).

Kirstein, T., D. Cottet, J. Grzyb and G. Troster, Wearable computing systems – Electronic textiles. In Tao, X. (ed.), *Wearable Electronics and Photonics*. Woodhead Publishing in Textiles, Houston, TX, pp. 177–197 (2005).

Knigge, A., M. Zorn, H. Wenzel, M. Weyers and G. Trankle, High efficiency again-based 650 nm vertical-cavity surface-emitting lasers, *Electron Lett.*, 37, 1222–1224 (2001).

Koncar, V., E. Deflin and A. Weill, Communication apparel, optical fibre fabric display. In Tao, X. (ed.), *Wearable Electronics and Photonics*, Woodhead Publishing in Textiles, Houston, TX, pp. 155–176 (2005).

Krebs, F.C., M. Biancardo, B. Winther-Jensen, H. Spanggard and J. Alstrup, Strategies for incorporation of polymer photovoltaics into garments and textiles, *Solar Energ. Mater. Solar Cells*, 90, 1058–1067 (2005).

Lam Po Tang, S., Recent developments in flexible wearable electronics for monitoring applications, *Trans. Inst. Measure. Control*, 29(3), 283–300 (2007).

Lauterbach, C., M. Strasser, S. Jung and W. Weber, Smart clothes self-powered by body heat, In *Avantex Symposium*, Frankfurt, Germany, pp. 83–87 (2002).

Linz, T. and C. Kallmayer, Embroidering electrical interconnects with conductive yarn for the integration of flexible electronic modules into fabric, In *Proceedings of IEEE ISWC*, Osaka, Japan, pp. 28–41 (2005).

Lorussi, F., W. Rocchia, E.P. Scilingo, A. Tognetti and D. De Rossi, Wearable, redundant fabric-based sensor arrays for reconstruction of body segment posture, *IEEE Sens. J.*, 4(6), 807–818 (2004).

Lorussi, F., E.P. Scilingo, M. Tesconi, A. Tognetti and D. De Rossi, Strain sensing fabric for hand posture and gesture monitoring, *IEEE Trans. Inf. Technol. Biomed.*, 9(3), 372–381 (2005).

Lukowicz, P., H. Junker, M. Stäger, T. von Büren and G. Tröster, Wear NET: A distributed multi-sensor system for context aware wearable's, In *Proceedings of the Fourth International Conference on Ubiquitous Computing*, Sweden, pp. 361–370 (2002).

Mackenzie, K., D. Hudson, S. Maule, S. Park and S. Jayaraman, A prototype network embedded in textile fabric, In *Proceedings of the International Conference on Compilers, Architecture, and Synthesis Embedded Systems*, New York, pp. 188–194 (2001).

Malloy, H.M. and J.J. Hoffman, Home apnea monitoring and sudden infant death syndrome, *Preventive Med.*, 25, 645–649 (1996).

Mann, S., Smart clothing: The wearable computer and wear cam, *Person. Technol.*, 1, 21–27 (1997).

Marculescu, D., R. Marculescu and P. Khosla, Challenges and opportunities in electronic textiles modeling and optimization, In *Proceedings of the ACM/IEEE 39th Design Automation Conference*, New Orleans, LA, pp. 175–180 (2002).

Matsuoka, T., T. Ito and T. Kaino, First plastic optical fibre transmission experiments using 520 nm LEDs with intensity modulation/direct detection, *Electron Lett.*, 36, 1836–1837 (2000).

Mayol, W.W., B. Tordoff and D.W. Murray, Wearable visual robots, *Person. Ubiquitous Comput.*, 6, 37–48 (2002).

Mazzoldi, D., D. De Rossi, F. Lorussi, E.P. Scilingo and R. Paradiso, Smart textiles for wearable motion capture systems, *AUTEX Res. J.*, 2(4), 54–68 (2002).

Mei, T., Y. Ge, Y. Chen, L. Ni, W. Liao, Y. Xu and W.J. Li, Design and fabrication of an integrated three-dimensional tactile sensor for space robotic applications, *IEEE MEMS*, 14, 112–117 (1999).

Michahelles, F., P. Matter, A. Schmidt and B. Schiele, Applying wearable sensors to avalanche rescue: First experiences with a novel avalanche beacon, *Comput. Graph.*, 27, 839–847 (2003).

Milojicic, D., W. LaForge and D. Chauhan, Mobile objects and agents, In *Proceedings of the USENIX Conference Object-Oriented Technologies and Systems*, California, pp. 1–14 (1998).

Moore, S.K., Just one word – Plastics. Organic semiconductors could put cheap circuits everywhere and make flexible displays a reality, *IEEE Spectrum*, 39, 55–59 (2002).

North, M.M., S.M. North and J.R. Coble, Virtual reality therapy: An effective treatment for phobias, *Stud. Health Technol. Inform.*, 58, 112–119 (1998).

Ockerman, J.J., L.J. Najjar and J.C. Thompson, FAST: Future technology for today's industry, *Comput. Indust.*, 38, 53–64 (1999).

Ohsuga, M. and H. Oyama, Possibility of virtual reality for mental care, *Stud. Health Technol. Inform.*, 58, 82–90 (1998).

Pacelli, M., G. Loriga, N. Taccini and R. Paradiso, Sensing fabrics for monitoring physiological and biomechanical variables: E-textile solutions, In *Proceedings of the Third IEEE-EMBS*, New York, pp. 17–24 (2006).

Park, S., C. Gopalsamy, R. Rajamanickam and S. Jayaraman, The wearable motherboard: An information infrastructure or sensate liner for medical applications, In Lymberis A. and DeRossi D. (eds.), *Studies in Health Technology and Informatics*, vol. 62, IOS Press, Amsterdam, the Netherlands, pp. 252–258 (1999).

Park, S. and S. Jayaraman, Adaptive and responsive textile structures, In Tao, X. (ed.), *Smart Fibers, Fabrics and Clothing: Fundamentals and Applications*. Woodhead, Cambridge, U.K., pp. 226–245 (2001).

Park, S. and S. Jayaraman, E-textiles, *Ind. Fabric Prod. Rev.*, 87(11), 64–75 (2002).

Park, S. and S. Jayaraman, Smart textiles: Wearable electronic systems, *MRS Bull.*, 28(8), 585–591 (2003).

Park, S., K. Mackenzie and S. Jayaraman, The wearable motherboard: A framework for personalized mobile information processing (PMIP), In *ACM/IEEE 39th Design Automation Conference*, New Orleans, LA, pp. 114–119 (2002).

Parker, R., R. Riley, M. Jones, D. Leo, L. Beex and T. Milson, STRETCH – An E-textile for large-scale sensor systems, In *International Interactive Textiles for the Warrior Conference*, Cambridge, MA, pp. 77–81 (2002).

Perumalraj, R., B.S. Dasaradan and V.R. Sampath, Study on copper core yarn conductive fabrics for Electromagnetic shielding, In *Proceedings of National Conference on Functional Textiles and Apparels*, PSG Tech, Coimbatore, India, pp. 187–203 (2007).

Pessa, M., M. Guina, M. Dumitrescu, I. Hirvonen, M. Saarinen, L. Toikkanen and N. Xiang, Resonant cavity light emitting diode for a polymer optical fibre system, *Semicond. Sci. Technol.*, 17, R1–R9, (2002).

Post, R. and M. Orth, Smart fabric, wearable clothing, In *Proceedings of the First International Symposium on Wearable Computers*, Cambridge, U.K., IEEE, pp. 167–168 (1997).

Post, E.R., M. Orth, P.R. Russo and N. Gershenfeld, Embroidery: Design and fabrication of textile-based computing, *IBM Syst. J.*, 39(3), 840–860 (2000).

Prystowsky, J.B., G. Regehr, D.A. Rogers, J.P. Loan, L.L. Hiemenz and K.M. Smith, A virtual reality module for intravenous catheter placement, *Am. J. Surg.*, 177(2), 171–175 (1999).

Rajamanickam, R., S. Park and S. Jayaraman, A structured methodology for the design and development of textile structures in a concurrent engineering environment, *J. Textile Inst.*, 89(3), 44–62 (1998).

Rantanen, J. and M. Hannikainen, Data transfer for smart clothing: Requirements and potential technologies, In Tao, X. (ed.), *Wearable Electronics and Photonics*, Woodhead Publishing in Textiles, Cambridge, U.K., p. 198 (2005).

Rantanen, J., T. Karinsalo, M. Mkinen, P. Talvenmaa, M. Tasanen and J. Vanhala, Smart clothing for the arctic environment, In *Proceedings of the International Symposium on Wearable Computers*, Arlington, VA, pp. 15–23 (2000).

Riemann, O. and J. Krause, The use of polymer optical fibres for in-house-networks, advantages of 520 nm LED transmission systems, In *Proceedings of the 24th European Conference on Optical Communication*, Madrid, Spain, pp. 381–382 (1998).

Riva, G., M. Baccetta, M. Baruffi, S. Rinaldi and E. Molinari, Experiential cognitive therapy: A VR based approach for the assessment and treatment of eating disorders, *Stud. Health Technol. Inform.*, 58, 120–135 (1998).

Rooman, M., R. Kuijk, R. Windisch, G. Vounckx, S. Borghs, A. Plichta, M. Brinkmann et al., Interchip optical interconnects using imaging fiber bundles and integrated CMOS detectors, In *Proceedings of the 27th European Conference on Optical Communication*, Amsterdam, the Netherlands, pp. 296–297 (2001).

Roycroft, B., M. Akhyer, P. Maaskant, P. De Mierry, S. Fernandez, F.B. Naranjo, E. Calleja, T. McCormack and B. Corbett, Experimental characterization of GaN-based resonant cavity light emitting diodes, *Phys. Stat. Sol.*, 192, 97–102, (2002).

Rudenko, A., P. Reiher, G.J. Popek and G.H. Kuenning, The remote processing framework for portable computer power saving, In *Proceedings of the ACM Symposium on Applied Computing*, San Antonio, TX, pp. 365–372 (1999).

Samuelson, L., F. Bruno, J. Kumar, R. Gaudiana and P. Wormser, Conformal solar cells for the soldier, In *International Interactive Textiles for the Warrior Conference*, Cambridge, MA, pp. 104–107 (2002).

Satava, R.M., Virtual reality surgical simulator: The first steps, *Surgical Endosc.*, 7, 111–113 (1993).

Satava, R.M. and S.B. Jones, Virtual reality and telemedicine: Exploring advanced concepts, *Telemed. J.*, 2(3), 195–200 (1996).

Schubert, E.F., Y.H. Wang, A.Y. Cho, L.W. Tu and G.J. Zydzik, Resonant cavity light emitting diode, *Appl. Phys. Lett.*, 60, 921–923 (1992).

Scilingo, E.P., A. Gemignani, R. Paradiso, N. Taccini, B. Ghelarducci and D. De Rossi, Performance evaluation of sensing fabrics for monitoring physiological and biomechanical variables, *IEEE Trans. Inf. Technol. Biomed.*, 9(3), 345–352 (2005).

Seyam, A.M., Electrifying opportunities, *Textile World*, 2, 33–37 (2003).

Shen, C.-L., T. Kao, C.T. Huang and J. Lee, Wearable band using a fabric based sensor for exercise ECG monitoring, *IEEE*, 1-4244-0598-X/06 (2006).

Snyder, W.E. and J. Clair, Conductive elastomers as sensor for industrial parts handling equipment, *IEEE Trans. Instrum. Measure.*, 27(1), 94–99 (1978).

Stylios, G.K. and L. Luo, The concept of interactive, wireless, smart fabrics for textiles and clothing, In *Proceedings of the Fourth International Conference Innovation and Modeling of Clothing Engineering Processes – IMCEP*, Maribor, Slovenia, pp. 9–11 (2003).

Swallow, S.S. and A.P. Thompson, Sensory fabric for ubiquitous interfaces, *Int. J. Hum. Comput. Interact.*, 13, 147–159 (2001).

Tao, X., Introduction, In Tao, X. (ed.), *Wearable Electronics and Photonics*, Woodhead Publishing in Textiles, Houston, TX, pp. 1–12 (2005).

Teller, A., A platform for wearable physiological computing, *Interacting Via Comput.*, 16, 917–937 (2004).

Van Laerhoven, K. and O. Cakmakci, What shall we teach our pants?, In *Proceedings of the Fourth International Symposium on Wearable Computers*, Atlanta, GA, pp. 77–83 (2000).

Van Laerhoven, K., A. Schmidt and H. Gellersen, Multi-sensor context aware clothing, In *Proceedings of the Sixth International Symposium on Wearable Computers*, Montreux, Switzerland, pp. 49–56 (2002).

Vigneswaran, C., L. Ashok Kumar and T. Ramachandran, Design development of signal sensing equipment for measuring light transmission efficiency of POF fabrics, *Indian J. Fibre Textile Res.*, 35, 317–323 (2010a).

Vigneswaran, C., L. Ashok Kumar and T. Ramachandran, Development of signal transferring fabrics using plastic optical fibres for defense personnel and study their performance, *J. Indust. Textiles*, 17(4), 91–97 (2010b).

Vigneswaran, C., P. Kandhavadivu and T. Ramachandran, Application of wireless communication device integrated apparel and bed linen, *Int. J. Eng. Sci. Technol.*, 2(10), 5970–5976 (2010c).

Vigneswaran, C. and T. Ramachandran, Design and development of copper core conductive fabrics for smart textiles, *J. Indust. Textiles*, 39, 81–93 (2009).

Winchester, R.C.C. and G.K. Stylios, Designing knitted apparel by engineering the attributes of shape memory alloy, *Int. J. Clothing Sci. Technol.*, 15, 359–366 (2003).

Wirth, R., C. Karnutsch, S. Kugler, W. Plass, W. Huber, E. Baur and K. Streubel, Resonant cavity LEDs for plastic optical fiber communication, In *Proceedings of the 10th International Plastic Optical Fibres Conference*, Amsterdam, the Netherlands, pp. 89–95 (2001).

Yang, B.H. and S. Rhee, Development of the ring sensor for health-care automation, *Robot. Autonom. Syst.*, 30, 273 (2000).

4

Product Development Using Wearable Electronic Integrated Fabrics

4.1 Introduction

This chapter discusses the development of wearable electronics products using cotton-wrapped nichrome yarn, POF yarn–based nichrome fabric for communication, illuminated and teleintimation. Apart from this, specialized miniaturized circuits were developed based on the applications that can be integrated with the fabrics to produce wearable electronic products. While developing the products using wearable electronic integrated fabrics, the following aspects are considered: physical environment, sensor behaviour, human body and motion, motion/draping of clothing, manufacturability, networking, power consumption and software execution. The warp and weft used in the wearable electronic integrated fabrics may also act as either sensors or conducting wires or as a power-generating source. Since the product development consists of cutting, stitching, folding, embroidery, etc., the positioning of sensors, conducting wires, processors and topology of the network were planned without any damage or disturbance for the proper functioning of the wearable electronic products. In this chapter, specialty garment manufactured using nichrome wires and POF fabrics are discussed and also plug-in method of integrating miniaturized electronic circuits in the garment is elucidated.

4.2 Heating Garment

The cotton-wrapped nichrome fabric is integrated with electronic circuits to develop heating garment. Heating garment is developed from the nichrome fabric for the military applications and commercial purposes; heating garments are used to comfort the soldier by regulating the temperature and by supplying the necessary conditions to provide technical superiority over the enemies and to monitor the soldier's health. The product will not only serve as a lightweight

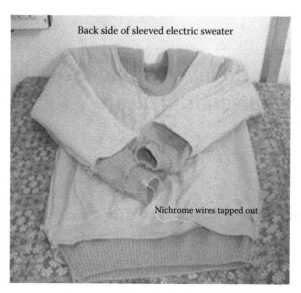

FIGURE 4.1
Heating garment sleeve style.

jacket but also enable soldiers in the warfront to combat cold weather conditions. The heating garment consists of two temperature sensors: one of the sensors will directly come underneath the armhole of the person wearing the garment to monitor the body temperature, and the other sensor monitors the temperature of the heating coil. The temperature of the heating coil is not to exceed 60°C and hence needs to be controlled. Power is supplied to the heating coils through a 6 V relay. The relay is triggered by a microcontroller. Both the temperature sensors are connected to the microcontroller to monitor the temperature. The microcontroller is programmed to record the temperature of both temperature sensors every second. Whenever the body temperature falls below a certain limit, the microcontroller will trigger the relay, thereby giving electric supply to nichrome wires. The supply will generate heat in wires. The heat thus generated permeates to the inner body. Whenever the temperature exceeds the upper limit, the power supply will be cut off. Figure 4.1 shows the heating garment connected with heating elements developed using sleeve nichrome fabric pad. Figure 4.2 shows the heating garment connected with heating elements developed using sleeveless nichrome fabric pad. Figure 4.3 shows the complete flowchart of the working of heating garment. The heating garment fabricated has been tested for its performance in real time by conducting trials in an air-conditioned room and was found to work satisfactorily.

4.2.1 Performance Analysis of Heating Garment

The performance of the heating garment has been analyzed by conducting tests at different ambient temperature. The wearer body temperature was

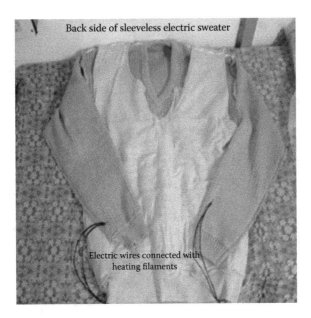

FIGURE 4.2
Heating garment sleeveless Style.

measured to be of normal 37°C and the body temperatures were measured at various room temperatures with 8:1 distance to spot ratio. The time the heating garment has taken to heat the coil was measured using gun type MTX-9 infrared non-contact thermometer of METRAVI make having temperature range of 18°C–1000°C. For study purpose, the minimum temperature to switch on the relay to heat up the nichrome fabric was set up to 18°C, and the maximum temperature to stop the process was set up to 40°C. Their corresponding test results are given in Table 4.1, and the test results show that the heating garment responds to the room temperature and the wearer temperature instantly and heats up the nichrome fabric with a time delay of less than 3 s.

4.3 Development of Communication Garment Using Copper Core Conductive Fabric

The communication garment has been developed using copper core conductive fabric and it is integrated with mobile phone charging circuit and temperature measurement circuit for charging the mobile phone and measuring the wearer temperature, respectively. Figure 4.4 shows the communication garment developed using copper core conductive fabric for body temperature measurement and mobile phone charging.

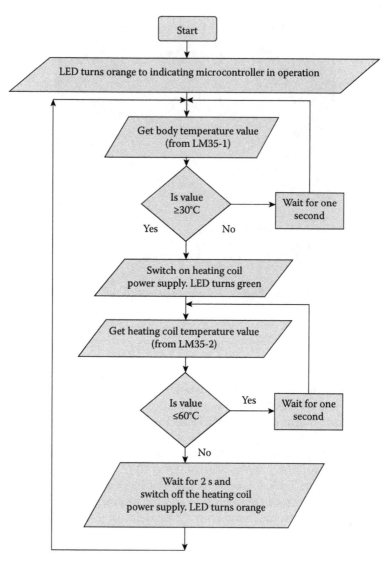

FIGURE 4.3
Flowchart for the working principle of heating garment.

4.3.1 Performance Analysis of Communication Garment Using Copper Core Conductive Fabric

The garment has been designed and embedded with temperature sensors to measure the body temperature. The temperature sensor is connected to the LCD unit through the copper core conductive fabric. The LCD can be placed anywhere in the garment depending upon the user's requirements. Figure 4.5 shows the core copper conductive fabric used for temperature measurement.

TABLE 4.1

Test Results of Heating Garment

S. No.	Test Condition	Body Temperature		Room Temper-ature (°C)	Garment Temper-ature (°C)	Heating Garment Temper-ature (°C)	Time (s)
		Before Operation	After Operation				
1.	AC room	36	40	18	20	40	240
2.	Normal room	38	43	33	30	42	180
3.	Open-air day time	38	44	35	32	42	183
4.	Open-air night time	36	38	20	21	40	236

FIGURE 4.4
Communication garment for temperature measurement and mobile phone charging. (From Vigneswaran, C. et al., *Indian J. Fibre Text. Res. (IJFTR)*, 35, 317, 2010a.)

The performance of the communication garment integrated with temperature measurement circuit has been tested by measuring the body temperature of the wearer in open-air at daytime, and the test results are compared with a standard clinical thermometer as shown in Table 4.2. The test results show that the value of the body temperature measured through copper core conductive fabric is similar to that of the temperature measurement through the standard clinical thermometer.

The performance of the copper core conductive fabric is also analyzed by integrating the fabric with mobile phone–charging circuit for charging a Nokia mobile phone. The supply to the mobile phone charger circuit is given

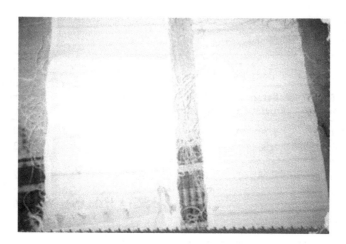

FIGURE 4.5
Fabric used for temperature measurement. (From Vigneswaran, C. et al., *Indian J. Fibre Text. Res. (IJFTR)*, 35, 317, 2010a.)

TABLE 4.2

Temperature Measurement Using Copper Core Conductive Fabric

S. No.	Temperature Measurement Using	Output Temperature (°C)
1.	Copper core conductive fabric	41
2.	Clinical thermometer	39

through AC mains of 230 V, and this voltage is stepped down to 5 V by a step-down transformer. The output of the transformer, 5 V, is given to mobile phone–charging circuit through the copper core conductive fabrics. The time it takes to charge the fully drained mobile phone battery is taken and the current values are also noted down. The test results are compared with a standard Nokia mobile phone charger as given in Table 4.3. The result shows that the copper core conductive fabric is charging the mobile phone almost as the same time as that of the time taken by the conventional mobile phone–charging circuit. The copper core conductive fabric is tested by charging the mobile phone as shown in Figure 4.6.

TABLE 4.3

Mobile Phone Charging Using Copper Core Conductive Fabric

S. No.	Mobile Phone–Charging Circuit	Voltage Applied (V)	Current Taken (μA)	Time Taken to Charge the Battery (h)
1.	Through conventional circuit	5	50	5
2.	Through copper core conductive fabric	5	55	4

FIGURE 4.6
Mobile phone charging using copper core conductive fabric. (From Vigneswaran, C. et al., *J. Ind. Text.*, 1, 91, 2010b.)

4.3.2 Development of Communication Garment Using Optical Core Conductive Fabric

The communication garment developed using optical core conductive fabric can be used for signal transmission as well as to locate the number and the place of bullet wounds. The optical core conductive fabric used for signal transmission is as shown in Figure 4.7.

4.3.2.1 Performance Analysis of Communication Garment Using Optical Core Conductive Fabric

The performance of the communication garment developed using optical core conductive fabric has been tested using signal-transferring circuit developed to test the signal-transferring capability of the fabric at various

FIGURE 4.7
Fabric used for signal-transferring using optical fibre. (From Vigneswaran, C. et al., *J. Ind. Text.*, 1, 91, 2010b.)

light sources. Figure 4.8a shows that the input source, white LED, is given at one end of the optical core conductive fabric, and Figure 4.8b shows that in the other end, light is received, showing no break in the optical core conductive fabric. The signal-transferring efficiency of the communication garment using optical core conductive fabric is analyzed using AT89C52 microcontroller designed with photodiode. The signal-transferring efficiency of the optical core conductive fabric is analyzed using AT89C52 microcontroller as shown in Figure 4.8. Figure 4.9 represents the signal

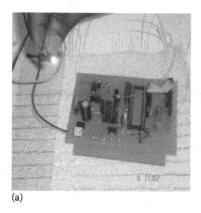
(a)

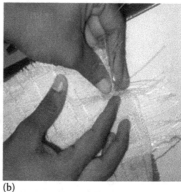
(b)

FIGURE 4.8
(a) Input source white LED connected to the signal-transferring circuit; (b) output of LED at the other end of the POF. (From Ashok Kumar, L. et al., *Indian J. Fibre Text. Res.*, 31, 577, 2006.)

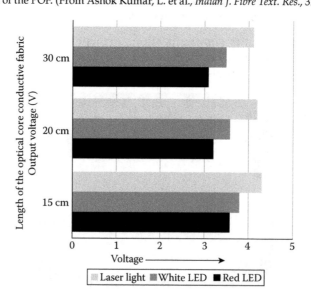

FIGURE 4.9
Signal-transferring capability of optical core conductive fabric. (From Ashok Kumar, L. et al., *Indian J. Fibre Text. Res.*, 31, 577, 2006.)

received in terms of output voltage for the optical core conductive fabric at different lengths such as 15, 20 and 30 cm using the red LED, white light LED and laser light LED with input voltage of +5 V. From the signal values measured in various places of fabrics at 15, 20 and 30 cm, cumulative average of optical length is taken and their signal output voltages at the inputs sources of red LED, white LED and laser light are 3.3, 3.6 and 4.2 V, respectively, for optical core conductive fabrics. Test results show that the optical core conductive fabric can be used for applications like communication purposes and for the development of teleintimation garment. The input light sources used at one end of the optical core conductive fabric were red LED, white LED and laser light; in this laser light there is maximum output voltage, resulting in minimum signal loss in the optical core conductive fabrics.

4.4 Development of Illuminated Garment

The fabric produced using polymethylmethaacrylate (PMMA) is used to develop illuminated garment. The plastic optic fibre is selected based on the requirements of size (length and width) of the strip. The prepared filaments are placed side by side above the base fabric. The filaments are laid parallel by simple combing. A small quantity of adhesive resin is applied across the strip below the top and bottom portions. The whole assembly is left undisturbed exposed to the atmosphere for 1 h for fixation to take place. Now the strip is ready for the assembly work. The colours of the light-emitting diodes (LEDs) are selected based on this requirement: they can be soldered with suitable resistance. All the LEDs are grouped together at the top and bottom portions separately. Both top and bottom groups of LEDs are connected with a 9 V DC battery in parallel. The LEDs are inserted with the corresponding sleeves. The fabric sample was embedded into the clothing system to get the illuminated effect. The detailed specifications of the sewing machine are given in Table 4.4.

The size of the design area is marked on the fabric, where the glowing system is required. The monogram portion like police and duck is designed by

TABLE 4.4

Sewing Machinery Details

Type of sewing machine	Flat bed single needle lock stitch
Type of stitch	Class 300 lock stitch
Feeding system	Drop Feed
Needle system	DB X 1
Type of seam	Superimposed and flat seam

using die cutting. To avoid unravelling, the edges of the letters are finished with embroidery work; embroidery thread colours are chosen based on the requirement. After completion of the embroidery work, the strip is attached on the garment's rear by using Velcro. By using a thin scratch film, the letter portions of the optic fibres are rubbed until the jacket (outer surface) gets removed. The edges of the filaments on both the sides are grouped into three equal parts and combined together. An assembly of panels of garment components is constructed by using a high-speed single needle lock stitch sewing machine. The standard seam types of flat and superimposed seam are used throughout the garment construction. Garment construction is made in such way that the illuminated system can be easily removed and attached whenever required. Figure 4.10 shows the illuminated garment with designs created using PMMA fabric.

The illuminated garment is designed and developed for safety and protection. It is possible to produce the woven fabric using PMMA to get the glowing effect with slight modification in the weave geometry. The sateen weave shows better illumination when compared to twill fabric; the structure of sateen weave has more weft float. Since the weft bending angle is less, it causes more illumination when compared to twill fabric. The illumination effect on clothing system is also influenced by the different colours of LEDs; the wavelength of different colours influences the illumination. The colour red LED has the minimum Lux value, which is due to the maximum wavelength of the red colour. The subjective evaluation results show that there is tremendous interest in the research work and future of the lighting system in various applications. The subjective evaluation is carried out for the end user; the specific attributes are selected and given the rating scale. The user should wear the garment and rate the attribute; these values are consolidated

FIGURE 4.10
Illuminated garment designs using PMMA POF. (From Ashok Kumar, L. et al., *Indian J. Fibre Text. Res.*, 31, 577, 2006.)

TABLE 4.5

Rating of Subjective Evaluation

S. No.	Features/Attributes	Average Rating
1.	Opinion about the garment	2.4
2.	How do you rate the practical usage of this garment?	2.6
3.	Do you feel comfortable while wearing?	2.6
4.	How do you feel the lighting system in the garment?	2.6
5.	Whether it fulfils the safety purpose during night time?	2.6

for the final scale rating. The final scale ratings are given in Table 4.5. The average attributes show positive feedback from the end user of the wearer.

With this development of the research work, illuminated garment can be implemented for applications such as transferring images or text from a mobile phone to the optical fibre screen of the garment to bring an animated effect. It is possible to vary the brightness of the optic fibre screen according to the sound environment using low-frequency sensitive circuit. The light may get brighter each time the wearer claps his or her hands or stamps the feet. Further work can also be carried out in medical textile by using this material as a surgical tool for phototherapy system.

4.5 Development of Teleintimation Garment

The teleintimation garment for defence personnel is made with POF fabric, which can sense the number and place of bullet wounds from a soldier's body. The POF is woven in weft and warp direction in the garment to form a matrix format. This matrix format allows the exact location of the bullet wound in the soldier's body. The POF is integrated with the microcontroller-based circuits to detect the bullet wound signals. These circuits were fabricated using flexible PCBs to give comfort and ease to the soldiers during combat situation. The bullet wound signals are transmitted to the soldier-monitoring station at the remote end, where an observer will monitor the soldier's status from the specially developed application software. The software contains the soldier's physical, biological data and the bullet wound status.

While designing the teleintimation garment, various aspects are considered: lightweight, breathability, comfortable (form-fitting), easy to wear and take off and be able to provide easy access to wounds. These are critical requirements in combat conditions so that the soldier's performance is not hampered by the protective garment. The teleintimation garment consists of the following building blocks or modules that are totally integrated to create a garment (with intelligence) that feels and wears like any ordinary undershirt. The modules are

1. A comfort component to provide basic comfort properties that any typical undergarment would provide to the user
2. A penetration-sensing component to detect the penetration of a projectile
3. An electrical sensing component to serve as a databus to carry the information from/to the sensors mounted on the user or integrated into the structure
4. A form-fitting component to ensure the right fit for the user

The schematic of one variation of the teleintimation garment prototype is shown in Figure 4.11. The POF is spirally integrated into the structure during the fabric production process without any discontinuities at the armhole or the seams. There is no need for the cut and sew operations to produce a garment from a two-dimensional fabric developed using handloom with detachable circuits. The width of the fabric has been woven in handloom and power loom to the chest measurement (38–40 in.) of the teleintimation garment. The developed garment is as shown in Figure 4.12. The POF fabric is also produced using rigid rapier loom with major modification in the machine as mentioned in Chapter 4. The developed teleintimation garment using this fabric is integrated with the flexible PCB as shown in Figure 4.13.

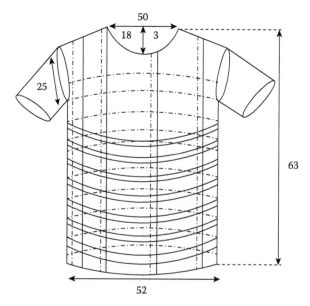

FIGURE 4.11
Teleintimation garment prototype.

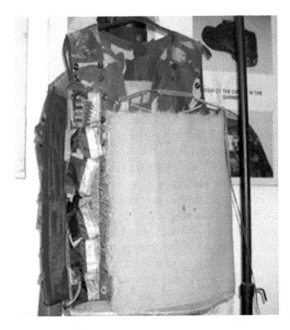

FIGURE 4.12
Teleintimation garment (handloom fabric) with detachable circuit. (From Ashok Kumar, L. et al., *Indian J. Fibre Text. Res.*, 31, 577, 2006.)

FIGURE 4.13
Flexible PCB and teleintimation garment (rapier loom fabric). (From Ashok Kumar, L. et al., *Asian Text. J.*, 6, 48, 2010.)

To facilitate integration of embedded systems to a flexible garment, it is proposed to use 0.5 mm of POF systems as a suitable mechanism to detect and locate penetrations in the garment, and to interface other potential sensors. Each row and column is scanned (via microcontroller) for end-to-end connectivity. When a penetration occurs, the POF path of the thread is broken and detected by the microcontroller, which in turn, provides grid location in (X–Y coordinates), indicating where the penetration physically occurred relative to garment and anthropomorphic features. Furthermore, the microcontroller allows detection of multiple penetrations with a resolution of 1–2 mm particle size.

The garment is designed to accommodate vertical and horizontal section of circuits so as to detect the bullet wound and location in the matrix format. The white bright LED is used at one end of the POF integrated garment. POF fibres from the garment are pulled and tied at one end. Another end of the POF is kept untied to connect to the photo-diode. PCB is used to connect the white bright LEDs serially. The tied POF line from the garment is connected to the light sources at the vertical end as well as light sources at the horizontal end. The POF from the garment is being inserted into the photo-diode. The photo-diodes are properly sleeved with heat shrinkable rubber tube and wire sleeves to keep the POF attached firmly with the sensors. After connecting the POF, each and every photo-diode output is checked for its proper sensing of the optical signals. The circuits are connected serially so that the row and column information from the matrix format will be displayed at the remote end server. The presence or absence of the bullet information is collected from the sensor circuitry that is being collected by the microcontroller and sent to the remote end server.

The penetration sensitivity and the controller capabilities of the soft circuit concept are tested using standard test rig. The developed teleintimation garment has been used for the following two unique applications: (1) The POF fabric is woven in the warp and weft direction with modifications in the rigid rapier loom and also the fabric is made using handloom for penetration detection and location of the bullet wound. (2) The POF fabric has been interfaced with the microcontroller and to the remote unit to monitor the status of the soldier.

Specifications of Teleintimation Garment Prototype	
Chest circumference	104 cm
Full length	63 cm
Shoulder point to point	50 cm
Half bottom	52 cm
Neck open	18 cm
Front neck drop	7 cm
Arm hole	25 cm
Back neck drop	3 cm

4.5.1 Detection of Bullet Wound in Teleintimation Garment

The bench-top set-up for testing the penetration sensing capability is shown in Figure 4.14. The laser or white LED has been used at one end of the POF to send pulses that 'lit up' the structure, indicating that teleintimation garment was armed and ready to detect any interruptions in the light flow that might be caused by a bullet. At the other end of the POF, a photo-diode connected

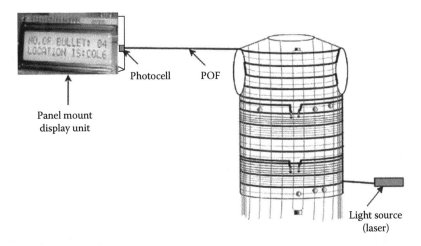

FIGURE 4.14
Bench-top setup for projectile penetration. (From Ashok Kumar, L. et al., *Asian Text. J.*, 6, 48, 2010.)

to a power-measuring device measured the power output from the POF. The penetration of teleintimation garment resulting in the breakage of POF was simulated by cutting the POF with a pair of scissors; when this happens, the power output at the other end on the measuring device falls to zero, the location of the actual penetration in the POF can be determined by a test rig. In the war field, the vital signs will be transmitted from the teleintimation garment to soldier's status monitor by the communication media to the remote unit. The soldier's vital signs can be continuously monitored by the triage unit and, if a penetration alert is received, when the soldier is shot, medical assistance can be provided immediately to the wounded soldier.

4.5.2 Performance Analysis of Teleintimation Garment Using DAQ Card

The performance analysis of teleintimation garment can be analyzed using USB-1208FS DAQ card, which consists of an eight analogue inputs, two 12-bit analogue outputs, 16 digital I/O connections and one 32-bit external event counter. The signals received from the microcontroller circuits are being captured using a data acquisition card for analysis of the data. The block diagram shown in Figure 4.15 is the typical setup to connect the DAQ hardware to the microcontroller circuit. The data is captured before the TTL to Serial RS 232 conversion logic. Hence, the data is captured and viewed in the DAQ application window. The DAQ card is a useful tool used to acquire the signals from the teleintimation garment. When there is no bullet wound, there will not be any transition in the graphical representation. When the POF is broken by a bullet penetration, the graphical representation will produce transitions equal to that of the number of bullet wound. Hence, the bullet wound information can be found by viewing the graphical representation.

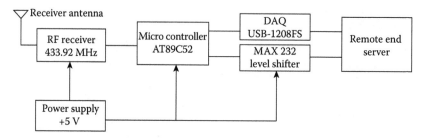

FIGURE 4.15

Connection setup of DAQ with the microcontroller. (From Ashok Kumar, L. et al., *Asian Text. J.*, 6, 48, 2010.)

4.5.3 Testing of Circuits Developed for 64 × 64 Matrix Format

The number of bullet counts was detected using the data acquisition card. The graphical representation in Figure 4.16 shows the bullet information from channel 0. Transition of the signal from channel 0 gives the information about the bullet count. Figure 4.16 shows the graphical view when there is no bullet. The acquiring values timing could be changed from 100 days to 1 μs. The data from the microcontroller is acquired with the time base value as 5 s. The data acquired in channel 0 shown indicates the data acquired from the POF integrated garment. In X-axis is the amplitude of the measuring signal

FIGURE 4.16

Strip chart data with no bullet wound. (From Ashok Kumar, L. et al., *Asian Text. J.*, 6, 48, 2010.)

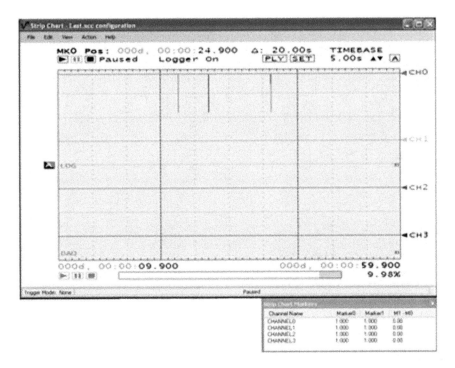

FIGURE 4.17
Strip chart data with three bullet wounds. (From Ashok Kumar, L. et al., *Asian Text. J.*, 6, 48, 2010.)

and in Y-axis is given time in seconds. The graph in Figure 4.16 shows that there is no transition in channel 0, which means no bullet has been detected. Whenever there is a bullet, the transition will occur according to the baud rate with which the data is being transmitted. The data captured from the teleintimation garment when bullets were detected is shown in Figures 4.17 and 4.18; the transitions were plotted in channel 0. The transition information contains the data to be shown in the LCD.

4.5.4 Testing of Circuits Developed Using PIC Microcontroller with Detachable PCBs

The number of bullet counts was represented using a data acquisition card where the data were captured and shown in graphical representation. The strip chart has two markers that can be moved to any location on the strip chart to view the values of specific data points. The markers are identified on the strip chart as MK 0 and MK 1. The value of the data point at each marker's position, and the amplitude difference between the markers for each channel, is shown on the strip chart markers window. By default, the markers have a vertical orientation. The left marker is labelled MK O, and the right marker is labelled MK 1. Figure 4.19 shows the marker 0 is in position 1000 m and marker 1 is in position 375 m and also it represents the bullet hit detection.

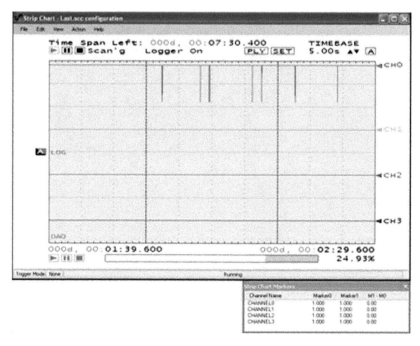

FIGURE 4.18
Strip chart data with multiple bullet wounds. (From Ashok Kumar, L. et al., *Asian Text. J.*, 6, 48, 2010.)

4.5.5 Testing of Circuits Developed Using PIC Microcontroller with Flexible PCBs

The signals received from the microcontroller are captured using a data acquisition card, and it is represented as a graphical values. These data are captured in various stages such as when there is no bullet hit, one bullet hit is made, two bullet hits are made and five bullet hits are made. The graphical representation of such conditions is shown in Figure 4.20. In this circuit, the communication technique used is GSM. Hence, it is made to transmit the soldier's unique ID and bullet wound information for every 10 s. The transitions shown in Figure 4.20 are due to the soldier ID, which is being transmitted every time when the bullet information is transmitted. Hence, more the bullet wound, more transitions will be given in the output window. The data captured from the earlier measurement could be saved to make analysis about the bullet information. Hence, the data captured during combat situations will be very much useful for further analysis. In real-time scenario, this information provides backup data where the graphical values are stored. It has a playback option where the bullet wound information could be exactly predicted. The playback option provides a recap of the signals acquired from the teleintimation garment, which will help to get the bullet wound data.

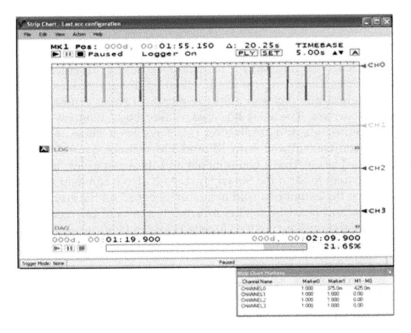

FIGURE 4.19
Strip chart data with bullet wounds. (From Ashok Kumar, L. et al., *Asian Text. J.*, 6, 48, 2010.)

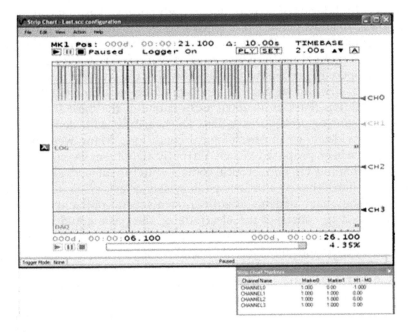

FIGURE 4.20
Strip chart data with multiple bullet wounds. (From Ashok Kumar, L. et al., *Asian Text. J.*, 6, 48, 2010.)

4.6 Smart Shirt

The Smart Shirt can be used for medical monitoring, athletics and military uses. Thus, they can provide information about the desired parameters throughout the day and alert the person if some anomaly in these parameters occurs. The vital body signs like body temperature, blood pressure, pulse rate and respiration rate have been measured, and the data has been transmitted with RF and FM technology and monitored in real-time. The electronic components and the sensor have been placed into the garment to form the Smart Shirt. And also it can help a physician to determine the extent of a soldier's injuries based on the strength of his temperature and heart beat. This information is vital for accessing who needs first assistance during the so-called Golden Hour in which there are numerous causalities.

4.6.1 Measurement of Vital Signs

4.6.1.1 Wearable Electrocardiograph

Wearable electrocardiograph (ECG) monitoring systems today use electrodes that require skin preparation in advance, and require pastes or gels to make electrical contact to the skin. Moreover, they are not suitable for subjects at high levels of activity due to high noise spikes that can appear in the data. To address these problems, a new class of miniature, ultralow noise, capacitive sensor that does not require direct contact to the skin, and has comparable performance to gold standard ECG electrodes, has been developed. Experimental results show that the wireless interface will add minimal size and weight to the system while providing reliable, untethered operation (Park et al. 2006). ECG is one of the most widely used biomedical sensing procedures to date. The heartbeat is the definitive indicator for a wide range of physiological conditions. Many wearable ECG systems have been proposed to date. Virtually all of them use some form of electrodes that must make electrical contact with the subject's skin surface. This necessitates the use of sticky pads, pastes or gel. While this method works for stationary patients, it suffers from several problems. First, the material used to construct the electrode or the paste could cause skin irritation and discomfort, especially if the subject is performing rigorous physical exercise and may be sweating. Another problem is that, during motion, the electrodes may become loose, breaking electrical contact and causing high noise spikes in the data. Paste/gel-free resistive contact ECG sensors have been developed. Many of them still suffer from similar noise levels to *wet* electrodes, and the contact can still cause irritation problems as well as being more sensitive to motion. Recent breakthroughs have been made in the form of insulated bio-electrodes (IBEs). They can measure the electric potential on the skin without resistive electrical contact and with very low capacitive coupling. This has been made possible by a combination of circuit design and the use of a new, low dielectric material. These IBEs enable through-clothing measurements, and results over 40 subjects have shown them to be capable of over 99% correlation with gold standard conventional electrodes.

Several wireless ECG monitoring systems have been proposed, all of them use conventional *wet* ECG sensors. For data sampling and wireless transmission, they use either existing standard wireless interfaces or general-purpose wireless sensor nodes. This combination results in many system-level drawbacks such as big form factor, low transmission speed, short battery lifetime and lack of wearability. In this section, first review the previous works and discuss their shortcomings. Researchers at Imperial College developed their own wireless sensor node, called the BSN node. It measures 28(L) × 37(W) × 12(H) mm^3 (w/a sensor board and w/o a battery). They used this platform to design a wireless ECG monitoring system. This system also used the 802.15.4 radio and conventional ECG sensors. Most systems use the 802.15.4 radio, even though it was originally developed for event-detection applications rather than real-time monitoring ones. There are two exceptions. The first one uses a CC1050 transceiver (Ashok Kumar and Ramachandran 2010c) (at 76.8 kbps max), which is similar to MICA2's transceiver. The other one uses a Bluetooth interface (721 kbps max). Neither radio interface was originally designed for real-time monitoring applications.

It is unlikely that dry electrodes will ever replace the adhesive, wet Ag/AgCl for in-hospital use. Standard electrodes adhere well to the body, are robust, inexpensive and simple. Properly used, wet electrodes provide an excellent signal. Dry or noncontact electrodes offer little advantages for the majority of hospital applications, while adding cost and complexity (such as the for active electrode circuitry). It is worth noting, however, that for situations where patients have extremely sensitive skin (i.e. burn units, neonatal care), dry and non-contact electrodes may be desirable. At a basic level, the polar heart rate monitor is one well-known example, although it comes under a non-clinical, of a dry electrode–based system for cardiac monitoring. The basic theme of a wearable, dry-contact chest strap/harness has been demonstrated by several authors and at least one known medical device company (Monebo). They provide a very easy way to continuously obtain a 1-lead ECG. Given the right analysis and wireless clinical infrastructure, dry-contact chest straps may prove to be a viable tool for long-term cardiac monitoring. With non-contact sensors, it is also possible to build a strap/harness that can be worn on top of a T-shirt, with electrodes placed in approximate positions to provide a derived 12-lead ECG. Motion artefacts and chest tightness, however, remain a difficulty with wearable, non-contact systems. Small bandage-like patches are even more convenient than chest straps for long-term, mobile monitoring. Recent advancements in microelectronics electronics have made it possible to integrate entire ECG monitoring systems within a small patch. Yoo et al. (2009) present a inductively powered ECG chest patch based on a single integrated circuit mounted on a fabric substrate. A few commercial offerings are also now on the market, in a somewhat larger form-factor (Corventis, iRythm, Proteus). Unfortunately, the short electrode-to-electrode distance makes it impossible to obtain the same waveform as a standard 1-lead ECG, although the QRS complex is readily visible in the most cases. These devices have potential to be highly useful for basic long-term cardiac monitoring, such as arrhythmia detection. Besides mobile wearable devices, non-contact electrodes have been used for rapidly obtaining chest body surface potential maps (BSPMs). In fact, the first demonstration of non-contact electrodes was for a chest array. Newer versions have been developed, mounted on a standard tablet PC. Non-contact electrodes have a distinct advantage, since it can be taken through clothing without any preparation. However, it is not clear what the clinical advantages are for non-contact BSPMs, especially in light of the noise and frequency responses of non-contact electrodes. A contact version, perhaps embedded within a tight garment, could prove useful, provided the extra information over a 12-lead ECG is clinically relevant.

Clinical ECG monitoring devices have traditionally required patients to wear a device on the body. With the exception of an implantable monitor, all of these systems require some degree of patient intervention and compliance. The advent of non-contact electrode technology has made it possible to integrate cardiac monitoring devices unobtrusively in the environment. Several attempts have been made to integrate electrodes in beds, chairs and even bathtubs and toilets. Obtaining signs of cardiac activity through an air gap (40 cm) is also possible. Unfortunately, signal quality from these devices is typically quite poor and riddled with motion artefact, noise and interference problems. At present, nothing has progressed beyond a basic proof-of-concept. More detailed clinical studies are required to find out if the degree of monitoring provided by beds and chairs is clinically useful.

4.6.1.1.1 Ultra-Wearability

Wearability is the most crucial issue in designing a wireless ECG monitoring system. However, to the best of our knowledge, none of the existing miniature sensing systems can be considered truly wearable in the strict sense, not just because they are still bulky but also because conventional ECG sensors can cause skin irritation. Therefore, using QUASAR's innovative ECG sensor and an ultra-compact wireless sensor node are specially designed for wearable applications.

4.6.1.1.2 System Design

In order to achieve the design goals described in the previous section, developing a new ECG monitoring system that takes Figure 4.21. System Architecture of ECG Monitoring System (Figure 4.22a) top view and (b) attached on T-shirts. Figure 4.22a QUASAR ECG Sensors and Figure 4.22b

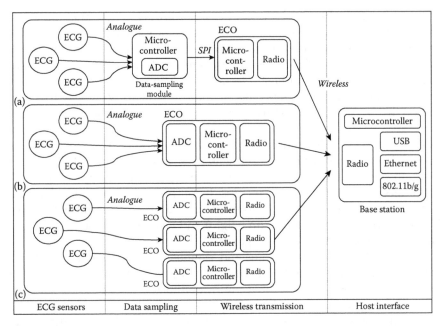

FIGURE 4.21
System architecture of ECG monitoring system. (From Park, C. et al., *Biomed. Circuit Syst.*, 21(1), 241, 2006.)

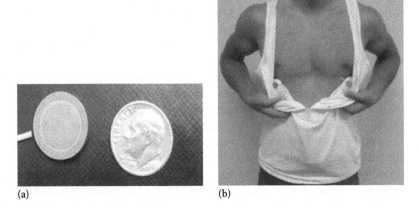

FIGURE 4.22
(a) QUASAR ECG sensors and (b) sensors attached on T-shirts and worn on a human body. (From Park, C. et al., *Biomed. Circuit Syst.*, 21(1), 241, 2006.)

sensors attached on a T-shirts and worn on a human body advantage of QUASAR's ECG sensors and Eco wireless sensor nodes. The QUASAR sensor is a wearable, tiny, low-power ECG sensing device, and Eco is an ultra-compact, low-power wireless sensor node. They are similar in size, power and are well matched in terms of data rate. This section first shows various system architectures for an ECG monitoring system that integrates these two technologies. Next, examine the specification of the QUASAR ECG sensor, Eco node and the base station in detail. ECG monitoring system can be functionally divided into four subsystems: ECG sensors, data sampling, wireless transmission and host interface. ECG signals are first digitized by analogue-to-digital converters (ADCs) and transmitted wirelessly to a base station that interfaces with a host computer via USB, Fast Ethernet or 802.11b.

ECG sensor: QUASAR's sensor (Figure 4.22) is a compact ECG sensor that does not require skin preparation, gels or adhesives. It includes not only a sensing device, but also signal conditioning circuitry such as low noise amplifiers and voltage reference chips. Its output signal range is adjustable from differential (−4.5 to 4.5 V) to single-ended (0–4.5 V). It measures only 15 mm (in diameter) × 3.8 mm (in height) and weighs 5 g. Also, it consumes only 1 mW active power on average.

Unlike ECG, which has a long and established clinical practice of out-patient monitoring systems, the difficulty in preparing a patient and data interpretation has largely limited brain monitoring to in-hospital settings. With the exception of an EEG counterpart to the 24/48 h ECG Holter device, mobile clinical EEG devices are still rarely used today. However, there does exist a strong need and interest for EEG monitoring for medical conditions such as sleep apnoea, epilepsy and traumatic brain energy. Thus, a novel applications and uses are even more critical for the success of dry electrode technology for EEG. Thus, it is expected that these wireless, outpatient EEG-based neural monitoring systems will become much more commonplace in the near future. A robust and patient-friendly dry electrode system will be a significant contribution to this field. At this time, there exists no clinical dry/non-contact EEG device in the market.

Several commercial offerings have been made mostly with a consumer focus for entertainment (Neurosky), sleep/wellness (Zeo) and marketing (Emsense). However, there has been significant activity with using dry EEG systems for research use. Sullivan et al. presented an architecture for high-density, dry-electrode EEG based on the concept of integrating the entire signal processing (amplification, filtering and digitizing) chain on top of a dry MEMS electrode. This enables electrodes to be easily daisy-chained and expanded with only one common-wire, significantly reducing the clutter associated with a conventional EEG system. It is easily wearable and provides access to the forehead locations without gels or other preparation. Monitoring of user attention or alertness is another area that has been explored as a candidate for dry-contact EEG. Several headsets have been developed with this in mind. Tseng et al. (2009) have presented a detailed study of using dry-contact EEG sensors to monitor for driver drowsiness. At present, dry-contact EEG systems are limited to only reliably acquiring forehead signals, due to the vastly varying thickness of human hair. Thus, for clinical applications to be truly viable, clinical procedures and validation must be established for this limited set of EEG signals first.

From an electronics perspective, nearly almost all of the circuit design issues are now well understood. In essence, a modern FET-input amplifier configured in unity-gain will be more than sufficient to buffer signals from virtually any electrode. Achieving sufficiently high input impedance is not a problem for the majority of cases. Input offsets are problematic, but DC-coupled instrumentation with very low gains (0 dB) and high-resolution ADCs (24-bit) can tolerate large electrode offsets. Except for esoteric applications, such as ECG sensing through a large air-gap, it is unlikely any circuit innovation directly at the electrode will be highly useful. It goes without saying, however, that there is much room for circuit/electronics innovation at the system level for building integrated, wearable and wireless biopotential sensors. Resolving the difficulties with motion artefacts remains the unsolved challenge in mobile, wearable ECG/EEG sensor systems. Unlike circuit characterization which involves standard, easily simulated and readily measured parameters like noise, gain and power consumption, motion artefacts are ill-defined and subject to human variability. In addition, different types of electrodes suffer from artefacts from distinct sources. The lack of quantifiable merits compounded with the difficulty in obtaining measurements has resulted in less attention in this area. In addition, fully understanding and characterizing the origin of skin-electrode noise is another under-addressed area in this field. It is well known that that the noise level, while strongly correlated with skin impedance, far exceeds the amount predicted by thermal noise at low frequencies. It has been theorized that the redox reaction at the electrode accounts for the $1 = f$ characteristic with wet Ag/AgCl electrodes. It has not been established that electrochemical noise contributes to capacitively coupled non-contact sensors, since redox reactions do not take place across the interface. Theory and experimental observations have shown that for non-contact electrodes, the thermal noise model is more accurate, and provides some clear guidelines for design considerations in the electrode interface. Again the lack of standard measurement methods combined with human variability makes an objective comparison scarce and difficult. An establishment of a clear measurement protocol followed by detail, objective and comparisons of the noise behaviour of all electrode types will be highly illuminating. Overall, there has to be a greater emphasis on the materials, packaging, signal processing and systems level. The ultimate solution will likely be a combination of some circuit design, but even more a matter of innovative mechanical construction and signal processing. Efforts directed in that direction are expected to yield significant returns for this field. The lead was disrupted at $t = 5$ s. It takes more than 15 s for the trace to recover, showing the problem with recovery time for AC coupled instrumentation. The input was designed to have a cut-off of 0.05 Hz in line with ECG standards.

4.6.1.2 ECG Shirt

The ECG shirt consists of the ECG module on flexible substrate, its encapsulation, a rechargeable flat battery, three ECG body contact electrodes and embroidered 'wiring' and interconnections. A T-shirt has been developed that measures an ECG signal. This work is different to other research in the field as it focuses more on advanced interconnection and integration technologies for electronics in textiles rather than on the ECG shirt as functionality. It is the first application using an interconnection technology based on embroidery of conductive yarn that has been developed recently (Linz et al. 2005). Until now, most wearable computers in the sense of real clothes exist only in laboratories. Great market success has been announced since many years. The reason for this slow start of wearable computing is the lack of real wearability of most prototypes. Yet this is essential for a broad market success. Therefore, Fraunhofer IZM has developed miniaturization and low-cost integration technologies that allow textile typical treatment like draping and washing. In earlier papers, the Fraunhofer IZM has shown different approaches to such technologies. The integration into woven structures has been reported. Temporary contacts and embroidered circuitry with conductive yarn have been introduced. Building embroidered electrical interconnection with a simple embroidery machine has been studied. Unlike earlier research (Post et al. 2000) that concentrated on the creation of conductive embroidered circuits, this work focused on the interconnection process (Figure 4.23). The usefulness of textile-integrated electronics stands out in applications where a distributed sensing, computing or energy sourcing is an indispensable part of the system. A sensing T-shirt is such an application that cannot be realized with a small device like a mobile phone. Information like ECG/EMG

FIGURE 4.23
Flexible electronic test module connected with embroidered conductive yarn.

signals, pulse oximetry (SpO$_2$), body posture and temperature can only be picked up locally on different parts of the body.

The high demands on comfort in a T-shirt and the need for reliable and precise measurements pose a challenge for the miniaturization and integration. The prototype development of a sensing T-shirt that measures an ECG signal is the objective and this version may later be enhanced to include other sensors like acceleration, SpO$_2$, EMG, etc. The work has been shared among Fraunhofer IIS who designed the circuit and Fraunhofer IZM who layout, miniaturized and integrated the circuit into a T-shirt using flexible substrate, flip chip technology and embroidery.

4.6.1.2.1 ECG Module

The ECG module's analogue part consists of a differential signal amplifier followed by an analogue second-order Butterworth low pass filters ($f_g = 125$ Hz). It amplifies the signal between the right shoulder and the lower left rib using the left shoulder as a reference. The AD converter is tunable to overcome signal variations, which occur naturally on the body in the order of second's duration. The flip chip microcontroller processes the data and sends it to the Bluetooth module, which transmits the data to a mobile phone or PDA. Data can be sent in continuous mode for real-time data or burst mode to save energy. A 2.5 mm thin battery with 550 mAh and 3.7 V allows a burst mode operation over 24 h. The battery is attached with snap fasteners for easy replacement and recharging and easy removal before the T-shirt is washed (Figures 4.24 through 4.26).

Integration on the T-Shirt: A commercially available tight fitting stretch T-shirt serves as a basis for the electronics and the embroidery. The tight fit is important for the reduction of movement artefacts and good signal quality.

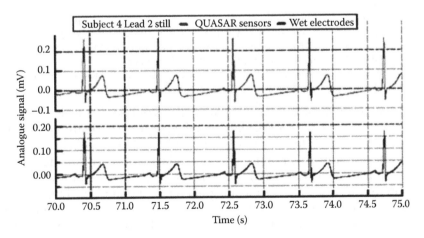

FIGURE 4.24
ECG data comparison. (From Park, C. et al., *Biomed. Circuit Syst.*, 21(1), 241, 2006.)

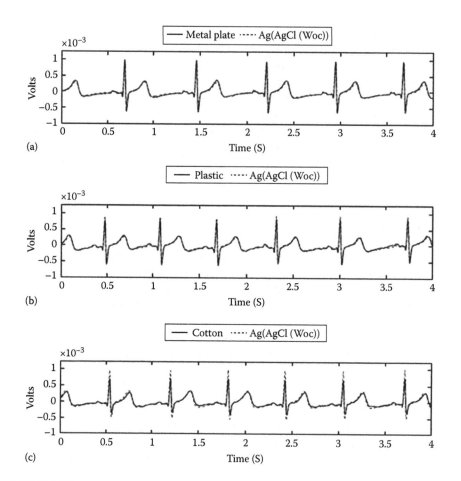

FIGURE 4.25

ECG samples taken from the various dry-contact and non-contact test electrodes (metal contact, thin-film insulation and cotton non-contact), plotted against the signal taken simultaneously from a wet Ag/AgCl electrode. The data is shown from a 0.7 to 100 Hz bandwidth without a 50/60 Hz notch. The increased noise floor of the plastic and cotton electrodes are not readily visible at ECG scales. Signal distortion can be seen on the R-wave for the cotton electrode due to the increased source impedance.

All components have been chosen in the smallest available size and assembled on a 50 μm thin flexible polyimide substrate (Figure 4.27). The dimensions of the module are 27 × 27 mm (not counting the connector, which can be cut off after programming). The thickness of 2.1 mm is due to the Bluetooth module, which could not be redesigned with reasonable amount of effort for this research. For the integration into the shirt, the ECG module has been equipped with metallized contact areas. The embroidery machine penetrates them with the needle. Both the conductors as well as the electrodes were embroidered with conductive yarn produced by *Shieldex*. The so-called *Statex 117/17 twine* is a silver-coated polyamide multifilament yarn.

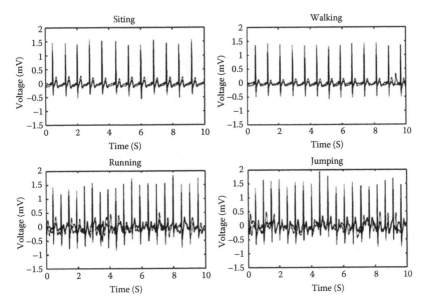

FIGURE 4.26
A 10 s comparison of noise and drift from wet Ag/AgCl (red trace) versus non-contact electrodes (black trace) during various activities, inducing motion and friction. The non-contact electrodes were fixed in a tight wireless chest band on top of a cotton shirt. (From Chi, Y.M. et al., *IEEE Rev. Biomed. Eng.*, 3, 106, 2012.)

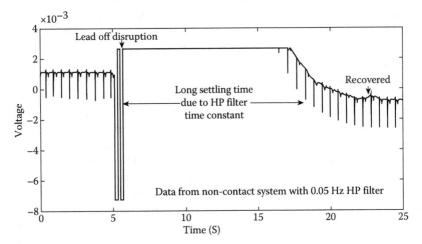

FIGURE 4.27
High impedance input node settling time.

Its resistance of around 500 Ω/m is very high. Nonetheless, this is acceptable for the sensing of the ECG signal, as there is very low current anyway. For the energy transport between battery and ECG module, it was critical to keep the distance short and to embroider the distance multiple times. The choice fell on *Statex* material, because it is conductive and embroider able.

FIGURE 4.28

27 × 27 mm ECG module on flex substrate. (From Linz, T. et al., Fully integrated EKG shirt based on embroidered electrical interconnections with conductive yarn and miniaturized flexible electronics, *Proceedings of the International Workshop on Wearable and Implantable Body Sensor Networks [BSN'06]* IEEE, 2006.)

The layout was embroidered with a semi-professional embroidery machine – Bernina artista 200 which can be programmed with special CAD software. With embroidery on a stretch T-shirt, it is important to ensure that the thread is not under mechanical tension when the shirt is worn. Therefore, it was necessary to stretch the T-shirt during the embroidery and during the encapsulation. A further improvement in this sense can be achieved by a zigzag layout of the long conductors to the ECG electrodes (as shown in Figure 4.28).

The interconnection technology with embroidery requires that the conductive thread is conductive on the surface and is not isolated. However, this application (as most others) demands isolated conductors between the ECG module and the electrodes. Therefore, the conductors have to be isolated after the integration using flexible polymer coating, which is a common process in the textile industry. Embroidery, which is essentially similar to sewing, relies on two threads: a needle thread, in our case on the outside of the shirt and a conductive bobbin thread, in our case on the inside (body side). For this application, both these treads were chosen to be conductive which reduced the resistance. All the conductors were embroidered six times, that is the substrate was penetrated three times at each contact pad, as this proved to be most conductive and most reliable in earlier tests. Furthermore, the conductors to the ECG electrodes were designed with double redundancy (see Figures 4.29 and 4.30). Ultimately, each path between an electrode and the EKG module shows a resistance of around 13 Ω. The conductors to the snap fasteners for the battery have around 1.3 Ω each.

4.6.1.2.2 Embroidered ECG Electrodes

The electrodes have been embroidered in one step together with the conductors which lead to them. This eliminates an unnecessary contact, which

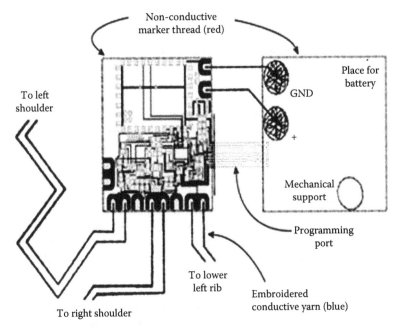

FIGURE 4.29
Schematics of the ECG module with embroidery and contacts for the battery. (From Linz, T. et al., Fully integrated EKG shirt based on embroidered electrical interconnections with conductive yarn and miniaturized flexible electronics, *Proceedings of the International Workshop on Wearable and Implantable Body Sensor Networks [BSN'06]* IEEE, 2006.)

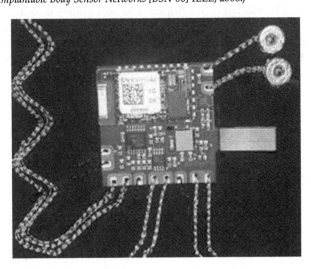

FIGURE 4.30
Zigzag-embroidered conductors for maximum stretch ability. (From Linz, T. et al., Fully integrated EKG shirt based on embroidered electrical interconnections with conductive yarn and miniaturized flexible electronics, *Proceedings of the International Workshop on Wearable and Implantable Body Sensor Networks [BSN'06]* IEEE, 2006.)

always comes with an additional contact resistance and contact noise. On the body side, they are enhanced with a solid gel pad-like substance to reduce movement, artefacts and to improve the contact between the ionic conductor (skin) and the electronic conductor (silver-coated threads).

4.6.1.2.3 Encapsulation

To protect the electronic components and their interconnections as well as the embroidered interconnects, encapsulation is indispensable. In a glob top encapsulation has been developed for a single chip module of a textile transponder. For the much larger module in this work, moulding is more promising as it offers a more defined geometry and better protection due to higher filler content of the material (see Figure 4.31). The moulding process requires that the fabric is very flat and neatly processed; otherwise, mould compound may be pressed outside the form. The curing temperature of typically 160°C–185°C is well below the melting point of *Statex* (PA66 melts at 260°C); however, it changes the characteristics of the spandex in the shirt to some extent. The material gets less elastic around the module. As this area is quite small, this may still be acceptable. Nonetheless, this issue will be addressed in further research. For this work, LOCTITE Hysol GR 9800 moulding compound with a curing temperature of 165°C has been applied.

This encapsulation method was first tested in and proved reliable under lab conditions. This ECG T shirt is the first real world application of mould encapsulation of electronics on textiles. In this work, an extremely miniaturized ECG module has been developed and integrated into a T-shirt using technologies, which make it very comfortable for the wearer. So far, the hardware has been tested in earlier state and showed good results. In a separate test, the electrodes showed similar results as standard electrodes with some

FIGURE 4.31

Flexible modules moulded on jeans fabric. (From Linz, T. et al., Fully integrated EKG shirt based on embroidered electrical interconnections with conductive yarn and miniaturized flexible electronics, *Proceedings of the International Workshop on Wearable and Implantable Body Sensor Networks [BSN'06]* IEEE, 2006.)

movement artefacts. In the coming weeks, the software has to be adapted to the slightly altered module that was used in the shirt and loaded onto it. Subsequently, intensive functionality tests of the T-shirt will be carried out. After that, the reliability of the shirt will be investigated more closely. The ECG electrodes may need improvement. Furthermore, the functionality may be enhanced with more sensors as mentioned in the introduction (Figures 4.32 and 4.33).

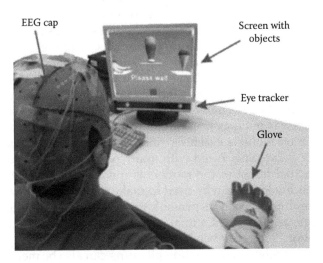

FIGURE 4.32
Physical experiment set-up. (From Novak, D. et al., *Biomed. Eng.*, 60(9), 2645, 2013.)

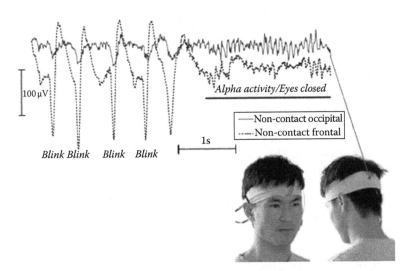

FIGURE 4.33
Non-contact EEG headband and data from both frontal and occipital electrodes. (From Chi, Y.M. et al., *IEEE Rev. Biomed. Eng.*, 3, 106, 2012.).

4.6.1.3 Electroencephalography (EEG)

Rapid recognition of voluntary motions is crucial in human–computer interaction, but few studies compare the predictive abilities of different sensing technologies. The performance of different technologies when predicting targets of human reaching motions is measured in terms of: EEG, electro-oculography (EOG), camera-based eye tracking, electromyography (EMG), hand position and the user's preferences. Eye tracking is robust and exhibits high accuracy at the very onset of limb motion. The information could aid human–computer interaction designers in selecting and evaluating appropriate equipment for their applications. Rapid recognition of voluntary motions is crucial in human–computer interaction and a number of different sensing technologies have been studied (Novak et al. 2013). For example, neuroprostheses can be controlled with brain–machine interfaces, exoskeletons use EMG to augment movement and rehabilitation robots use kinematics to assist patients who cannot complete a motion by themselves. Among voluntary motions, reaching motions towards one of several objects are especially interesting. Faced with many possible targets, the goal is to predict the actual target of the motion as quickly and accurately as possible. Many technologies can be used to make the prediction; when choosing among multiple objects, a person's gaze gradually shifts towards the eventually selected item, and a final gaze fixation precedes the start of the reaching motion. This gaze behaviour can be measured with eye trackers. Brain activity associated with motion planning can also be measured using EEG. Once a decision is made, autonomic nervous system responses may precede the motion and electrical activity of the involved muscles indicates the intended movement direction before the arm is moved. Finally, even before the user's hand touches the object, the direction of motion indicates the target. All these possibilities have been extensively studied, and some studies have combined multiple sensing technologies. For example, Corbett et al. (2012) used eye tracking to predict the reaching target and then planned the motion trajectory using EMG. Furthermore, Huang et al. (2012) combined EMG and mechanical sensors for locomotion mode classification in neuroprosthetics. They found that combining the two modalities not only improves accuracy, but also allows locomotion mode transitions to be detected earlier. The aforementioned studies combine two sensing modalities, with one of them also measuring how early a prediction can be made with each modality. However, there is still a lack of studies that would directly compare several sensing technologies in a single setting with regard to both how accurate they are and how early they can predict the target of a reaching motion. Novak et al. (2013) have made an attempt to address this issue by predicting targets of reaching motions using EEG, EOG, camera-based eye tracking, EMG, hand position and the user's preferences. The two primary goals of the study are as follows: (1) to quantitatively compare individual modalities' abilities to predict the target at different points in time before and during limb motion in a single setting; (2) to evaluate the degree to which accuracy can be improved by combining multiple modalities.

Eye tracking is accurate at motion onset, handles multiple targets well and generalizes well to new users. However, it predicts only the target, not the motion dynamics. It is thus useful if we wish to impose a motion (e.g. the user looks at the target, but the motion is made by a robot) or bring the target object closer to the user. The user's preferences can predict the target well before the motion and can augment any measurements, but could be difficult to obtain in real-world settings. Novak et al.'s (2013) approach with periods of interest and prior probabilities has proven effective at fusing multimodal data. Combining EEG and EOG results in a higher accuracy than using either individual modality and is convenient, since the two are often measured together anyway.

4.6.1.3.1 Wireless EEG System

These requirements can be translated into technical challenges, which are easily explained by looking at the different components that make up the complete signal chain, as depicted in Figure 4.34.

1. *Dry Electrodes*: Dry electrodes are the preferred solution for home use of EEG. Electrodes with contact gel (or other liquids) are not suited for long-term use, because evaporation will reduce contact properties over time. Moreover, the gel cannot be removed from the hair without washing, creating discomfort for the patient. For EEG, a structure that penetrates the hair to (almost) touch the surface of the skin is typically required.

2. *Analogue Readout Circuit*: Several challenges are associated to the analogue readout circuit. Several sources of interferences will lower the signal quality, which becomes even more challenging with the use of dry electrodes. Careful amplifier design and the use of active electrodes or active shielding will overcome some of these problems.

3. *Digital Signal Processing*: In medical applications, the patient can be asked to sit quietly, and to limit eye movement and blinking as much as possible. If the patient is monitored in his daily life environment, several artefacts will contaminate the EEG signal. When using dry electrodes,

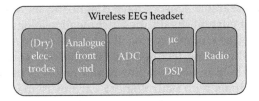

FIGURE 4.34
Components of biopotential signal acquisition unit.

we will see an increase in the severity of motion artefacts, on top of the biological artefacts (EOG and EMG). Digital signal-processing algorithms, such as ICA, are widely used to attenuate these artefacts.

4. *Headset Design*: Some (mechanical) challenges will have to be addressed in the headset, in which aesthetics, comfort and pressure on the electrodes are key for a successful application. Different application paradigms will typically come with a unique set of requirements (such as sensor positions), and hence, it is likely that headset designs will be application specific.

4.6.1.4 Future Research Scope: EEG and ECG

The first decade of research in the field of wearable technology was marked by an emphasis on the engineering work needed to develop wearable sensors and systems; recent studies have been focused on the application of such technology towards monitoring health and wellness. Also, a great deal of emphasis on surveying studies focused on the deployment of wearable sensors and systems in the context of concrete clinical applications, with main focus on rehabilitation. The interest of researchers and clinicians for pursuing applications of wearable sensors and systems has caused a shift in the field of wearable technology from the development of sensors to the design of systems. Consequently, it witnessed a great deal of work towards the integration of wearable technologies and communication as well as data analysis technologies so that the goal of remote monitoring individuals in the home and community settings could be achieved. Besides, when monitoring has been performed in the home, researchers and clinicians have integrated ambient sensors in the remote monitoring systems.

Patel et al. (2012) have summarized the recent developments in the field of wearable sensors and systems that are relevant to the field of rehabilitation. The growing body of work focused on the application of wearable technology to monitor older adults and subjects with chronic conditions in the home. The clinical applications of wearable technology is currently undergoing assessment rather than describing the development of new wearable sensors and systems. A short description of key enabling technologies (i.e. sensor technology, communication technology and data analysis techniques) that have allowed researchers to implement wearable systems is followed by a detailed description of major areas of application of wearable technology. Wearable sensors have diagnostic, as well as monitoring applications. Their current capabilities include physiological and biochemical sensing, as well as motion sensing. It is hard to overstate the magnitude of the problems that these technologies might help solve. Physiological monitoring could help in both diagnosis and ongoing treatment of a vast number of individuals with neurological, cardiovascular and pulmonary diseases such as seizures, hypertension, dysrthymias and asthma. Home-based motion sensing might assist in falls prevention and help maximize an individual's independence and community participation.

4.6.1.4.1 Remote Monitoring Systems

Remote monitoring systems have the potential to mitigate problematic patient access issues. Nearly 20% of those in the United States live in rural areas, but only 9% of physicians work in rural areas. Access may get worse over time as many organizations are predicting a shortfall in primary care providers as health care reform provides insurance to millions of new patients. There is a large body of literature that describes the disparities in care faced by rural residents. Compared to those in urban areas, those in rural areas travel two to three times farther to see a physician, see fewer specialists and have worse outcomes for common conditions such as diabetes and heart attack. Wearable sensors and remote monitoring systems have the potential to extend the reach of specialists in urban areas to rural areas and decrease these disparities. A conceptual representation of a system for remote monitoring is shown in Figure 4.35. Wearable sensors are used to gather physiological and movement data, thus enabling patient's status monitoring. Sensors are deployed according to the clinical application of interest. Sensors to monitor vital signs (e.g. heart rate and respiratory rate) would be deployed, for instance when monitoring patients with congestive heart failure or patients with chronic obstructive pulmonary disease undergoing clinical intervention.

Recent developments in the field of microelectronics have allowed researchers to develop miniature circuits entailing sensing capability, front-end amplification, microcontroller functions and radio transmission. The flexible

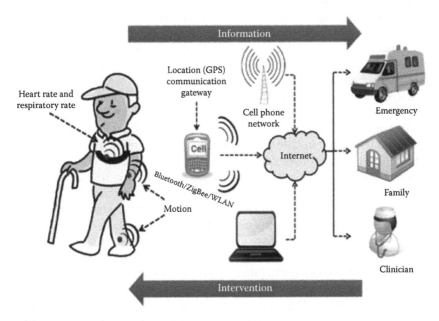

FIGURE 4.35
Illustration of a remote health monitoring system based on wearable sensors. (From Patel, S. et al., *J. NeuroEng. Rehabil.*, 9, 21, 2012.)

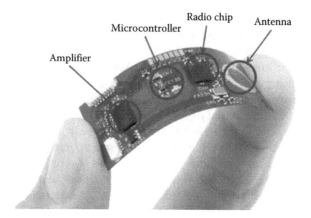

FIGURE 4.36
Flexible wireless ECG sensors with a fully functional microcontroller by IMEC.

circuit shown in Figure 4.36 is an example of such technology and allows one to gather physiological data as well as transmit the data wirelessly to a data logger using a low-power radio. Particularly relevant to applications in the field of rehabilitation are advances in technology to manufacture micro-electromechanical systems (MEMS). MEMS technology has enabled the development of miniaturized inertial sensors that have been used in motor activity and other health status–monitoring systems. By using batch fabri-cation techniques, significant reduction in the size and cost of sensors has been achieved. Microelectronics has also been relied upon to integrate other components, such as microprocessors and radio communication circuits, into a single integrated circuit, thus resulting in system-on-chip implementations.

Advances in material science have enabled the development of e-textile based systems. These are systems that integrate sensing capability into gar-ments. Figure 4.37 demonstrates how sensors can be embedded in a garment to collect, for instance electrocardiographic and EMG data by weaving elec-trodes into the fabric and to gather movement data by printing conductive elastomer-based components on the fabric and then sensing changes in their resistance associated with stretching of the garment due to subject's move-ments. Rapid advances in this field promise to deliver technology that will soon allow one to print a full circuit board on fabric. Health-monitoring applications of wearable systems most often employ multiple sensors that are typically integrated into a sensor network either limited to body-worn sen-sors or integrating body-worn sensors and ambient sensors. In the early days of body-worn sensor networks (often referred to as '*body sensor networks*'), the integration of wearable sensors was achieved by running 'wires' in pockets created in garments for this purpose to connect body-worn sensors. Recently developed wearable systems integrate individual sensors into the sensor net-work by relying on modern wireless communication technology. Embedded sensors provide one with the capability of recording electrocardiographic

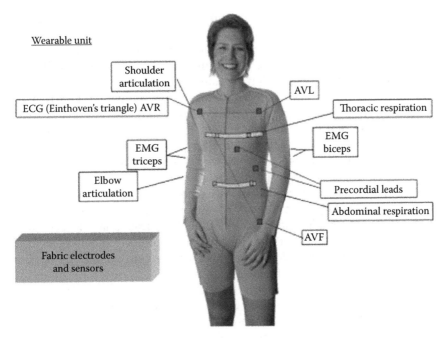

FIGURE 4.37
Example of e-textile system for remote, continuous monitoring of physiological and movement data. (From Patel, S. et al., *J. NeuroEng. Rehabil.*, 9, 21, 2012.)

data (ECG) using different electrode configurations as well as electromyographic (EMG) data. Additional sensors allow one to record thoracic and abdominal signals associated with respiration and movement data related to stretching of the garment with shoulder movements.

4.6.1.4.2 Mobile Phone Technology

Mobile phone technology has had a major impact on the development of remote monitoring systems based on wearable sensors. Monitoring applications relying on mobile phones are becoming commonplace. Smart phones are broadly available. The global smart phone market is growing at an annual rate of 35% with an estimated 220 million units shipped in 2010. Smart phones are preferable to traditional data loggers, because they provide a virtually 'ready to use' platform to log data as well as to transmit data to a remote site. Besides being used as information gateways, mobile devices can also function as information-processing units. The availability of significant computing power in pocket-sized devices makes it possible to envision ubiquitous health monitoring and intervention applications. In addition, most mobile devices now include an integrated GPS tracking system, thus making it possible to locate patients in case of an emergency. Also, as storage and computation becomes more and more cloud based, health-monitoring systems can become low-cost, platform-independent, rapidly deployable and universally accessible. Monitoring devices can become simpler and cheaper as the computation is pushed to the cloud. This enables users to buy off-the-shelf devices and access customized monitoring applications via cloud-based services. Cloud-based systems can prove especially useful for bringing health-care services to rural areas. In addition, monitoring applications deployed via the cloud can be easily updated without requiring that the patient to install any software on his/her personal monitoring device, thus making system maintenance quicker and cost-effective. Finally, the massive amount of data that one can gather using wearable systems for patient's status monitoring has to be managed and processed to derive clinically relevant information. The Android-based mobile application allows low-power ECG sensors to communicate wirelessly with the phone. With increasing computational and storage capacity and ubiquitous connectivity, smart phones are expected to truly enable continuous health monitoring (Figure 4.38). Data analysis techniques such as signal processing, pattern recognition, data mining and other artificial intelligence-based methodologies have enabled remote monitoring applications that would have been otherwise impossible.

Recent advances in smart phone technology have led to their use in fall detection systems. Often, these systems combine fall detection with localization of the person who fell via a GPS-based method. Schwickert et al. (2013) have developed a fall detection system that relied upon the accelerometers available

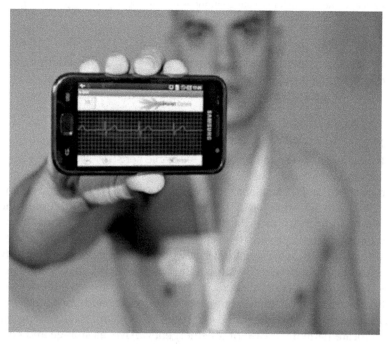

FIGURE 4.38
Smart phone–based ECG monitoring system.

in smart phones and incorporated different algorithms for robust detection of falls. Their implementation leveraged the characteristics of the Android 2 operating system. The authors developed advanced signal-processing techniques to achieve high accuracy of falls detection. Besides, the system provided subject's location using Google maps. Using this approach, a warning about the fall and the location of the subject undergoing monitoring is transmitted to a caregiver or family member via SMS, email and Twitter messages. Ongoing research is geared towards the prevention of fall-related injuries. Numerous systems have been developed by leveraging airbag technology. These systems rely upon wearable accelerometers and gyroscopes to trigger the inflation of the airbag when a fall is detected. Although these systems can potentially help to prevent fall-related injuries, further development is needed to miniaturize the airbag system that provides protection to the subject before an impact occurs. Individuals with movement impairments require more specific approaches to detect or prevent falls. Bächlin et al. (2009) have developed a system to detect freezing of gait (FOG), a commonly found gait symptom in Parkinson's disease that is highly related to falls. The system was designed to provide subjects with a rhythmic auditory signal aimed to stimulate the patient to resume walking when a FOG episode is detected. Smith and Bagley (2010) have developed a system to be used in children with difficulty in walking, which is known to be associated with frequent falls. They collected tri-axial accelerometer data and digital video recordings for over 50 h from children with cerebral palsy and 51 control subjects. The dataset was used to develop algorithms for automatic real-time processing of the accelerometer signals to monitor a child's level of activity and to detect falls. Sposaro and Tyson (2009) have focused their attention on older adults with dementia. These subjects require frequent caregivers' assistance to accomplish standard activities of daily living. The authors relied upon an Android application, iWander, which uses GPS and communication functions available via the smart phone, to provide tracking of subjects' location and assistance when needed. The system was shown to improve functional independence among dementia patients while decreasing the stress put on caregivers. Another application of wearable sensors and systems that has received a great deal of attention among researchers and clinicians is the detection of epileptic seizures. Primary and secondary compulsive epileptic crises (EC) cause a sudden loss of consciousness. These events are accompanied by stereotypical movements that one can observe in association with characteristic changes in the electroencephalogram (EEG). During the acute phase, the subject is completely unable to interact with the environment. To detect EC, systems and methods relying upon wearable sensors have been proposed and evaluated. Electroencephalographic (EEG) sensors, 3D accelerometers on a wrist, combination of EMG and accelerometers, and electrodermal activity (EDA) have been used to develop methods to distinguish EC from normal motor activities. Nokia N810 and the SHIMMER platform were used as wearable sensors to detect seizure events. An interesting recent development is the integration of various sensors and systems in a network for comprehensive safety monitoring and smart home health-care applications. AlarmNet is an example of such systems. It collects and analyzes various data streams to monitor a resident's overall wellness, known medical conditions, activities of daily living and emergency situations. The whole project deals not only with wearable sensing technology but also with security/privacy issues in patient's data transfer, and real-time data streaming. A major contribution towards the development of new solutions in the field of wearable and ambient sensors and their integration in comprehensive safety monitoring and smart home health-care applications is provided by the European AALIANCE (Ambient Assisted Living Innovation Alliance). AALIANCE is an active project that includes many research institutes, companies and universities in Europe. The project's aim is to define the necessary future R&D steps towards developing Ambient Assisted Living (AAL) solutions. The project builds infrastructure for practical applications of wearable technologies such as telemonitoring of patient's status and self-management of chronic diseases.

4.6.2 Electronic Shirt

Australia's Common Wealth Scientific Industrial and Research Organization (CWSIRO) has developed the wearable instrument shirt (WIS), which uses custom sensors to turn body movements into music. Users can play 'air guitar' by moving one arm to pick chords and the other to strum the imaginary strings. The battery-powered WIS consists of a wearable sensor interface embedded in a conventional shirt that uses custom software to map gestures with audio samples. It recognizes and interprets arm movements, and then, it wirelessly relays this data to a computer for audio generation. No cables or microphones are necessary. 'It's easy to use virtual instrument that allows real-time music making', says Richard Helmer, leader of the team of researchers that developed the WIS. By customizing the software, the team also has tailored the technology to create an air tambourine and an air guiro, a percussion instrument. 'The technology, which is adaptable to almost any kind of apparel, takes clothing beyond its traditional role of protection and fashion into the realms of entertainment and a wide angel of other applications including the development of clothes which will be able to monitor physiological changes', Helmer says. CSIRO's researchers believe electronic textiles have many practical uses in applications that require easy human–computer interfacing (Figure 4.39). They expect the technology to progress from entertainment uses to sports and then rehabilitation and medicine.

4.6.2.1 Smart Shirt: Temperature Measurement

Current trends in e-textiles and smart fabrics try to find a symbiosis of electronics and clothing. Augmenting a fabric with an intrinsic sensing function is an elegant way to measure the environment. This approach supports the

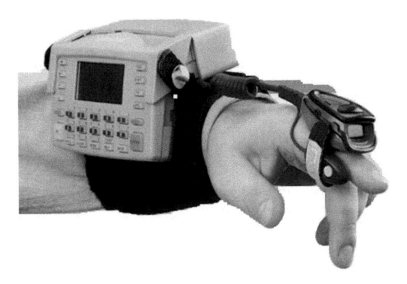

FIGURE 4.39
Sensor wear.

idea of a smooth combination of the two worlds of electronics and clothing. Our hybrid fabric senses temperature and estimates the temperature profile along the surface. We explain the measurement principle and show how a measurement accuracy of 0.5 K with a 10 cm long piece of fabric can be achieved. We also point out potential applications of our fabric temperature sensor. The realization of our vision of continuous health care and assistance during daily life necessitates an unobtrusive embedding of sensors and electronics in our outfit without affecting wearing comfort. Sensors are distributed all over the body, often requiring direct contact to the skin. Smart fabrics with intrinsic sensing functionality offer an optimal solution to satisfy these requirements. Moreover, anything that utilizes fabrics from clothing until car interior could be made *smart*. In this chapter, we present a smart fabric system with temperature-sensing capability. The goal is to achieve a measurement accuracy of 0.5 K over a temperature range of 10°C–60°C that is comparable to commercial products such as thermistors (e.g. Pt100) and thermocouples. The woven structure of the hybrid fabric itself represents an array of temperature sensors that can be utilized to measure the temperature profile of a surface. To our knowledge, such a smooth embedding of temperature-sensing capability into a fabric is novel. Our hybrid fabric is not only capable of temperature sensing, but also for signal transmission. Currently, many research labs concentrate on the side of signal transmission. Temperature measurement is one option, but fabrics have the capacity to integrate other sensing functionality. Lorussi et al. (2004) work on fabric-based strain sensors in order to track body postures. Companies such as Softswitch, Eleksen and ITP investigate fabric pressure sensors.

4.6.2.2 Measurement Concept

In this section, we briefly present the physical principle for our temperature measurements. In our method, we utilize the temperature dependence of the electrical resistance of metals according to Equation 4.1.

$$R_{wire} = \rho_{metal}(1 + \alpha_{metal} \times DT)\frac{l_{wire}}{A_{wire}} \; (\Omega) \tag{4.1}$$

where
   ρ_{metal} is the specific resistance
   α_{metal} is the temperature coefficient
   A_{wire} denotes the cross-section area
   l_{wire} is the length of the metal wire

α_{metal} is similar for all metals and features about 0.4%/K for copper. The length elongations as well as the quadratic coefficient of resistance (due to temperature variations) are negligibly small for our intended measurement range from 0°C to 60°C. We utilize thin copper wires with about 14.16 Ω/m at 20°C for our measurements. The wires are embedded in a fabric; the intended temperature profile estimation requires a grid of sensors. In a demo application, we applied a 10 × 10 cm² hybrid fabric patch as measurement array with a spatial resolution of 1 cm.

4.6.2.3 Hybrid Fabric Sensor

The hybrid fabric is fabricated by Sefar Inc. It consists of woven polyester yarn (PET) with diameter of 42 µm and copper alloy wires with diameter 50 ± 8 µm. The hybrid fabric with a mesh opening of 95 ± 10 µm and an opening area of 44% is shown in Figure 4.40. Each copper wire itself is coating

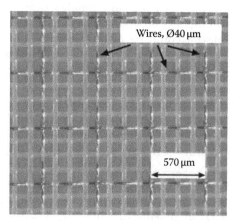

FIGURE 4.40
Fabric with embedded copper wires as temperature sensors (Sefar Inc.).

with a polyurethane varnish as electrical insulation. The copper wire grid in the textile features a spacing of 0.57 mm (mesh count in warp and in weft is 17.5 cm^{-1}). The combination of PET yarn and copper wires requires a special weaving technology, which includes two yarn systems in warp and weft direction (three PET wires and one copper wire) with separate tensioning systems. The hybrid fabric with its weight of 74 g/m^2 is positioned as interlining. Its application field is therefore very versatile. The fabric represents a compromise between preserving textile properties and copper wire density. To our knowledge, such a precise hybrid fabric consisting of PET yarn and copper wire is new to the market. Due to the low mass of a wire (123 dtex = 12.3 µg/m) and the high thermal conductivity of copper, such a temperature sensor reacts quickly towards thermal changes.

4.6.2.4 Electronic Measurement Circuit

Since the resistances of the thin copper wire as well as its temperature coefficient α_{Cu} are small, the electronics must measure very precisely to achieve the targeted accuracy of 0.5 K. Thus, we utilize the four-wire resistance measurement principle. Secondly, we propose a new difference measurement method in order to eliminate circuit offsets.

4.6.2.5 Measurement Principle

The principle measurement circuit is shown in Figure 4.41. By imposing a constant current through the copper wire and measuring the resulting voltage drop, the wire resistance can be extracted and, therefore, the temperature according to Equation 4.2. $R_{wire\ 0}$ defines the reference resistance of the wire at a certain temperature, that is 20°C. A constant current through the wire is ensured by the operational amplifier (OA), which regulates its output such that its (+) and (–) inputs feature the same potential. The instrumental amplifier (IA) boosts the differential signal such that a measurement resolution of 0.4 K/LSB is achieved for our demonstrator. The low-pass filter reduces noise and avoids aliasing by limiting the signal bandwidth to f_c = 160 Hz. This bandwidth is a compromise between noise reductions and measurement speed. The ADC features a resolution

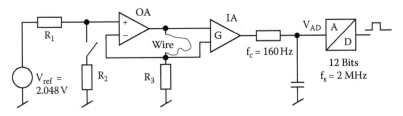

FIGURE 4.41
Temperature measurement principle.

of 12 Bits, which translates to a least significant bit (LSB) of 0.5 mV using a reference voltage of 2.048 V. R_2 can be switched to the circuit that allows applying different currents through the copper wire. This procedure further improves measurement accuracy.

$$\Delta T = \left(\frac{R_{wire}}{R_{wire0}} - 1 \right) \cdot \frac{1}{\alpha_{Cu}} \ (K) \tag{4.2}$$

Formula given in Equation 4.3 gives the wire resistance in relation to the measured voltages V_{ADh} and V_{ADl} at the ADC. The two voltages correspond to the two currents that are generated by switching R_2 on and off the circuit. Due to subtraction of the voltages, constant circuit offsets (e.g. amplifier offsets) are eliminated. Putting Equation 4.3 into Equation 4.2, the constants R_1 – R_3, G and V_{ref} cancel out, leading to formula given in Equation 4.4. As a result, the accuracies of these constants become unimportant when using a reference wire resistance R_{wire0} determined by calibration. In fact, the voltages in formula given in Equation 4.4 can directly be replaced with the bits$_x$ corresponding to the binary values of the ADC. This then leads to formula given in Equation 4.5.

$$R_{wire} = \frac{R_3}{G} \cdot \frac{R_1 + R_2}{R_2} \cdot \frac{V_{ADh} - V_{ADl}}{V_{ref}} \ (\Omega) \tag{4.3}$$

$$T = \left(\frac{V_{ADh} - V_{ADl}}{V_{AD_{refh}} - V_{AD_{refl}}} - 1 \right) \frac{1}{\alpha_{Cu}} + T_0 \ (^\circ C) \tag{4.4}$$

$$T = \left(\frac{bits_h - bits_l}{bits_{refh} - bits_{refl}} - 1 \right) \frac{1}{\alpha_{Cu}} + T_0 \ (^\circ C) \tag{4.5}$$

R_{wire0}, that is bits$_{refh}$ and bits$_{refl}$, can be determined with a single point calibration. However, α_{Cu} determines the slope of the resistance change by temperature. The value of α_{Cu} is not precisely known, since literature gives values ranging from 0.0039 to 0.00404 K^{-1}. Secondly, the wires in our fabric consist of a copper alloy whose α differ from a pure copper wire. Therefore, a two-point calibration is necessary at least.

4.6.2.6 Theoretical Temperature Accuracy

The total measurement tolerance depends on different factors. On the one hand, accuracy improves the longer a measurement wire within the fabric is. A longer wire directly corresponds to a higher resistance. On the other hand, location information about temperature changes decreases the longer a wire is. For instance a temperature change could be spread along the entire wire as well as just locally at a single point. This behaviour resembles the uncertainty principle. Secondly, all measurements are noisy up to the bandwidth of 160 Hz and all electronic components are temperature dependent as well. We model our measurement system including parasitic for a worst-case scenario in the following way (Equation 4.6)

$$V_{AD} = f(T_{wire}) + V_{ofs} + V_{noise} + V_{drift} + V_{tol} \ (V) \tag{4.6}$$

Whereas the circuit offsets V_{ofs} are cancelled by the difference measurement described. The circuit noise V_{noise} (thermal noise) can statically be reduced by multiple measurements N with an accuracy improvement following the $1/\sqrt{N}$-law. The reference voltage source V_{ref}, the OA and the IA identify the biggest noise sources, though they are rather constant over temperature. V_{drift} describes the temperature dependence of the electronic circuit, whereas $R_1 - R_3$ feature the biggest influence with a datasheet value of 15 ppm/K. Thirdly, the ADC with its accuracy of 0.19 K/LSB according to the datasheet introduces a quantization noise of about 0.22 K. Figure 4.42 depicts the different inaccuracy sources of the circuit as a function of temperature. Circuit noise contributes most to the error, but it can be minimized by multiple measurements. Therefore, an accuracy of about 0.5 K can be achieved with single measurement.

4.6.2.7 Temperature Measurements

In this section, measurement verification of our demonstrator is described. Note that our reference thermometer (Fluke DMM with thermocouple) itself

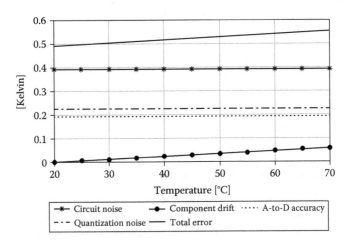

FIGURE 4.42
Contributions to temperature measurement inaccuracy.

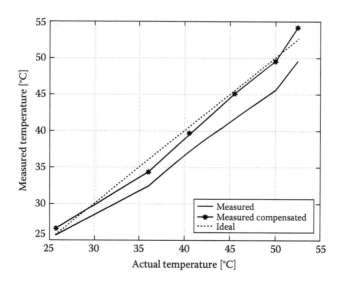

FIGURE 4.43
Temperature measurement linearity.

has an accuracy of only $\alpha \pm 1\%$ (+3°C). Figure 4.43 shows the temperature linearity neglecting the tolerances of the reference thermometer. Since we do not know α_{metal} exactly, the slope of the measured curve with an assumed $\alpha = 0.00404$ K^{-1} (thin solid line in the figure) is too flat. However, we approximated the true α_{metal} with the given measurements by means of Formula 4.7, resulting in an $\alpha_{true} = 0.0035$ K^{-1}. The corresponding α-compensated curve is depicted as thick solid line in the figure. The measurement points of the fitted curve feature a standard deviation of $\sigma_e = 1.2$ K to an ideal curve (straight line).

$$\alpha_{true} = \alpha_{Cu} \frac{(T_{meas} - T_0)^T (T_{meas} - T_0)}{(T_{act} - T_0)^T (T_{meas} - T_0)} \; (K^{-1}) \tag{4.7}$$

The approximation (Equation 4.7) was computed using the least mean square error approach shown in Formula 4.8.

$$\frac{\partial}{\partial \alpha_{true}} \sum_i \left[T_{act_i} - T_0 - (T_{meas_i} - T_0)\frac{\alpha_{Cu}}{\alpha_{true}} \right]^2 = 0 \tag{4.8}$$

4.6.2.8 Temperature Measurement Accuracy

In order to determine the measurement accuracy and repeatability, we repeatedly measured the same temperature (e.g. 500 times) using a climate chamber. Thus, we achieved a small standard deviation of $\alpha = 0.39$ K at a temperature of about 50°C. The standard deviation behaved similarly for different temperatures. The temperature distribution is depicted in Figure 4.44.

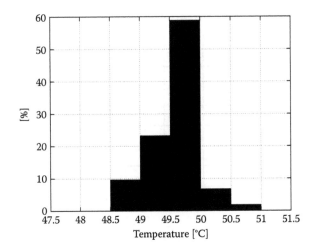

FIGURE 4.44
Measurement distribution over 500 iterations at 50°C.

4.6.2.9 Surface Temperature Profile

The grid of copper wires in the hybrid fabric (Section 4.3) allows measurement of temperature along a surface. For our demonstrator, we utilized a 10 × 10 cm² hybrid fabric patch as temperature sensor array. A principle schematic is given in Figure 4.45. Due to the wire topology, location information of *hot spots* can be extracted by measuring wire resistance in x- and y-direction. Instantly, such a fabric seems to be attractive to measure temperature distribution across a surface. However, a measurement returns an average temperature along the wire whereby *hot spots* get smeared out over the entire wire length. From a mathematical point of view, the system is under-determined,

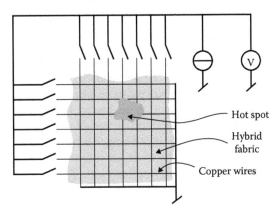

FIGURE 4.45
Temperature measurement array.

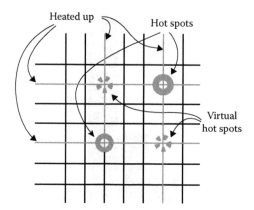

FIGURE 4.46
Measurement ambiguity due to grid structure.

since we only have 2N measurements in an N × N grid with N^2 cross points for temperature measurement. This fact leads to spatial ambiguity. Unless all the 'hot spots' are aligned in a row or a column, additional virtual 'hot spots' occur. For example the measurement of two actual *'hot spots'* on the fabric results in detection of two additional virtual 'hot spots' as can be seen in Figure 4.46. Three actual 'hot spots' result in an additional 6 virtual 'hot spots', etc. A restriction to single 'hot spot' detection relaxes the problem such that no ambiguity occurs. The temperature distribution estimation of such a single 'hot spot' is shown in Figure 4.47, where a finger touched the fabric in the rear corner. The estimated distribution of the temperatures is computed by taking the outer product of the temperatures along the row and column wires according to Equation 4.9.

$$\text{Test} = \sqrt{T_{row} \cdot T_{col}^{T}} \ (°C) \tag{4.9}$$

4.6.2.10 Measurement Ambiguity Avoidance

By shrinking the patches to a size where ambiguity does not matter anymore, for example an area of 2 × 2 cm^2, and by applying several of these patches in an array, overall ambiguity can be eliminated. Such patches can be designed by fabric interconnection techniques. An array of patches is depicted in Figure 4.48. The connections guarantee a longer wire such that the targeted temperature resolution can be achieved.

The fabric temperature sensor was developed with an accuracy of about 1 K when using a two-point calibration. This accuracy can further be improved to 0.5 K by recording the entire curve over the temperature range such that non-linearity can be compensated. Additionally, more sophisticated electronics with temperature compensation and less noisy components improve performance as well. Secondly, we have shown how

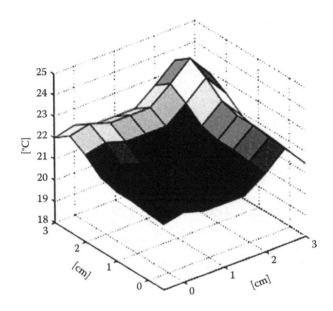

FIGURE 4.47
Temperature distributions along fabric.

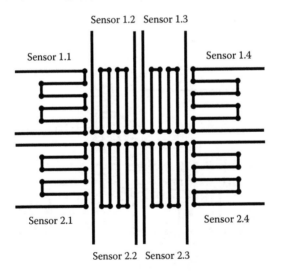

FIGURE 4.48
Array of temperature measurement patches.

spatial resolution can be improved without ambiguity. Such fabric temperature sensors can smoothly be integrated into clothing providing a platform for temperature profile measurement along the human body. Many applications are feasible, for example monitoring of temperature to which fire-fighters are exposed during their missions. Other applications could lie in the medical field. A new idea is using sensor array to map the food intake over the day by measuring skin temperature close to the liver. The sensors do not necessarily need to be integrated into clothing, for example a car seat equipped with these sensors could measure skin temperature as well (Figures 4.49 through 4.52).

FIGURE 4.49
Flat battery can be removed for washing and charging. (From Linz, T. et al., Fully integrated EKG shirt based on embroidered electrical interconnections with conductive yarn and miniaturized flexible electronics, *Proceedings of the International Workshop on Wearable and Implantable Body Sensor Networks [BSN'06]* IEEE, 2006.)

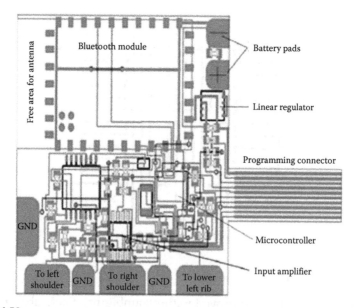

FIGURE 4.50
Largest and most energy-consuming part of the ECG module is the Bluetooth unit. (From Linz, T. et al., Fully integrated EKG shirt based on embroidered electrical interconnections with conductive yarn and miniaturized flexible electronics, *Proceedings of the International Workshop on Wearable and Implantable Body Sensor Networks [BSN'06]* IEEE, 2006.)

FIGURE 4.51
ECG shirt with electronics, snap fasteners for the battery and embroidered electrodes and conductors. (From Linz, T. et al., Fully integrated EKG shirt based on embroidered electrical interconnections with conductive yarn and miniaturized flexible electronics, *Proceedings of the International Workshop on Wearable and Implantable Body Sensor Networks [BSN'06]* IEEE, 2006.)

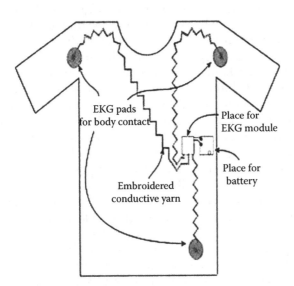

FIGURE 4.52
Illustration of the shirt with electronics, snap fasteners for the battery and embroidered electrodes and conductors. (From Linz, T. et al., Fully integrated EKG shirt based on embroidered electrical interconnections with conductive yarn and miniaturized flexible electronics, *Proceedings of the International Workshop on Wearable and Implantable Body Sensor Networks [BSN'06]* IEEE, 2006.)

4.7 E-Textile Craft

The blossoming research field of electronic textiles (or e-textiles) seeks to integrate ubiquitous electronic and computational elements into fabric. One of the most challenging aspects of the design and construction of e-textile prototypes is: engineering the attachment of traditional hardware components to textiles. New techniques for attaching off-the-shelf electrical hardware to e-textiles are: (1) the design of fabric PCBs or iron-on circuits to attach electronics directly to a fabric substrate; (2) the use of electronic sequins to create wearable displays and other artefacts and (3) the use of socket buttons to facilitate connecting plug-gable devices to textiles. The blossoming research field of electronic textiles (or e-textiles) focuses on ubiquitous computing as realized by a particular type of material: e-textile researchers seek to integrate pervasive electronic and computational elements into cloth. The ultimate goal of e-textiles is to develop technologies that are entirely fabric based and though terrific progress has been made, this vision is still a way off; at least for the present, researchers are forced to incorporate traditional electronics like microcontrollers and light-emitting diodes (LEDs) into their designs. One of the most challenging aspects of creating e-textile prototypes is: engineering the attachment of traditional hardware components to textiles. Three new techniques for attaching off-the-shelf electrical hardware to e-textiles were studied. In exploring this subject, we have tried to limit ourselves (if that is the right expression) to techniques that are accessible and materials that are easily obtained. A set of methods was developed that will make e-textile technology available to interested amateurs of all sorts. In the longer term, we hope to provide means for crafters and designers of a variety of ages and skill levels to create robust, affordable e-textile prototypes without postulating radically new hardware: to initiate a library of materials and techniques that form the foundation of 'do-it-yourself' electronic textiles. Area originates in a desire to integrate research in ubiquitous or pervasive computing with the best traditions of educational computing traditions that emphasize student creativity, empowerment and self-expression. E-textiles offer an especially promising field for creating educational and expressive artefacts that are at once intellectually challenging, aesthetically appealing and personally meaningful. Indeed, in such a young discipline, it is especially helpful for researchers to draw on a wide range of techniques geared towards the creation of affordable, robust prototypes. Moreover, investigating new techniques of incorporating electronics into fabrics could conceivably inform the next generation of electronics package design, since it is likely that in the near future, components will be manufactured specifically for wearable computing and e-textile applications. In the remainder of this chapter, we describe our techniques, situating them in the larger research areas of e-textiles and, more generally, ubiquitous computing. The following (second) section summarizes major trends in e-textile research as background to our work. In the third and central section, we present our three techniques: (1) the design of fabric PCBs or iron-on circuits to attach electronics directly to a fabric substrate; (2) the use of electronic sequins to create wearable displays and other artefacts, and (3) the use of socket buttons to facilitate connecting pluggable devices to textiles. We will conclude the third section with a discussion of a variety of materials and techniques we use to insulate conductive materials in e-textile prototypes and a presentation of the results of preliminary washing studies for each of our techniques. In the fourth section, we use our techniques as the basis for a wider-ranging discussion on the opportunities for do-it-yourself e-textiles, and we argue that this is an especially important and significant area for research. While the discussion in this section draws on our interests in educational computing, we maintain that the issues raised by do-it-yourself e-textiles are important to other areas of ubiquitous computing research as well. The fifth and final section discusses our ongoing work and outlines promising areas for future research.

Wearable computing explores technologies that are portable and attached to or carried on the body; head mounted displays, cell phones and PDAs, for example are 'wearable' computing devices. E-textile research is a closely related field, but has a slightly different focus: investigating electronic and computational technology that is embedded into textiles. E-textiles are often clothing, but can also be wall hangings, pillows, rugs and other pervasive fabric artefacts. Rather than focusing on existing (hard) electronic devices, e-textile researchers strive to build things that are as soft and flexible as traditional cloth. E-textiles can be beautiful examples of Weiser's ideas put into practice; at their best, they truly, unobtrusively 'weave themselves into the fabric of everyday life'. A good deal of e-textile research has been undertaken to advance the medical and military realms. For example in one groundbreaking early project, Park and Jayaraman (2003) built the 'Georgia Tech Wearable Mother-board'. This vest-like garment utilized woven optical fibres and conductive yarns in conjunction with integrated electronics to detect bullet wounds and monitor physiological signs like heart rate and temperature. This and other similar early e-textile research efforts have since borne fruit in the commercial arena. There are now several commercial devices primarily in medicine and sports that make use of e-textile research. One excellent example is the Life Shirt, developed by VivoMetrics. This shirt, similar in many ways to the Georgia Tech Wearable Motherboard, continuously monitors and records the heart rate, respiration rate and posture of its wearer. This data can then be 'downloaded' from the garment and analyzed by doctors and researchers, giving them a comprehensive portrait of the wearer's physiological patterns. Another genre of e-textiles research has focused less on applications of the technology and more on materials and techniques. In another example of groundbreaking early research, Post, Orth and their colleagues developed both simple and sophisticated e-textile engineering methods including: the embroidery of conductive yarns to act as data and power busses, resistors and capacitors on fabric; the development of capacitive sensing cloth touch pads and the use of gripper snaps in e-textile applications. Other innovations in e-textile techniques have focused on assessments of specific materials. For example Edmison et al. (2002) have described how to utilize piezoelectric materials in e-textiles, and in conjunction with researching new transistor materials, several groups are researching techniques for embedding transistors in fabric, taking the first steps towards realizing entirely fabric-based computation. Other practitioners have explored playful applications of e-textiles in fashion and other aesthetically driven applications. Post et al. (2000) and Berzowska et al. (2005a) are two of the more prominent researchers investigating this area. Post and Orth became well known in part for using thermo-chromic materials and resistive heating techniques to weave beautiful, controllable and non-emissive textile displays. Berzowska and her colleagues have embedded electronics like LEDs, sensors and speakers into fabric to create fanciful fashions and wall hangings. It is worth noting that the community of people developing artistic e-textiles includes individuals from outside technical academia, such as artists, designers and hobbyists of various stripes. The main interests are two-fold: to build artefacts that will capture people's imagination and to develop tools and techniques that will allow them to experiment with e-textiles themselves tools and techniques for a 'do-it-yourself' e-textile community. In short, to bring e-textiles 'to the people' by supporting and encouraging e-textile craft.

4.7.1 Techniques for E-Textile Craft

This section will introduce three techniques and analyze the potential of each as a do-it-yourself method, assessing everyone in terms of the expense and availability of its required materials and how difficult it is to employ. Printed circuit boards (PCBs) allow for the precise placing of electrical components into small spaces. In prototyping and hobby contexts, a circuit board pattern is first etched out of copper-clad board; then, holes for hardware are drilled into the board and finally, components are soldered to the copper traces. An analogous technique for creating PCBs on cloth using conductive fabric and an iron-on adhesive and also detail laser-cut fabric PCBs focus on hand-cut fabric PCBs was described.

4.7.1.1 Laser-Cut Fabric PCBs

Laser cutters can cut a wide range of materials with astonishing precision and speed. The next few paragraphs will describe how one can make use of these wonderful devices to build complex circuits out of conductive cloth. There are several steps in this process, most of which are shown in Figure 4.53. A heat activated adhesive is attached to a conductive fabric (usually use a metallized fabric called 'Zelt', which has a Sn/Cu plating and a surface resistivity of less than 0.1 Xs/sq). One is left with a piece of conductive fabric that

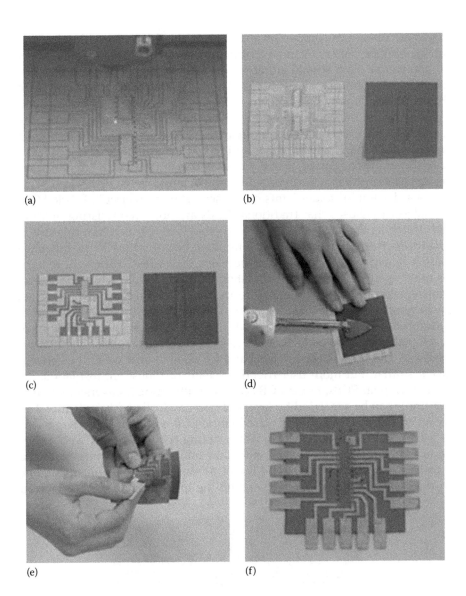

FIGURE 4.53
Steps to building a laser-cut fabric PCB. (a) The circuit is cut by a laser cutter. (b) The cut circuit and its backing fabric. (c) The paper underneath the circuit is removed. (d) The circuit is ironed onto its backing fabric. (e) The conductive fabric that is not part of the circuit is removed. (f) The completed circuit.

has a layer of adhesive covered with a layer of paper on one side. This fabric is placed, paper side up, into a laser cutter where a circuit pattern is etched into the fabric. The settings on the laser cutter should be adjusted so that the adhesive and paper backing are cut, but the fabric is only scored. Figure 4.53a shows a laser cutter etching a circuit, and Figure 4.53b shows the completed etched circuit and its companion substrate of blue fabric. Once the circuit is cut, the backing paper is removed from underneath the circuit – only where the conductive cloth should adhere to the baking fabric (Figure 4.53c). The circuit is then carefully aligned on its fabric substrate and ironed into place (Figure 4.53d). Finally, as is shown in Figure 4.53e, the circuit is separated from the rest of the conductive fabric. Note how the laser cutter scored the conductive fabric so that it comes apart easily at this stage, but remained together beforehand so that the circuit could be accurately placed. Figure 4.53f shows the completed circuit. Once created, a fabric PCB – with generous applications of flux – can be soldered like a traditional PCB. Figure 4.54 shows a close-up of solder joints on a laser-cut iron-on circuit. Fabric PCBs are subject to abuses that traditional PCBs are not – the twisting, folding and stretching of cloth – and solder joints inevitably break under this strain. To address this issue, each solder joint must be covered with an inflexible coating before the fabric PCB can be worn or washed. Variety of materials to encapsulate solder joints on fabric PCBs are used. Figure 4.55 shows joints encapsulated with an epoxy resin. Yet to find an easy way to protect solder joints using non-toxic materials, further research is on. Many other researchers, including Linz et al. (2005) and Kallmayer et al. (2003), have developed encapsulation techniques to protect various types of circuitry on e-textiles. The work is encouraging and we do hope that we will be able to invent similar, but more user-friendly, means of robust encapsulation. As with traditional PCBs, fabric PCBs can be multi-layered. Layers of conductive traces can be separated by layers of non-conducting fabric. Figure 4.56 shows an example of a multi-layered fabric PCB with two layers separated by purple insulating fabric. It is worth noting that a hobbyist who etches his or her own circuit boards cannot make multi-layered circuits: he or she is restricted to etching the circuit out of, at most, two layers of copper, one on the bottom of the board and one on the top. Traditional multi-layered

FIGURE 4.54
Solder joints on a fabric PCB.

FIGURE 4.55
Solder joints encapsulated in epoxy.

FIGURE 4.56
Multi-layered fabric PCB.

PCBs must be ordered from commercial PCB manufacturers. In this arena, fabric PCBs has a distinct advantage over the traditional PCBs made by hobbyists. Laser-cut fabric PCBs exhibit many of the advantages of traditional PCBs while preserving the essential qualities of cloth. As can be seen in Figure 4.56, the traces are flexible and can be sewn as well as soldered. In a typical use of the technique, solder IC sockets or microcontrollers to the traces and stitch other components.

4.7.1.2 Hand-Cut Fabric PCBs

Laser-cut fabric circuits are powerful e-textile tools, but they are challenging to build and require access to expensive equipment. How then, does this technique fit into the do-it-yourself or crafting category? In fact, iron-on traces need not be etched by a laser cutter; they can be cut by hand. The process to building hand-cut fabric PCBs is similar to that of building laser-cut circuits. One starts by attaching an iron-on adhesive to a piece of conductive fabric. Then, one can draw out a design on the adhesive's paper backing, cut the design out with scissors or a mat knife and iron it onto a backing fabric. Again, electrical components can be soldered or stitched to these traces. Figure 4.57 shows a hand-cut trace with an LED sewn to it. As is hinted at in the image, hand-cut traces (and laser-cut ones) can function as lovely decorative elements in e-textiles.

The hand-cut iron-on circuit technique is very accessible. Though youngsters might need assistance working with irons, with help even they should be able to design and cut fabric PCBs by hand. After familiarizing themselves with this basic technique, users will be comfortable with the possibility of employing the more sophisticated laser-cutting method. Laser cutters are expensive, *high-tech* machines, but they are becoming increasingly popular and are often available in places like high schools. This trend will continue and it can envision laser-cut fabric PCBs being part of a high school course on e-textiles in a few years. Even with access to a laser cutter, there are challenges to working with fabric PCBs. In particular, they can be difficult to solder. Since most fabrics are flammable, heat is an obvious concern; the backing fabric for the circuit should thus be chosen for properties of heat resistance. (Thin polyesters, e.g. are a bad choice for a backing fabric.) The encapsulation of solder joints is likely to present another challenge to novices. More research needs to be done to find user-friendly means of encapsulation before fabric PCBs can be easily used by novices. While these features of the medium present issues that users and researchers will have to grapple with, they do not believe that any of them present insurmountable barriers. Furthermore, iron-on circuits need not be soldered; they can also be employed in situations where all components are sewn. The fabrics that we use were developed for electromagnetic field shielding applications and can be purchased online from several different sites. Conductive fabric is expensive relative to other fabrics, but is quite reasonable in the small quantities that are needed to make iron-on circuits. The other components of the technique, iron-on adhesives, are inexpensive and available in craft and fabric stores.

FIGURE 4.57
Hand-cut iron-on circuit with LED and power supply stitched using stainless steel thread.

4.7.1.3 Electronic Sequins

In early experience of designing e-textiles, the lack of suitable electronic packages, LEDs and other components were available that could be sewn to cloth, like sequins or beads. Electronic sequins evolved from efforts to create such a package. In first crude attempt at realizing this goal, it twisted the leads of through-hole LEDs to create LEDs with looped leads that could be sewn down with conductive thread. Figure 4.58 shows one of these stitchable LEDs. As it requires no soldering, it could be easily employed by novices. Still, as an LED package design, it left much to be desired: it was bulky and ugly, and the leads were prone to breaking off during twisting. Creating a sewable LED package was considerably more successful. Silver crimping beads to the leads of surface mount LEDs with lead-free solder. The resulting 'LED sequins' could be stitched to fabric with conductive thread in much the same manner as traditional beads. Figure 4.58 shows LED sequins before and after being stitched. This sequin package is stable, because almost all of the stress of flexing fabric is allowed and forgiven by the thread moving inside the bead. Very little strain is forced onto the solder joints and these joints remain intact as they are used. (Section 3.4 will present the results of preliminary washing tests for LED sequins.) The sequin package can be used for other electrical components in addition to LEDs. Almost any two-lead component can be attached to beads or devices that achieve the same effect. Since the realization of the initial LED sequin, it has created switch, capacitor, resistor, sensor and battery sequins. Figure 4.59 shows a picture of switch and battery sequins. Electronic sequins are being used in a number of e-textile prototypes. Figure 4.60 shows one application of the LED sequin, a wearable display in which 140 LED sequins were stitched into a shirt to create a wearable display that is capable of displaying low-resolution animations like scrolling text and cellular automata. LED sequins have proven to be useful, robust and – most importantly – important in the way that they suggest a wide variety of

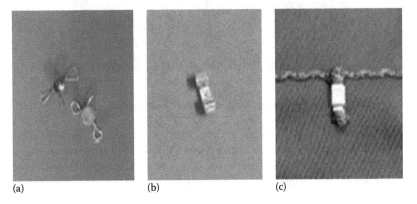

(a)   (b)   (c)

FIGURE 4.58
(a) A stitchable LED; (b) LED sequin and (c) LED sequin stitched into fabric.

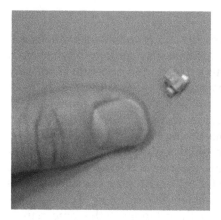

FIGURE 4.59
Switch and battery sequins.

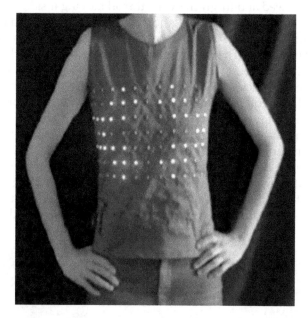

FIGURE 4.60
Wearable LED display matrix: a shirt embellished with LED sequins.

(initially unanticipated) designs. Figure 4.61 shows an example of one such unforeseen design, a beaded bracelet woven out of LED sequins. In this case, prompted to use the LED sequins just as one might traditional beads, weaving the bracelet from glass beads, LED sequins, standard thread and conductive thread on a bead loom. Much like the shirt in Figure 4.60, the bracelet can also be used to display scrolling text, cellular automata and other low-resolution animations. It is worth noting that much of the electrical wiring for

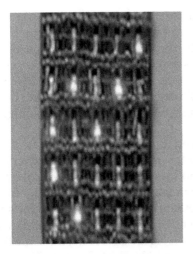

FIGURE 4.61
Wearable LED display matrix: a bracelet woven out of LED sequins and glass beads.

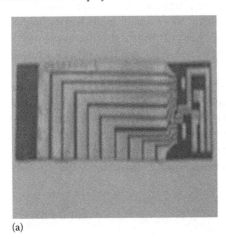

(a)   (b)

FIGURE 4.62
(a) The first layer of the fabric PCB that powers the bracelet. (b) A flat view of the bracelet.

the bracelet was built out of a multi-layered fabric PCB. One can get a sense of this layout by referring to Figure 4.62. This highlights yet another aspect of the techniques described here – namely, that they are complementary in their functions. That is while LED sequins and fabric PCBs are separate techniques for e-textile craft, they can be profitably used in conjunction.

4.7.1.4 Socket Buttons

The iron-on circuits was described; while extremely handy, they do not have universal applicability. In particular, not all users will be able to cut out the precision circuits that are required to attach components like micro-controllers to fabric or to solder iron-on circuits. When attempting to attach a microcontroller or other device requiring densely packed traces to fabric, an user may encounter other problems as well. Sometimes, for instance the stiffness of solder joints is undesirable, or a user may be working with a fabric that needs to stretch. In these situations, we have developed another means to attach IC sockets, and thus microcontrollers and other 'pluggable' components, to fabric.

Sockets with through holes can be sewn onto fabric like buttons to create sockets. Each hole in a socket can be stitched onto a fabric backing with con-ductive thread. One can use this thread to continue a trace across the fabric, or simply to make a connection between the socket and an existing trace on the fabric. When a microcontroller or other device is plugged into the socket, it makes contact with the traces on the textile via the stitching on the socket. An example of a socket button is shown in Figure 4.63. Here, the same thread was used to stitch out traces on the garment (the shirt shown in Figure 4.60) and sew on the IC socket. The socket can hold a 40-pin microcontroller; the result is that complex programmes can be plugged into (o unplugged from) the shirt; one can effectively reprogramme the display by inserting a new microcontroller into its socket button. Socket buttons might also serve as dif-ferent types of plugs, facilitating connections to devices like power supplies and computers.

While the socket button technique has the disadvantage of being time-consuming, it provides a powerful and important benefit: it allows users to build sophisticated e-textile prototypes using only a needle and thread. A circuit can be sewn in conductive thread; stitchable electronics can be stitched to these traces for decorative or other purposes and finally, socket buttons allow designers to attach controlling computer chips to their

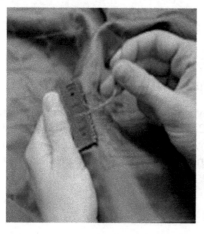 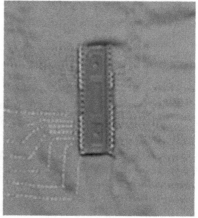

FIGURE 4.63
Socket button used to hold the microcontroller that powers the shirt.

textiles. The socket button technique is quite accessible, requiring only IC sockets and conductive thread. IC sockets are cheap and available from electronics retailers and, as was mentioned earlier, conductive thread is also easily obtainable for reasonable prices. The delicate stitching that the technique requires may be difficult for young people or those lacking fine motor skills, but most patient teenagers and adults should have no problem in employing it.

4.7.1.5 Insulators

Textiles bend and fold during use. This presents a challenge for e-textile design – electrical traces may come in contact with one another as a fabric flexes. Exposed traces are acceptable on circuit boards that are not only stiff, but also usually encased in hard protective boxes. Traces on textiles, however, must be insulated and protected to prevent short circuits. Furthermore, the insulation must preserve the qualities of the textile; it should be soft, flexible and even stretchy if necessary. It will focus on three techniques in turn: the use of stitching, the use of iron-on patches and the use of fabric paints. A 'couching' embroidery stitch can be used to insulate traces sewn in conductive thread. The couching stitch is a stitch that covers what is underneath it with a densely packed, zigzagging layer of thread. Figure 4.64 shows a picture of two traces – one exposed and one covered with a couching stitch. The couching method allows the insulator to become part of the fabric with the advantages that that implies: stitches wear wonderfully, and they simply look like natural parts of a textile.

FIGURE 4.64
Exposed trace sewn in stainless steel thread and a similar trace covered with a couching stitch.

The second insulating technique examined is the use of non-conducting iron-on fabric patches. To employ this technique, substitute a traditional fabric for the conductive cloth. Figure 4.65 shows an example of how this technique can be fruitfully employed: a simple hand-cut iron-on circuit is applied to cloth and then a matching iron-on insulator is applied over it. If additional support is desired, the iron-on insulator can be stitched down with non-conducting thread after it is ironed. Iron-on insulators do not interfere with the conductivity of the materials to which they are applied. The conductivity of sewn traces and iron-on circuits before and after iron-on insulators were applied, and no increase in the resistance of the traces was detected. It should be noted that we employed a commercial digital multimeter for these tests; it may be the case that the iron-on adhesive does interfere with the conductivity of yarns and fabrics very slightly but the meter used was incapable of detecting the change. A justification for measuring technique is detailed in the next section. As is suggested by Figure 4.65, the iron-on insulator method can be used particularly successfully in conjunction with iron-on circuits.

A disadvantage of these insulators is that they stiffen the area to which they are applied. Though the patches will soften with use, they will always remain less flexible than the rest of the textile, and multi-layered areas will stay particularly stiff. Also, like the iron-on circuits, they cannot be applied to stretch fabric. The iron-on patch technique is quite easy to employ. One needs only adhesive, an iron and a piece of fabric to make use of it. As discussed earlier, these materials are cheap and easy to find, and novice e-textile practitioners could make use of iron-on insulators. Perhaps the most useful and robust class of insulators we have investigated is paint-on materials.

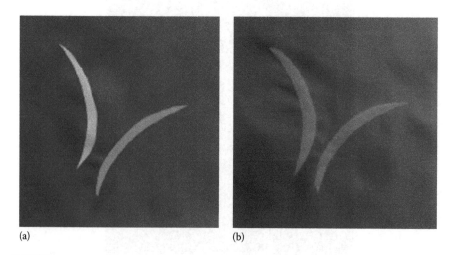

(a)   (b)

FIGURE 4.65
(a) A hand-cut iron-on circuit consisting of two traces in a copper plated fabric and (b) iron-on insulators have been applied to the traces.

It was experimented with a variety of paint-on substances including latex, acrylic gel mediums and an assortment of fabric paints. The most robust insulators found so far are 'puffy' fabric paints. Latex rubber also works well, but because latex allergies are not uncommon, it will focus on fabric paint in this discussion. Figure 4.66 shows a picture of two traces covered with puffy fabric paint. The image illustrates how effective fabric paints are as insulators. One trace was sewn out and painted, then a second trace was stitched over the first and painted. The fabric paint separates the first trace from the second one, providing a robust insulation.

Fabric paint does not interfere with the conductivity of yarns and fabrics. Again, the conductivity of traces with a digital multi-meter before and after paint was tested and applied, and no change in the resistance of the traces was detected. There are some noteworthy drawbacks in using fabric paint as an insulator. Fabric paint can change the look and feel of a textile. To get good insulating coverage, the thread or fabric must be completely covered, and this necessitates a raised area of paint. Also, the paint does not provide as elegant a covering as couching stitches or iron-on patches, being a conspicuously non-fabric material. Furthermore, it can be difficult to match the colour of the paint to the colour of the background cloth and thus difficult to camouflage painted traces. And finally, paint stiffens fabric – while bendable, painted areas are slightly hardened. Though fabric paints are not the ideal insulator, they have several benefits. In particular, they are non-toxic, familiar to many hobbyists, and quite easy to obtain and use. They are cheap and come in a variety of colours. What is more, like the other insulating techniques have examined, they can be creatively employed to decorate as well as insulate.

FIGURE 4.66
Two criss-crossing traces prevented from contacting one another by applications of puffy fabric paint.

4.8 Future Research Scope

E-textiles represent one particular genre of ubiquitous computing; but it is a genre that offers tremendous opportunities for creative contributions by students, hobbyists and amateurs. From own viewpoint – as researchers interested in e-textiles and educational technology – the ability of students and children to design, decorate, programme and customize their own fabric-based artefacts is especially powerful, since clothing and fabrics occupy such a central and meaningful aspect in young people's lives. Unlike so many other efforts in educational technology, e-textile crafts offer students a natural combination of aesthetics, challenging engineering and design and personal expression. Moreover, as argued, such activities represent an introduction to still other, and broader, realms of electronics, programming and mathematics. The techniques described in this chapter should thus be understood in the context of this larger project of nurturing a popular, democratized and even child-friendly genre of ubiquitous computing. They represent early steps in this project, but they are useful steps. Their utility is, ideally, not only as specific techniques for e-textile crafting, but also as illustrations of a more general belief in the value of widely available and diffused expertise. To put it another way: as computing moves further and further from the desktop, it is important that it does not become the exclusive province of a rarefied caste of specialists. Computer science has deep historical roots in the world of hobbyists, garage tinkerers and other 'amateurs'. One can trace back innumerable important ideas in the field – in home computing, in interface design, in computational graphics and music – to the work of exuberantly obsessed young men and women. The e-textile research community (and the ubiquitous computing community more broadly) should likewise take care to empower their youngsters, and to nurture the participation of the 'amateurs' from whom so much of the professional community has sprung.

4.9 Summary

In this chapter, specialty garment manufactured using nichrome wires and POF fabrics were discussed, and also plug-in method of integrating miniaturized electronic circuits in the garment are elucidated. Heating garment is developed from the nichrome fabric for the military applications and commercial purposes. The performance of the heating garment has been analyzed by conducting tests at different ambient temperature. The wearer body temperature was measured to be of normal 37°C and the body temperatures were measured at various room temperatures with 8:1 distance to spot ratio.

The time heating garment has taken to heat the coil was measured using gun type MTX-9 infrared non-contact thermometer of METRAVI make having temperature range of 18°C–1000°C. The test results show that the heating garment responds to the room temperature and the wearer temperature instantly and heats up the nichrome fabric with time delay of less than 3 s. The communication garment has been developed using copper core conductive fabric, and it is integrated with mobile phone–charging circuit and temperature measurement circuit for charging the mobile phone and measuring the wearer temperature, respectively. The performance of the communication garment integrated with temperature measurement circuit has been tested by measuring the body temperature of the wearer in open-air at daytime, and the test results are compared with the standard clinical thermometer. The test results show that the value of the body temperature measured through copper core conductive fabric is similar to that of the temperature measurement through standard clinical thermometer. The performance of the copper core conductive fabric is also analyzed using integrating the fabric with mobile phone–charging circuit for charging the Nokia mobile phone. The result shows that the copper core conductive fabric is charging the mobile phone, almost at the same time as that of the time taken by the conventional mobile phone charging circuit. The communication garment developed using optical core conductive fabric is used for signal transmission as well as to locate the number and the place of the bullet wound. From the test results it has been found that the optical core conductive fabric shows the maximum output voltage results that have minimum signal loss. The input light sources used at one end of the optical core conductive fabric were red LED, white LED and laser light, in this laser light is showing. The fabric produced using PMMA is used to develop illuminated garment. The subjective evaluation is carried out for the end user; the specific attributes are selected and given the rating scale. The user should wear the garment and rate the attribute; these values are consolidated for the final scale rating. It is possible to vary the brightness of optic-fibre screen according to the sound environment, using low-frequency sensitive circuit.

The teleintimation garment for defence personnel is made with POF fabric, which can sense the number and place of bullet wound from the soldier's body. The POF is woven in weft and warp direction in the garment to form a matrix format. This matrix format allows exactly locating the bullet wound in the soldier body. The POF is integrated with the microcontroller-based circuits to detect the bullet wound signals. The performance of the teleintimation garment has been tested using standard test rig developed for assessing the signal loss percentage. The performance analysis of teleintimation garment circuits has been assessed using USB-1208FS DAQ card. The Smart Shirt has been developed for medical monitoring, athletics and military uses. In the Smart Shirt, temperature of the wearer can be determined by placing the sensor in the armhole of the garment. Three different temperatures measurements are taken at different conditions, and the test results are compared with the standard clinical thermometer and it shows that the developed circuit results are almost same as that of the standard measuring instrument. The pulse rate of the person was measured when the person is under normal, after running, after fast walking and walking conditions. The test results were obtained from the pulse rate module and these results were compared with the standard clinical pulse oximeter of StarPlus LT make and it showed that the developed circuit is almost giving the same values as that of the standard testing instrument. The respiration rate measurement unit is designed to find the respiration rate of the patient. The test results were acquired from the respiration rate module and these results were compared with the standard clinical respiration rate measurement device of StarPlus LT make and the developed circuit results were almost same as that of the conventional measuring device.

Bibliography

Ashok Kumar, L., Stand-alone solar power generation system with constant current discharge, *Int. J. Emerg. Trends Electric. Electron.*, 5(1), 108–115 (2013), ISSN: 2320-9569.

Ashok Kumar, L. and T. Ramachandran, A methodology for product development of nonwoven wearable electronics, *Int. J. Nonwovens Tech. Text.*, 4(6), 25–28 (2010a).

Ashok Kumar, L. and T. Ramachandran, Emerging trends in flexible wearable electronics for versatile applications, *Indian Text. J.*, 7, 62–66 (2010b).

Ashok Kumar, L. and T. Ramachandran, Investigation on acquisition of signals from teleintimation garment for extreme health care and defence personnel, *Int. J. Biomed. Signal Process.*, 1(2), 49–57 (2010c).

Ashok Kumar, L. and T. Ramachandran, Remote utilization of textile testing equipments using microcontrollers with LabVIEW, *Pakistan Text. J.*, 11, 43–45 (2010d).

Ashok Kumar, L. and T. Ramachandran, Teleintimation garment for personnel monitoring applications, *Melliand Int.*, 53, 53–54 (2011a).

Ashok Kumar, L. and T. Ramachandran, Design and development of telemonitoring system using mobile network for medical applications, *Natl. J. Technol.*, 7(2), 1–7 (2011b).

Ashok Kumar, L. and T. Ramachandran, Communication architectures in smart grid, *Electric. India*, 53(2), 24–46 (2013a).

Ashok Kumar, L. and T. Ramachandran, Design and development of cost effective global positioning system general packet radio service based soldier tracking system, *J. Eng. Technol.*, 3(1), 36–40 (2013b).

Ashok Kumar, L., T. Ramachandran and I. Vennila, Development of bullet wound detection circuit using flexible PCB for teleintimation garment, *Natl. J. Technol.*, 6(1), 75–83 (2010a).

Ashok Kumar, L., A.N. Sai Krishnan and A. Venkatachalam, Design and development of illuminated clothing with PMMA for versatile applications, *Indian J. Fibre Text. Res.*, 31, 577–579 (2006).

Ashok Kumar, L. and M. Senthilkumar, Concept & applications of RFID in textile & apparel industry, *Indian Text. J.*, 4, 40–44 (2010), ISSN: 0019-6436.

Ashok Kumar, L., P. Siva Kumar, K. Sunderasen and T. Ramachandran, Development of wearable solar jacket for remote and military applications, *Asian Text. J.*, 6, 48–52 (2010b).

Ashok Kumar, L. and A. Venkatachalam, Implementation of MEMS accelerometer in looms for warp stop motion, *J. Inst. Eng. (India)*, 91, 16–20 (2010).

Bächlin, M., J.M. Hausdorff, D. Roggen, N. Giladi, M. Plotnik and G. Tröster, Online detection of freezing of gait in Parkinson's disease patients: A performance characterization, In *Proceedings of the Fourth International Conference on Body Area Networks*, Belgium, pp. 146–149 (2009).

Berry, P., G. Butch and M. Dwane, Highly conductive printable inks for flexible and low-cost circuitry, In *Proceedings of the 37th International Symposium on Microelectronics, IMAPS*, California (2004).

Berzowska, J., Electronic textiles: Wearable computers, reactive fashion, and soft computation, *Textile*, 3(1), 2–19 (2005a).

Berzowska, J., Electronic textiles: Wearable computers, reactive fashion, and soft computation, *J. Cloth Culture*, 3(1), 58–75 (2005b).

Body Media Product Literature, Bodymedia Incorporated, Pittsburgh, PA, http://www.bodymedia.com. Accessed 11 August, 2013.

Bonderover, E. and S. Wagner, A woven inverter circuit for e-textile, *Appl. IEEE Electron Device Lett.*, 25(5), 295–297 (2004).

Bonfiglio, A., D. De Rossi, T. Kirstein, I.R. Locher, F. Mameli, R. Paradiso and G. Vozzi, Organic field effect transistors for textile applications, *IEEE Trans. Inf. Technol. Biomed.*, 3(9), 319–324 (2005).

Buechley, L., DIY: Make your own electronic sewing kit and materials links. Available: http://www.cs.colorado.edu/_buechley. Accessed 5 November, 2013.

Buechley, L. and M. Eisenberg, Fabric PCBs, electronic sequins, and socket buttons techniques for e-textile craft, *Pers. Ubiquit. Comput.*, Springer-Verlag London Limited, 24, 57–63 (2007).

Buechley, L., N. Elumeze, C. Dodson and M. Eisenberg, Quilt snaps: A fabric based computational construction kit, In *International Workshop on Wireless and Mobile Technologies in Education (WMTE)*, Tokushima, Japan (2005).

Buechley, L., N. Elumeze and M. Eisenberg, Electronic/computational textiles and children's craft, In *Proceedings of Interactive Design and Children (IDC)*, Tampere, Finland (2006).

Carmo, J.P., P.M. Mendes, C. Couto and J.H. Correia, 2.4 GHz wireless sensor network for smart electronic shirts, *Smart Sens. Actuators MEMS Proc. SPIE*, 5836, 579–586 (2005).

Carter, G.S., M.A. Coyle and W.B. Mendelson, Validity of a portable cardio-respiratory system to collect data in the home environment in patients with obstructive sleep apnea, *Sleep Hypnosis*, 6(2), 85–92 (2004).

Chi, Y.M., T.-P. Jung and G. Cauwenberghs, Dry and non-contact biopotential electrodes, *IEEE Rev. Biomed. Eng.*, 3, 106–119 (2010).

Choi, S. and Z. Jiang, A novel wearable sensor device with conductive fabric and PVDF film for monitoring cardio respiratory signals, *Sens. Actuators A: Phys.*, 128, 317–326 (2006).

Corbett, E.A., E.J. Perreault and K.P. Körding, Decoding with limited neural data: A mixture of time-warped trajectory models for directional reaches, *J. Neural Eng.*, 9(3), 54–57 (2012).

Cottet, D., J. Grzyb, T. Kirstein and G. Troster, Electrical characterization of textile transmission lines, *IEEE Trans. Adv. Packaging*, 26(2), 182–190 (2003).

Cutecircuit. Galaxy dress. Available: http://www.cutecircuit.com/. Accessed 25 June, 2013.

De Rossi, D., F. Carpi, F. Lorussi, A. Mazzoldi, R. Paradiso, E.P. Scilingo and A. Tognetti, Electroactive fabrics and wearable biomonitoring devices, *AUTEX Res. J.*, 3(4), 21–25 (2003).

De Vries, J., J.H. Strubbe, W.C. Wildering, J.A. Gorter and A.J.A. Prins, Patterns of body temperature during feeding in rats under varying ambient temperatures, *Physiol. Behav.*, 53, 229–235 (1993).

Dhawan, A., A.M. Seyam, T.K. Ghosh and J.F. Muth, Woven fabric-based electrical circuits – Part I, Evaluating interconnect methods, *Text. Res. J.*, 74, 913–919 (2004).

Dunne, L.E., S. Brady, R. Tynan, K. Lau, B. Smyth, D. Diamond and G.M.P. O'Hare, Garment-based body sensing using foam sensors, In *Proceedings of the Australasian User Interface Conference*, Hobart, Australia, Vol. 50, pp. 165–171 (2006).

Edmison, J., M. Jones, Z. Nakad and T. Martin, Using piezoelectric materials for wearable electronic textiles, In *Proceedings of Sixth International Symposium on Wearable Computers, IEEE (ISWC)*, Seattle, WA, pp. 41–48 (2002).

Einsmann, C., M. Quirk, B. Muzal, B. Venkatramani, T. Martin and M. Jones, Virginia Polytechnic Institute and State University, Modeling a wearable full-body motion capture system, In *Proceedings of the Ninth IEEE International Symposium on Wearable Computers (ISWC'05)*, Osaka, Japan, pp. 144–151 (2005).

Elton, S.F., Ten new developments for high-tech fabrics & garments invented or adapted by the research & technical group of the defence clothing and textile agency, In *Proceedings of the Avantex, International Symposium for High-Tech Apparel Textiles and Fashion Engineering with Innovation-Forum*, Frankfurt am Main, Germany, pp. 27–29 (2000).

Fensli, R., E. Gunnarson and T. Gundersen, A wearable ECG-recording system for continuous arrhythmia monitoring in a wireless tele-home-care situation, In *Proceedings of the IEEE International Symposium on Computer-Based Medical Systems (CBMS)*, pp. 407–412 (2005).

Firoozbakhsh, B., N. Jayant, S. Park and S. Jayaraman, Wireless communication of vital signs using the Georgia Tech Wearable Motherboard, In *IEEE International Conference on Multimedia and Expo*, Vol. 3, pp. 1253–1256 (2000).

Gibbs, P.T. and H.H. Asada, Wearable conductive fiber sensors for multi-axis human joint angle measurements, *J. Neuroeng. Rehabil.*, 2(1), 7 (2005).

Gorlick, M.M., Electric suspenders: A fabric power bus and data network for wearable digital devices, In *Proceedings of the Third International Symposium on Wearable Computers*, San Francisco, CA, IEEE, pp. 114–121 (1999).

Gould, P., Textiles gain intelligence, *Mater. Today*, 38, 38–43 (2003).

Guler, M. and Ertugrul, S., Measuring and transmitting vital body signs using MEMS sensors, In *First Annual RFID Eurasia 2007*, IEEE, Kobe, Japan, pp. 1–4 (2007).

Hartmann, W.D., *High-Tech Fashion*, Heimdall, Witten, Germany (2000).

Huang, D., K. Qian, D.-Y. Fei, W. Jia, X. Chen and O. Bai, Electroencephalography (EEG)-based Brain–Computer Interface (BCI): A 2-D virtual wheelchair control based on event-related desynchronization/synchronization and state control, *IEEE Trans. Neural Syst. Rehabil. Eng.*, 20(3), 379–388 (2012).

International Fashion Machines. Electropuff. Available: http://www. ifmachines. com. Accessed 18 June, 2013.

Jung, S., C. Lauterbach, M. Strasser and W. Weber, Enabling technologies for disappearing electronics in smart textiles, In *IEEE International Solid-State Circuits Conference*, San Francisco, CA, Vol. 1, pp. 386–387 (2003).

Kallmayer, C., New assembly technologies for textile transponder systems, In *Electronic Components and Technology Conference*, New Orleans, LA (2003).

Kallmayer, C., T. Linz, R. Aschenbrenner and H. Reichl, System integration technologies for smart textiles, *MST News*, 2, 42–43 (2005).

Kallmayer, C., R. Pisarek, A. Neudeck, S. Cichos, S. Gimpel, R. Aschenbrenner and H. Reichlt, New assembly technologies for textile transponder systems, In *Electronic Components and Technology Conference*, New Orleans, LA (2003a).

Kallmayer, C., R. Pisarek, A. Neudeck, S. Cichos, S. Gimpel, R. Aschenbrenner and H. Reichlt, New assembly technologies for textile transponder systems, In *53rd Electronic Components and Packaging Conference (ECTC)*, New Orleans, LA, pp. 1123–1126 (2003b).

Kim, G.B.C., G.G. Wallace and R. John, Electroformation of conductive polymers in a hydrogel support matrix, *Polymer*, 41, 1783–1790 (2000).

Leckenwalter, R., Innovations in new dimensions of functional clothing, In *Proceedings of the Avantex International Symposium for High-Tech Apparel Textiles and Fashion Engineering with Innovation-Forum*, Frankfurt am Main, Germany (2000).

Lee, J. and Subramanian, V., Weave patterned organic transistors on fiber for E-textiles, *IEEE Trans. Electron*, 52(2), 269–275 (2005).

Lehn, D., e-TAGs: e-textile attached gadgets, In *Proceedings of Communication Networks and Distributed Systems (CNDS)*, San Diego, CA (2004).

Less EMF. Conductive fabrics. Available: http://www.lessemf.com/fabric.html. Accessed 27 June, 2013.

Linz, T. and C. Kallmayer, New interconnection technologies for the integration of electronics on textile substrates, In *Ambience 2005*, Tampere, Finland (2005).

Linz, T., C. Kallmayer, R. Aschenbrenner and H. Reichl, Embroidering electrical interconnects with conductive yarn for the integration of flexible electronic modules into fabric, In *Proceedings of the Ninth IEEE International Symposium on Wearable Computers*, IEEE ISWC, Osaka, Japan, pp. 86–89 (2005).

Linz, T., C. Kallmayer, R. Aschenbrenner and H. Reichl, Fully integrated EKG shirt based on embroidered electrical interconnections with conductive yarn and miniaturized flexible electronics, In *Proceedings of the International Workshop on Wearable and Implantable Body Sensor Networks (BSN'06) IEEE*, Aachen, Germany (2006).

Lo, B., S. Thiemjarus, R. King and G.Z. Yang, Body sensor network – A wireless sensor platform for pervasive healthcare monitoring, In *Adjunct Proceedings of the Third International Conference on Pervasive Computing*, Munich, Germany (2005).

Locher, I., T. Kirstein and G. Tröster, Routing methods adapted to e-textiles, In *Proceedings of 37th International Symposium on Microelectronics (IMAPS)*, California (2004).

Loriga, G., N. Taccini, D. De Rossi and R. Paradiso, Textile sensing interfaces for cardiopulmonary signs monitoring, In *Proceedings of the 2005, IEEE Engineering in Medicine and Biology 27th Annual Conference*, Shanghai, China, pp. 7349–7352 (2005).

Lorussi, F., W. Rocchia, E.P. Scilingo, A.T. and D. De Rossi, Wearable, redundant fabric-based sensor arrays for reconstruction of body segment posture, *IEEE Sens. J.*, 4(6), 807–818 (2004).

Lymberis, A. and D.E. De Rossi, *Wearable Health Systems for Personalized Health Management – State of the Art and Future Challenges*, IOS Press, Amsterdam, the Netherlands (2004), ISBN: 1 58603 449 9.

Maccioni, M., E. Orgiu, P. Cosseddu, S. Locci and A. Bonfiglio, Towards the textile transistor: Assembly and characterization of an organic field effect transistor with a cylindrical geometry, *Appl. Phys. Lett.*, 89, 143–515 (2006).

Mann, S., Smart clothing: The shift to wearable computing, *Commun. ACM*, 39(80), 23–24 (1996).

Mann, S., An historical account of the 'WearComp' and 'WearCam' inventions developed for applications in 'Personal Imaging', In *First International Symposium on Wearable Computers*, Boston, MA (1997).

Martinsen, O.G. and Grimnes, S., *Bioimpedance and Bioelectricity Basics*, Academic Press, London, U.K. (2000).

Mei, T., Y. Ge, Y. Chen, L. Ni, W. Liao, Y. Xu and W.J. Li, Design and fabrication of an integrated three-dimensional tactile sensor for space robotic applications, In *IEEE MEMS*, Orlando, FL, pp. 112–117 (1999).

Mitcheson, P.D., Architectures for vibration-driven micropower generators, *J. Microelectromechan. Syst.*, 13(3), 429–440 (2004).

Mitcheson, P.D., T.C. Green, E.M. Yeatman and A.S. Holmes, Microeletromech, *IEEE/ASME J. Syst.*, 13(3), 429–440 (2004).

Moore, S.K., Just one word-plastics, organic semiconductors could put cheap circuits everywhere and make flexible displays a reality, *IEEE Spectr.*, 55–59 (2002).

Mundt, C.W., K.N. Montgomery, U.E. Udoh, V.N. Barker, G.C. Thonier and A.M. Tellier, A multiparameter wearable physiologic monitoring system for space and terrestrial applications, *IEEE Trans. Inf. Technol. Biomed.*, 9, 382–391 (2005).

Novak, D., X. Omlin, R. Leins-Hess and R. Riener, Predicting targets of human reaching motions using different sensing technologies, *Biomed. Eng.*, 60(9), 2645–2654 (2013).

Orth, M., Defining flexibility and sewability in conductive yarns, *Proc. Mater. Res. Soc. Symp. (MRS)*, 736, 37–48 (2002).

Otto, C., A. Milenkovic, C. Sanders and E. Jovanov, System architecture of a wireless body area sensor network for ubiquitous health monitoring, *J. Mobile Multimedia*, 1(4), 307–326 (2006).

Pacelli, M., G. Loriga, N. Taccini and R. Paradiso, Sensing fabrics for monitoring physiological and biomechanical variables: E-textile solution, In *Proceeding of the Third IEEE EMBS International Summer School and Symposium on Medical Devices and Biosensors*, Boston, MA, pp. 1–4 (2006).

Paradiso, J. and M. Feldmeier, A compact, wireless, self-powered pushbutton controller, In *Ubicomp 2001: Ubiquitous Computing, LNCS 2201*, Springer-Verlag, Berlin, Germany, pp. 299–304 (2005).

Paradiso, R., G. Loriga and N. Taccini, Wearable health care system for vital signs monitoring, In *Mediterranean Conference on Medical and Biological Engineering*, Ischia, Italy (2004).

Paradiso, R., G. Loriga, N. Taccini, A. Gemignani and B. Ghelarducci, WEALTHY, a wearable health-care system: New Frontier on Etextile, *IEEE JTIT*, 4, 105–109 (2005).

Park, C., P.H. Chou, Y. Bai, R. Matthews and A. Hibbs, An ultra-wearable, wireless, low power ECG monitoring system, *Biomed. Circuit Syst.*, 21(1), 241–244 (2006).

Park, S. and S. Jayaraman, Adaptive and responsive textile structures, In X. Tao (ed.), *Smart Fibers, Fabrics and Clothing: Fundamentals and Applications*, Woodhead, Cambridge, U.K., pp. 226–245 (2001).

Park, S. and S. Jayaraman, Smart textiles: Wearable electronic systems, *MRS Bull.*, 28(8), 585–591 (2003).

Park, S., K. Mackenzie and S. Jayaraman, Wearable motherboard: A Framework for personalized mobile information processing (PMIP), In *Design Automation Conference (DAC)*, New Orleans, LA (2002).

Patel, S., H. Park, P. Bonato, L. Chan and M. Rodgers, A review of wearable sensors and systems with application in rehabilitation, *J. NeuroEng. Rehabil.*, 9, 21 (2012).

Post, E.R. and M. Orth, Smart fabric, wearable clothing, In *Proceedings of the First International Symposium on Wearable Computers*, Cambridge, MA, IEEE, pp. 167–168 (1997).

Post, E.R., M. Orth, P.R. Russo and N. Gershenfeld, E-Broidery: Design and fabrication of textile based computing, *IBM Syst. J.*, 39(3–4), 840–860 (2000).

Rantanen, J. and H. Ninen, Data transfer for smart clothing: Requirements and potential technologies, In Tao, X. (ed.), *Wearable Electronics and Photonics*, Woodhead Publishing in Textiles, England, p. 198 (2005).

Riley, R., S. Thakkar, J. Czarnaski and B. Schott, Power-aware acoustic beamforming, In *Proceedings of the Military Sensing Symposium Specialty Group on Battlefield Acoustic and Seismic Sensing, Magnetic and Electric Field Sensors*, Laurel, MD (2002).

Satava, R.M., Virtual reality and telepresence for military medicine, *Ann. Acad. Med.*, Singapore, 26(1), 118–120 (1997).

Schwickert, L., C. Becker, U. Lindemann, C. Maréchal, A. Bourke, L. Chiari, J.L. Helbostad, W. Zijlstra, K. Aminian and C. Todd, Fall detection with body-worn sensors, *Zeitschrift für Gerontologie und Geriatrie*, Springer, 46(8), 706–719 (2013).

Scilingo, E.P., A. Gemignani, R. Paradiso, N. Taccini, B. Ghelarducci and D. De Rossi, Sensing fabrics for monitoring physiological and biomedical variables, *IEEE Trans. Inf. Technol. Biomed., Special Sect. New Generat. Smart Wearable Health Syst. Appl.*, 9(3), 345–352 (2005a).

Scilingo, E.P., A. Gemignani, R. Paradiso, N. Taccini, B. Ghelarducci and D. De Rossi, Performance evaluation of sensing fabrics for monitoring physiological and biomechanical variables, *IEEE Trans. Inf. Technol. Biomed.*, 9(3), 345–352 (2005b).

Service, R., Technology: Electronic textiles charge ahead, *Science*, 201(5635), 909–911 (2003).

Shen, C.-L., T. Kao, C.-T. Huang and J.-H. Lee, Wearable band using a fabric-based sensor for exercise ECG monitoring, In *10th IEEE International Symposium on Wearable Computers, IEEE*, Montreux, Switzerland, pp. 143–144 (2006).

Shih, E., V. Bychkovsky, D. Curtis and J. Guttag, Continuous medical monitoring using wireless micro sensors, *Proc. Sen. Syst.*, 310, 325–328 (2004).

Slade, J., M. Agpaoa-Kraus, J. Bowman, A. Riecker, T. Tiano, C. Carey and P. Wilson, Washing of electrotextiles, *Proc. Mater. Res. Soc. Symp. (MRS)*, 736, 99–108 (2002).

Smith, W.D. and A. Bagley, A miniature, wearable activity/fall monitor to assess the efficacy of mobility therapy for children with cerebral palsy during everyday living, In *Engineering in Medicine and Biology Society (EMBC), Annual International Conference of the IEEE*, Argentina, Vol. 4, pp. 5030–5033 (2010).

Snyder, W.E. and J. Clair, Conductive elastomers as sensor for industrial parts handling equipment, *IEEE Trans. Instrum. Meas.*, 27(1), 94–99 (1978).

Sposaro, F. and G. Tyson, iFall: An android application for fall monitoring and response, *Conf. Proc. IEEE Eng. Med. Biol. Soc.*, 17(4), 6119–6122 (2009).

The Georgia Tech Wearable Motherboard: The intelligent garment for the 21st century. Available: http://www.smartshirt.gatech.edu. Accessed 17 August, 2013.

Thomas, B., K. Grimmer, J. Zucco and S. Milanese, Where does the mouse go? An investigation into the placement of a body-attached touchpad mouse for wearable computers, *Personal Ubiquit. Comput.*, 6(2), 97–112 (2002).

Toffoli, T. and N. Margolus, *Cellular Automata Machines: A New Environment for Modeling*, MIT Press, Cambridge, U.K. (1987).

Tognetti, A., F. Lorussi, R. Bartalesi, S. Quaglini, M. Tucson, G. Supine and D. De Rossi, Wearable kinesthetic system for capturing and classifying upper limb gesture in post-stroke rehabilitation, *J. NeuroEng. Rehabil.*, 2(8), 17–22 (2005).

Tröster, G., The agenda of wearable healthcare, *The IMIA Yearbook of Medical Informatics 2005: Ubiquitous Health Care Systems*, Schattauer, pp. 125–138 (2005).

Troster, G., T. Kirstein and P. Lukowicz, Wearable computing: Packaging in textiles and clothes, In *Proceedings of the 14th European Microelectronics and Packaging Conference (EMPC)*, Firedrichshafen, Germany (2003).

Tseng, S.-Y., C.-H. Tsai, Y.-S. Lai and W.-C. Fang, A wireless biomedical sensor network, *Life Sci. Syst. Appl.*, 2, 183–186 (2009).

Vigneswaran, C., Development of signal transferring fabrics using plastic optical fibres (POF), *Non woven Tech. Text.*, 1(2), 40–45 (2008).

Vigneswaran, C., L. Ashok Kumar and T. Ramachandran, Design development of signal sensing equipment for measuring light transmission efficiency of POF fabrics, *Indian J. Fibre Text. Res. (IJFTR)*, 35, 317–323 (2010a).

Vigneswaran, C., L. Ashok Kumar and T. Ramachandran, Development of Signal Transferring fabrics using Plastic Optical fibres for defence personnel and study their performance, *J. Ind. Text.*, 1, 91–97 (2010b).

Vigneswaran, C., P. Kandahavadivu and T. Ramachandran, Application of wireless communication device integrated apparel and bed linen, *Int. J. Eng. Sci. Technol.*, 2(10), 5970–5976 (2010c).

Vigneswaran, C. and T. Ramachandran, Design and development of copper core conductive fabrics for smart textiles, *J. Ind. Text.*, 39, 81–93 (2009).

Vivometrics. Wireless monitoring life shirt. Available: http://www.vivometrics.com. Accessed 27 June, 2013.

Wakita, A. and M. Shibutani, Mosaic textile: Wearable ambient display with non-emissive color-changing modules, In *Proceedings of the International Conference on Advances in Computer Entertainment Technology (ACE)*, Hollywood, CA (2006).

Weiser, M., The computer for the 21st century, *Sci. Am.*, 265(3), 94–104 (1991).

Westerterp Plantenga, M.S., L. Wouters and F. Ten Hoor, Deceleration in cumulative food intake curves, changes in body temperature and diet-induced thermogenesis, *Physiol. Behav.*, 48, 831–836 (1990).

Wijesiriwardana, R., T. Dias and S. Mukhopadhyay, Resistive fibremeshed transducers, In *Proceedings of the Seventh IEEE International Symposium of Wearable Computers*, New York, pp. 200–209 (2003).

Wijesiriwardana, R., K. Mitcham and T. Dias, Fibre-meshed transducers based real time wearable physiological information monitoring system, In *Proceedings of the Eighth IEEE International Symposium on Wearable Computers, ISWC*, New York, Vol. 1, pp. 40–47 (2004).

Wilson, P., Textiles from novel means of innovation, In McQuaid, M. (ed.), *Extreme Textiles: Designing for High Performance*, Princeton Architectural Press, New York, pp. 180–213 (2005).

Yao, J., R. Schmitz and S. Warren, A wearable point-of-care system for home use that incorporates plug-and-play and wireless standards, *IEEE Trans. Inf. Technol. Biomed.*, 9(3), 363–371 (2005).

Yoo, J., L. Yan, S. Lee, H. Kim and H.-J. Yoo, A wearable ECG acquisition system with compact planar-fashionable circuit board-based shirt, *Inf. Technol. Biomed.*, 13(6), 897–902 (2009).

Zheng, J.W., Z.B. Zhang, T.H. Wu and Y. Zhang, A wearable mobihealth care system supporting real-time diagnosis and alarm, *Med. Biol. Eng. Computer*, 45, 877–885 (2007).

5

Software Development for Wearable Electronics

5.1 Introduction

In this chapter, the software developed to monitor the wearer at the user end is noted besides the various communication methods used to interface the wearable electronic products to the remote end are also discussed. Through this research work, wearable electronics such as teleintimation garment and smart shirt have been developed. To operate the garment in an effective manner, various factors and data have to be collected, analysed and processed to operate the required sensors and processors. For this purpose, software has been developed using Visual Basic and LabVIEW known as Soldier's Status-Monitoring (SSM) Software for monitoring applications. This chapter also discusses the software developed to track the soldier based on global positioning system (GPS) – general packet radio service (GPRS) technology. The purpose of body sensor networks (BSNs) is to provide an integrated hardware and software platform for facilitating the future development of pervasive monitoring systems.

5.2 Biosensor Design and MEMS Integration

Recent advances in biological, chemical, electrical and mechanical sensor technologies have led to a wide range of wearable and implantable sensors suitable for continuous monitoring. For instance, a biosensor proposed a chemical glucose sensor for diabetic monitoring as well as an implantable blood pressure sensor cuff for tonometric blood pressure measurement. Interuniversity Micro-Electronics Centre (IMEC) has developed a Si-based DNA sensor that can directly detect hybridization without any fluorescent labelling. Today, miniaturized pH biosensors based on electrochemically modified electrodes have been developed. Use of implantable magneto-elastic sensors for remote sensing of temperature and glucose are being under research for biosensor development. The reliability of biosensor is often based on the interface between the sensor and tissue or blood, especially for implantable sensor; biocompatibility is the major issue to be considered in the sensor design. Sensor array was used in biosensor to ensure the reliability of the sensors, and is a critical point of product design and development. Recently, polymeric barrier membranes that can be used in enzyme-based electrochemical sensors have been introduced and are available in the market. Parallel advances in the micro-electromechanical systems (MEMS) technology have facilitated the development of physiological sensors such as the micro-needle array for drug delivery and glucose measurement and the piezoresistive shear stress sensor. Context-awareness sensors include 3D accelerometers and gyroscopes. In addition, technological advancement in photonics has enabled the realization of unobtrusive optical biosensor, such as the optical glucose sensor.

5.2.1 Pervasive Monitoring

In parallel to the development of sensing and monitoring devices, several research platforms are emerging. One approach is to incorporate physiological sensors into the garment by linking sensors to a wearable processing device, such as the knitted bioclothes developed by the EU Project Wealthy. Similarly, the EU project, MyHeart, aims to provide continuous monitoring of vital signs for cardiac patients. The concept of an intelligent biomedical cloth (IBC) is proposed where biosensors are embedded inside clothes for measuring physiological signals and to provide immediate diagnosis and trend analysis. Although embedding the sensors into the garment could provide a convenient wearable system for the patient, the design is not flexible for the addition or relocation of sensors. In addition, different sizes of clothes have to be designed for different persons, which can introduce a significant cost burden. An alternative approach is to use on-body sensors such as the Human++ research programme at IMEC and the EU-funded project Healthy Aims. Another research proposed a Bluetooth-based wearable ECG server integrated with a GPS sensor, which can report the location of the patient when an emergency event is detected. Other systems include CardioNet and MIThril for remote heart monitoring.

5.2.2 Body Sensor Networks

Thus far, most hardware platforms for pervasive health-care applications are proprietary designed. The lack of interoperability and standards has prohibited a common approach towards the development of pervasive sensing applications. The BSN architecture from Imperial College was developed in response to the significant research activities in this area. The basic concept of BSN is illustrated in Figure 5.1 where wireless sensors are either worn by or implanted into the patient, and the sensor data is gathered by a local

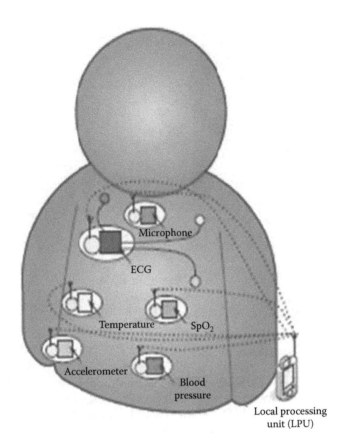

FIGURE 5.1
Basic design of the body sensor network.

processing unit (LPU), such as a PDA before it is further processed or transmitted to the central monitoring server. With regard to the BSN concept, the hardware platform, BSN node, is designed and developed. Despite providing the wireless communication and local processing capability, the BSN is designed to ease the integration of different sensors, such as ECG, SpO2 and other context-awareness sensors. In addition, by adopting the IEEE 802.15.4 standard, interoperability is assured between different sensor platforms.

5.2.3 BSN Architecture

Figure 5.2 illustrates the basic structure of the BSN node. The BSN node uses the Texas Instrument (TI) MSP430 16-bit ultra-low-power RISC processor with 60 kB + 256 B flash memory, 2 kB RAM, 12-bit ADC and six analog channels (connecting up to six sensors). The wireless module has a throughput of 250 kbps with a range over 50 m. In addition, 512 kB serial flash memory is incorporated in the BSN node for data storage or buffering. The BSN node

FIGURE 5.2
Pictorial illustration of the BSN node.

runs TinyOS by U.C. Berkeley, which is a small, open source and energy-efficient embedded operating system. It provides a set of modular software building blocks, from which designers could choose the components they require. The size of these files is typically as small as 200 bytes and thus, the overall size is kept to a minimum. The operating system manages both the hardware and the wireless network – taking sensor measurements, making routing decisions and controlling power dissipation. By using the ultra-low power TI microcontroller, the BSN node requires only 0.01 mA in active mode and 1.3 mA when performing computation-intensive calculation like an FFT. With a size of 26 mm, the BSN node is ideal for developing wearable biosensors. In addition, the stackable design of the BSN node and the available interface channels ease the integration of different sensors with the BSN node. Together with TinyOS, the BSN node can significantly cut down the development cycle for pervasive sensing development.

5.2.4 UbiMon

The DTI funded project, Ubiquitous Monitoring Environment for Wearable and Implantable Sensors (UbiMon), aims to provide a continuous and unobtrusive monitoring system for patients in order to capture transient but life-threatening events. Based on the BSN design, the basic framework of the UbiMon is illustrated in Figure 5.3. The system consists of five major components, namely, the BSN nodes, LPU, the central server (CS), the patient

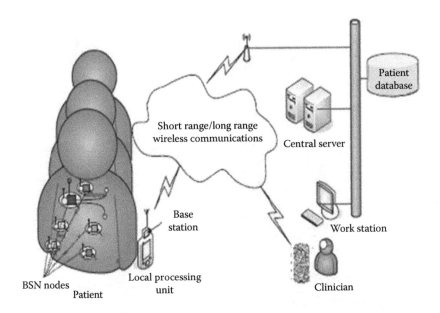

FIGURE 5.3
Schematic illustration of the UbiMon system architecture.

database (PD) and the workstation (WS). With the current UbiMon structure, a number of wireless biosensors including three-leads ECG, two-leads ECG strip and SpO2 sensors have been developed (Figure 5.4). To facilitate the incorporation of context information, context sensors including accelerometers, temperature and skin conductance sensors are also integrated to the BSN node. Furthermore, a compact flash BSN card is developed for PDAs, where sensor signals can be gathered, displayed and analysed by the PDA, as shown in Figure 5.4d.

Apart from acting as the local processor, the PDA can also serve as the router between the BSN node and the CS, where all sensor data collected will be transmitted through a WiFi/GPRS network for long-term storage and trend analysis. A graphical user interface (GUI) is developed at the WS for retrieving the sensor data from the database, as illustrated in Figure 5.5. To assist patient management, subjects with the highest risk are listed at the top of the patient table, which can be interactively interrogated. In addition, the historical record of the sensor readings can also be played back for any specific episodes.

5.3 Health-Care Monitoring System

Kańtoch et al. (2011) have studied the wearable ubiquitous health-care-monitoring system that integrates electrocardiogram (ECG device) and an accelerometer sensor with a mobile device in a Bluetooth-based BSN.

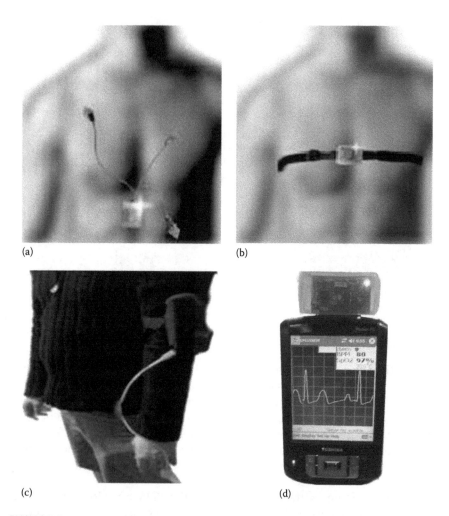

(a)

(b)

(c)

(d)

FIGURE 5.4
(a) Wireless 3-leads ECG sensor, (b) ECG strap (centre), (c) SpO2 sensor and (d) the PDA base station.

This research was focused on the right connection of the hardware units, combination of the detection of QRS complexes, calculation of heart rate (HR) and the detection of human falls. The main aim of this research was the early detection of abnormal situations (high/low HR, a fall) and the HR variability analysis. The human falls are very risky events that occur not only in the elderly people's daily living but also by epileptics and asthmatics. For these people, independent living is strictly forbidden. A wearable sensor-based monitoring system can inhibit serious injuries and allow those people live an independent life. Additionally, an SMS messaging module was integrated with the monitoring system. After detecting a non-standard situation, a short notice is sent. The implementation of the QRS complex detection algorithm that was based on the Tompkins formula was tested on the records from the MITBIH

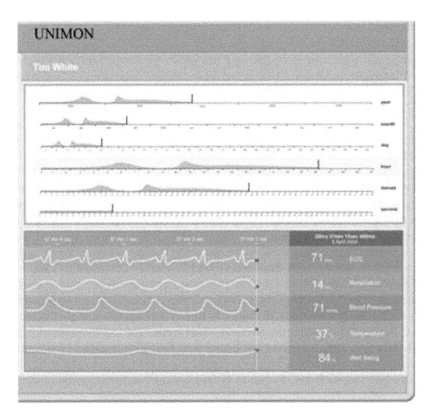

FIGURE 5.5
Workstation and database GUI for UbiMon.

database. Recent advances in biomedical engineering, wireless network and computer technologies have enabled the possibility of remote patient monitoring. Based on these technologies, it is possible to improve patient care, chronic disease management and promote lifelong health and well-being for the ageing population. ASPEKT 500 is a digital unit designated for wireless ECG signal transmission to PC or mobile device. The transmitter allows a free movement of the patients up to a distance of 10 m from the receiver. Small dimensions and weight make the examination more comfortable for a patient. ASPEKT 500 – Wireless ECG signal transmitter is equipped with 10 electrode cables. ASPEKT 500 – wireless ECG signal transmitter is equipped with ten electrode cables and it will receive 12-leads (Einthoven, Goldberger, Wilson). ECG data is sampled at 500 Hz frequency. Battery operated time is about 12 h.

5.3.1 System Architecture

The monitoring system prototype has been built and analysed software has been implemented in order to check if it is possible to monitor human activity remotely. The wireless ECG recorder transmits data to a mobile device via wireless network based on Bluetooth technology. The data are analysed by implemented software and forwarded using GPRS to Internet database, where medical data is accessible to a doctor. Implemented software is organized into four modules: ECG signal analysis, fall detection, HR variability and database. The most important module of the monitoring system is the ECG data analysis module, which analyses the incoming signal data from the ECG recorder. The ECG data analysis module executes the Pan and Tompkins algorithm. The output of this module is transmitted to the HR variability module (Figure 5.6). This part of the system calculates the HRV and sends the results to the database via GPRS.

The fall detection module is responsible for analysing data from the accelerometer and calculating the fall factor. If the fall is detected, the alarm module is switched on. The alarm module is a combination of two methods. The first one is a signal bell that can be switched off during 10 s. If it is not accomplished, a text message is sent to a trusted friend. The proposed health-care-monitoring system can help to monitor health conditions (heart rate and HR variability) and support elderly, sick and disabled people in their independent living. Taking electroencephalogram (EEG) out of the traditional, clinical environment into people's homes will require a paradigm shift in which a set of technological challenges would have to be tackled, from electrodes and analog readout to application and headset design. The challenges associated with the electronical and mechanical aspects of a wearable EEG acquisition unit are described. In clinical settings, biopotential signals such as EEG and ECG are measured as input for medical diagnosis. A new and evolving field is the continuous monitoring of bio-potential

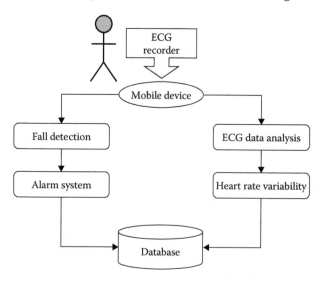

FIGURE 5.6
Monitoring system architecture. (From Kantoch, E. et al., Design of a wearable sensor network for home monitoring system, *Proceedings of the Federated Conference on Computer Science and Information Systems*, 2011, pp. 401–403.)

signals out of the hospital environment. Continuous monitoring could help in detecting early onsets of certain disorders, or to monitor recovery or disease progression. In some cases, it can even be used as integral part of treatment of brain-related disorders. However, recording bio-potential signals in an uncontrolled environment is very challenging (technical challenges) (Grundlehner et al. 2013).

5.3.2 Body Sensor Networks: Technical Challenges

Preventative care and chronic disease management to reduce institutionalization is a national priority for most western countries. This chapter outlines some of the key technical challenges related to pervasive health monitoring with BSNs. Issues concerned with the development of a common architecture with context-aware sensing are discussed. In order to achieve unobtrusive pervasive sensing that links physiological/metabolic parameters and lifestyle patterns for improved well-being monitoring, and early detection of changes in disease, we have demonstrated the basic structure and our experience of the UbiMon platform for continuous monitoring of patients under their natural physiological conditions.

The last decade has witnessed a rapid surge of interest in new sensing and monitoring devices for health care. One key development in this area is implantable *in vivo* monitoring and intervention devices. While the problem of long-term stability and biocompatibility is being addressed, several promising clinical prototypes are starting to emerge. For example in the case of managing patients with acute diabetes, the blood glucose level can be monitored continuously *in vivo*, which controls the insulin delivery from an implanted reservoir. For the treatment of epilepsy and other debilitating neurological disorders, there are already implantable, multi programmable brain stimulators available in the market which save the patient from surgical operations of removing brain tissue. In cardiology, the value of implantable cardioverter-defibrillator (ICD) has increasingly been recognized for the effective prevention of sudden cardiac death (SCD). In Europe, 900,000 patients die suddenly each year and about 90% of these deaths are caused by an arrhythmogenic event. Disturbingly, many arrhythmogenic deaths could be prevented if ICD implantation is done when the risk of SCD is identified. It is possible to envisage a large percentage of the population having permanent implants which would provide continuous monitoring of the most important physiological parameters: these parameters help in identifying in advance the precursors of major adverse cardiac events including sudden death. Such technological development echoes the social, industrial and clinical perspectives of future health-care delivery.

Technological developments of sensing and monitoring devices are reshaping the general practice in clinical medicine. Although extensive measurement of biomechanical and biochemical information is available in almost all UK hospitals, the diagnostic and monitoring utility is generally limited to brief periods of time and perhaps unrepresentative physiological states such as supine and sedated, or artificially introduced exercise tests. Transient abnormalities, in this case, cannot always be captured. For example many cardiac diseases are associated with episodic rather than continuous abnormalities such as transient surges in blood pressure, paroxysmal arrhythmias or induced or spontaneous episodes of myocardial ischaemia. These abnormalities are important, but their timing cannot be predicted and much time and effort is wasted in trying to capture an *episode* with controlled monitoring. Important and even life-threatening disorders can go undetected, because they occur only infrequently and may never be recorded objectively. High-risk patients such as those with end-stage ischaemic heart disease or end-stage myocardial failure often develop life-threatening episodes of myocardial ischaemia or ventricular arrhythmia. These episodes, if reliably detected, could lead to better targeting of potentially life-saving but expensive therapies. With the emergence of miniaturized mechanical, electrical, biochemical and genetic sensors, there is likely to be a rapid expansion of biosensor development over the next 10 years with corresponding reduction in size and cost. This will facilitate continuous wireless monitoring, initially of at-risk patients but eventually screening an increasing proportion of the population for abnormal conditions.

5.4 Wireless Sensor Networks for Health Monitoring

CustoMed is a fascinating and critical class of distributed embedded systems. The main attribute of CustoMed is its fast and easy customizability capability based on patients' needs. Firstly, the customization may be at the device level. The choice of the device, the placement of the device on the body and the interaction level of the device with the environment will be tailored to the individual and his/her needs. Secondly, the software downloaded onto the devices can be customizable. Depending on the gender, age, medical condition and other variables, the software downloaded onto the devices differs. The idea of customization has not been emphasized before, but is an important concern, if the system is made to be robust enough to handle many different needs, as well as unexpected needs that may arise. CustoMed can be easily custom-built for patients by non-engineering staff. The system will be quickly assembled from basic parts and configured for use. The vision is that in the doctor's office, in about 5 min, the appropriate devices and the correct number of them will be assembled and affixed to the patient. Other systems take months or even years to be built, and hence lack the adaptability of the system we propose. The physician will also pick from a wide range of code to download onto the devices. We developed such a tool which enables physicians to pick a specific variation of code for a particular application and download it to the system components of CustoMed. Furthermore, it works with the environment made available to the patient. For example in case of an emergency, the sensors can alert the security system in the house. In less urgent cases, an email can be sent across the Internet or a home appliance can be turned off or on.

Several *wearable* technologies exist to continually monitor the patient's vital signs, utilizing low-cost, well-established disposable sensors such as blood oxygen finger clips and ECG electrodes. The Smart Shirt from Sensatex is a wearable health-monitoring device that integrates a number of sensory devices onto the wearable motherboard from Georgia Tech. Several other technologies have been introduced by MIT called MIThril, e-Textile from Carnegie Mellon University and Wearable e-Textile from Virginia Tech. Furthermore, the Lifeguard project carried out at Stanford University is a physiological monitoring system comprised of physiological sensors (ECG/respiration electrodes, pulse oximeter, blood pressure monitor and temperature probe), a wearable device with built-in accelerometers (chronic pulmonary obstructive disease [CPOD]) and a base station (Pocket PC). The CPOD acquires and logs the physiological parameters measured by the sensors. The Assisted Cognition Project carried out by the University of Washington's Department of Computer Science is exploring the use of AI systems to support and enhance the independence and quality of life of Alzheimer's patients. Assisted cognition systems use ubiquitous computing and artificial intelligence technology to replace some of the memory and problem-solving abilities that have been lost by an Alzheimer's patient. Nevertheless, none of these projects/systems support the concept of being custom-built in minutes and reconfiguration to the extent that CustoMed does.

5.5 Wearable and Wireless Brain–Computer Interface

EEG is a powerful non-invasive tool widely used by for both medical diagnosis and neurobiological research as it provides high temporal resolution in milliseconds. Another important advantage of EEG is that it involves sensors light enough to allow near-complete freedom of movement of the head and body, making EEG the clear choice for brain imaging of humans performing normal tasks in real-world environments. However, the lack of portable and user-acceptable (e.g. comfortably wearable) sensors and miniaturized supporting hardware/software to continuously acquire and process EEG has long thwarted the applications of EEG 742 C.-T. However, the lack of portable and user-acceptable (e.g. comfortably wearable) sensors and miniaturized supporting hardware/software to continuously acquire and process EEG has long thwarted the applications of EEG 742 C-T (Lin et al. 2005). Recently, we developed and tested a prototype four-channel mobile and wireless EEG system incorporating a miniature data acquisition (DAQ) circuitry and dry MEMS electrodes with 400 ganged contacts for acquiring signals from non-hairy sites without use of gel or skin preparation. This study extends our previous work, NCTU BCI-cap, to a smaller, lighter, wearable and wireless brain–computer interface (BCI), NCTU BCI-headband. The NCTU BCI-headband features: (1) disposable dry MEMS electrodes;

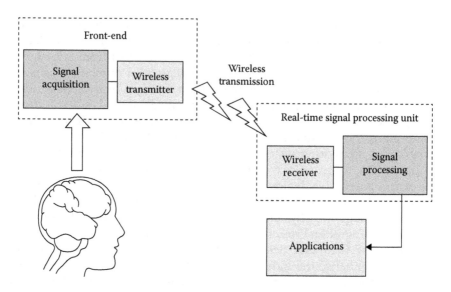

FIGURE 5.7

System diagram of a wearable and wireless brain–computer interface. (From Schmorrow, D.D., Why virtual?, In: J.Cohn & A.Bolton (Eds.), *Special issue on Optimizing Virtual Training systems in Theoretical issues of Ergonomics Science*, CRC Press, Boca Raton, FL, Vol. 3, pp. 257–258, 2009.)

(2) an eight-channel DAQ unit; (3) wireless telemetry and (4) real-time digital signal processing (DSP) implemented on a commercially available cell phone or a digital signal-processing module. The applicability of the NCTU BCI-headband to EEG monitoring in operational environments was demonstrated by a sample study: cognitive-state monitoring and management of participants performing normal tasks in real-world environments.

Figure 5.7 shows the system diagram of the mobile and wireless BCI. The front-end unit integrates (1) clip-on electrode holders for dry MEMS or commercially available wet EEG electrodes, (2) a DAQ unit and (3) wireless transmission circuitry into a quickly and easily donned and doffed headband that can acquire and transmit EEG signals from up to eight channels. The back-end unit integrates a wireless signal receiver and online DSP. EEG signals are first acquired by dry MEMS or commercially available electrodes, amplified by the preamplifier, converted to digital signals and then wirelessly transmitted to the data receiver. The DSP unit processes the EEG data and displays the results. The raw EEG data can also be wirelessly transmitted to a remote PC for further offline analysis and/or database collection.

5.5.1 Dry MEMS Electrodes and Electrode Holders

Use of MEMS technology to build a silicon-based spiked electrode array or so-called dry electrode is being used to enable EEG, EOG, ECG and EMG monitoring without conductive paste or scalp preparation. However, the connectors between the dry sensors and DAQ board were not very robust in the BCI cap. This study incorporated snap-on electrode holders to house dry electrodes or commercially available EEG sensors.

5.5.2 Data Acquisition Unit

The data acquisition unit integrated an analog preamplifier, a filter and an analog-to-digital converter (ADC) into a small, lightweight, battery-powered DAQ. EEG signals are sampled at 512 Hz with 12-bit precision, amplified by 6000 times and band-pass filtered between 1 and 50 Hz. Figure 5.8 shows the block diagram of the DAQ unit. To reduce the number of wires for high-density recordings, the power, clocks and measured signals are daisy-chained from one node to another with bit-serial output. That is adjacent nodes (electrodes) are connected together to (1) share the power, reference voltage and ADC clocks and (2) daisy-chain the digital outputs.

5.5.3 Wireless Transmission Unit

The wireless transmission unit consisted of a wireless module and a microcontroller. It used a Bluetooth module to send the acquired EEG signals to a

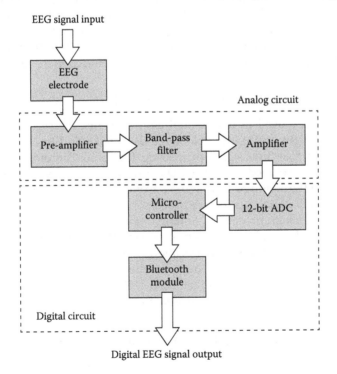

FIGURE 5.8
Block diagram of the data acquisition unit.

FIGURE 5.9
Picture of the wearable and wireless EEG system.

custom real-time DSP unit described in the following or a Bluetooth-enabled cell phone which was used as a real-time signal-processing unit. The dimension of the wireless transmission circuit was 40×25 mm^2 and integrated four-channel wireless EEG system. Reference and ground channels were also included in the system (not shown). The integrated circuitry can be embedded into a headband, NCTU BCI-headband, as shown in Figure 5.9. The power-consumption of the NCTU BCI-headband is very low (1100 mAh Li-ion battery can last over 33 h). It comprises four- or eight-channel snap-on electrode holders (plus a reference and a ground channels), miniature bio-amplifier, a band pass filter, an ADC and a Bluetooth module. All channels were referred to the left mastoid.

5.5.4 Real-Time Digital Signal Processing Unit

To be practically used in operational environments, the signal-processing unit must be lightweight, portable, have low power and have online data-receiving and real-time signal-processing function. Therefore, this study designed and developed a real-time DSP unit which used a Bluetooth module to receive the acquired EEG signals from the NCTU BCI-headband and process the EEG signals via its core processor in near-real-time. The core processor is the Black fin processor (Analog Device Incorporation, ADSP-BF533) which provided a high performance, power-efficient processor choice for demanding signal-processing applications. The dimension of the miniature DSP unit is about 65×45 mm^2. The maximum high processing performance of the BF533 core processor can reach 600 MHz. Furthermore, the following peripheral modules were also incorporated into the unit.

5.5.5 Data Logging and Digital Signal Processing on a Cell Phone

To demonstrate the application of the wearable and wireless BCI during long and routine recording in operational environments, we have developed and installed a data logging Java GUI on a Bluetooth-enabled cell phone. The Java program receives EEG signals from the NCTU BCI headband and plots them on the LCD screen. We have also implemented power spectrum density (PSD) estimation using a 512-point fast Fourier transform (FFT) on the cell phone. Lin et al. (2008) have reported a four-channel BCI-cap which measured EEG and transmitted it to a commercially available DSP kit by Texas Instruments. This study extended the EEG recording system into a truly mobile BCI which acquired and processed EEG signals in near-real-time. It is evident that, compared to BCI cap, BCI headband is lighter, smaller, more power-efficient and accommodates more channels with higher sampling rate and digitization precision.

Real-time alertness monitoring using NCTU BCI headband and a cell phone recently demonstrated the feasibility of using dry MEMS EEG electrodes, supporting hardware and commercially available TI DSP kit to continuously and accurately estimate the driving performance (putative drowsiness level) based on EEG data from four frontal non-hairy positions in a realistic VR-based dynamic driving simulator. Bluetooth-enabled cell phone reported the received EEG signals and processed them with the on-board processor. The cell phone delivered arousing feedback when the participants were drowsy. Figure 5.10a shows the flowchart of the signal processing implemented on the cell phone. Figure 5.10b shows the evident alpha activities when the subject was drowsy. This study demonstrated a truly portable, lightweight and readily wearable BCI that featured dry MEMS electrodes and a miniaturized DAQ, wireless telemetry and online signal processing. The main goal of the design and development of wearable and wireless BCI is to maximize their wearability, unconstrained mobility, usability and reliability in operational environments. The signal-processing module and the Bluetooth-enabled cell phone were programmed to assess fluctuations in individuals' alertness and capacity for cognitive performance based on the EEG signals. The BCI delivered arousing feedback to the driver to maintain optimal performance. The cell phone and DSP unit, however, can be programmed for many other brain–system interface applications. Truly portable and user-acceptable BCI will have enormous future impacts on clinical research and practice in neurology, psychiatry, gerontology and rehabilitation medicine.

5.5.6 Wet Electrodes

EEG electrodes are small metal plates that are attached to the scalp using a conducting electrode gel. They can be made from various materials. Most frequently, tin (Sn) and silver/silver-chloride (Ag/AgCl) electrodes are used, but there are gold (Au) and platinum (Pt) electrodes as well. Figure 5.11 shows electrodes wet electrodes used in biomedical applications. Correct EEG electrode placement is important to ensure proper location of electrodes in relation to cortical areas so that they can be reliably and precisely maintained on an individual basis. The International 10–20 System of Electrode Placement is the most widely used system which describes the location of scalp electrodes. The 10–20 system is based on the relationship between the location of an electrode and the

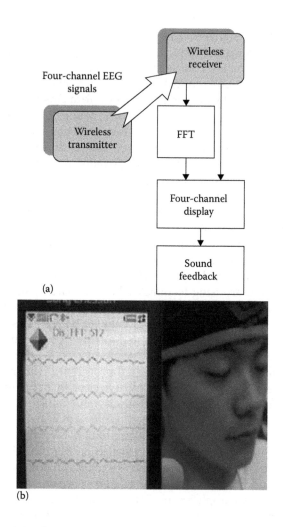

FIGURE 5.10

(a) The flowchart of signal processing on a cell phone; (b) four-channel EEG signals were displayed on the cell phone. (From Lin, C.T. et al., *IEEE Trans. Biomed. Eng.*, 55(5), 1582, 2008a.)

underlying area of cerebral cortex. Each site has a letter which identifies the lobe and a number or another letter to identify the location of hemisphere. Figure 5.12 shows the international 10–20 electrode placement system. The letters used are: 'F' – frontal lobe, 'T'–temporal lobe, 'C' – central lobe, 'P' – parietal lobe and 'O' – occipital lobe. Even numbers such as 2, 4, 6, 8 refer to the right hemisphere and odd numbers such as 1, 3, 5, 7 refer to the left hemisphere. Letter 'Z' refers to an electrode placed on the midline. The smaller the number, the closer is the position of electrode to the midline. 'Fp' stands for front polar. 'nasion' is the point between the forehead and nose. 'inion' is the bump at the back of the skull. The letters used are: F – frontal lobe, T – temporal lobe, C – central lobe, P – parietal lobe and O – occipital lobe 'Z' refers to an electrode placed on the midline.

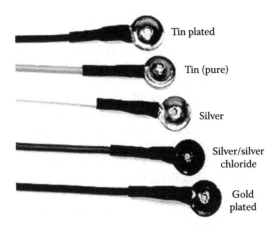

FIGURE 5.11
EEG electrodes.

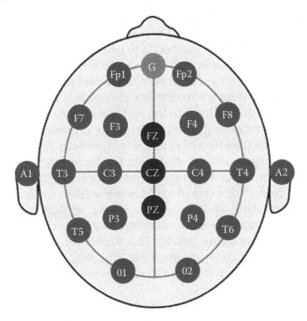

FIGURE 5.12
International 10–20 electrode placement system. (From Matthews, R. et al., *Proc. IEEE Annu. Int. Conf. Engineering Medicine Biology Soc.*, 17, 5276, 2007.)

5.5.7 Placement of Electrode

Mark the nasion and inion. Get the total. Mark 50% of the total. This is the first mark of CZ. Mark 10% up from the nasion and 10% up from the inion. This is the first mark of FPZ and OZ. Mark 20% up from either the first mark of FPZ or CZ. This will be the first mark of FZ. Measure the distance from preauricular point to preauricular point. Have the patient open his mouth slightly. Lightly run the finger up and down just anterior to the ear; the indentation above the zygomatic notch is easily identified. Obtain the total. Mark half of the total (this number should intersect with the first 50% mark). This is the true CZ mark 10% up from each preauricular point. These are the first marks of T3 and T4. Measure from the first mark of T3 to your CZ and obtain the total. Mark half of the total. This is the first mark of C3. Measure from the first mark of T4 to CZ and obtain total. Mark half of the total. This is the first mark of C4. Draw a cross-section on the first mark of FPZ. This is the true FPZ mark. Encircle the measuring tape all across the 10% mark, obtaining the total circumference. Mark half of the total. This will be the true mark of OZ. Mark 5% of the half to the left and right of OZ. These will be the true marks of O1 and O2, respectively. Mark 5% to the left and right of FPZ. These will be the true marks of FP1 and FP2, respectively. Other important electrodes are reference and neutral electrodes. The reference electrode is critical. It is the second input for the digital amplifiers for all head electrodes being recorded. The reference electrode is commonly placed half way between CZ and PZ or half way between CZ and FZ. Another method is to use two electrodes at A1 and A2 linked together. The principle is the same: to try to use a location that is as electrically silent as possible. It is a very important electrode while recording as if it comes off, data from all other channels is recorded as interference. A scalp electrode is affixed to the midline forehead or other relatively neutral site. The neutral or ground electrode is important as it is used by the system to reduce the effect of external interference. Its position is not so critical, if this electrode comes off during the recording; one may or may not have a problem depending on the environment. When more detailed EEG has to be recorded, extra electrodes are added, utilizing the spaces in between the existing 10–20 system as shown in Figure 5.13. This new electrode naming system is more complicated giving rise to the Modified Combinatorial Nomenclature (MCN). This MCN system uses 1, 3, 5, 7, 9 for the left hemisphere which represents 10%, 20%, 30%, 40%, 50% of the inion-to-nasion distance, respectively. The introduction of extra letters allows the naming of extra electrode sites.

Two issues limit the use of these systems. The first is that electrically conductive gel is required for a good contact between the sensors and the scalp. It takes a lot of time to apply sensors on the scalp; it may diffuse through the hair to create short circuits between the sensors and it tends to dry out, which limits the recording time. The second limiting issue with these EEG systems is that they are not mobile. Moreover, it consumes high power. For reduced application time and increased subject comfort, EEG cap can be used for recording multiple EEG channels. Electrodes are pre-positioned in the international 10/20 montage. So, even novice EEG researchers can minimize electrode placement errors. Figure 5.14 shows an EEG cap used to record multiple channels.

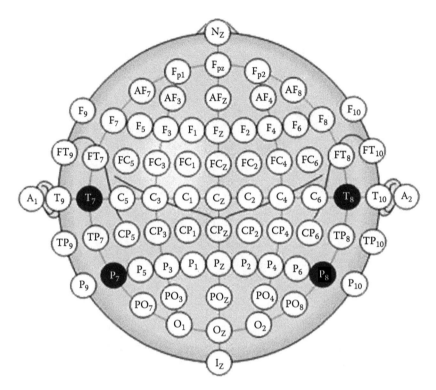

FIGURE 5.13
Advanced International 10–20 electrode placement.

5.5.8 MEMS-Based Dry Electrodes

One key problem researcher's face is how to measure the signals without applying sticky electrodes or using conductive gel. One such type of electrode used is MEMS-based dry electrodes that create a good contact with the skin without requiring gel. One application of MEMS is the manufacture of miniaturized mechanical and electro-mechanical devices and structures that are made by exploiting the technique of micro-fabrication. The functional elements of MEMS are miniaturized structures, sensors, actuators and microelectronics; the most notable (and perhaps most interesting) elements are the micro-sensors and micro-actuators. Micro-sensors and micro-actuators are appropriately categorized as 'transducers', which are defined as devices that convert energy from one form to another. Stratum basale and stratum spinosum together are called stratum germinativum. The dermis also varies in thickness, depending on the location of the skin. It is 0.3 mm on the eyelid and 3.0 mm on the back. The subcutaneous tissue is a layer of fat and connective tissue that houses larger blood vessels and nerves. The fabricated dry electrode is designed to pierce the stratum corneum into the electrically conducting tissue layer stratum germinativum, but to not reach

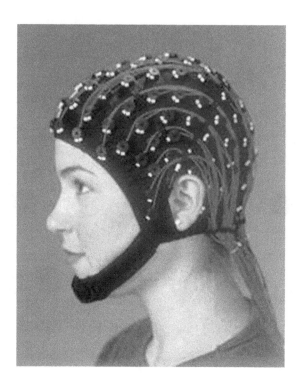

FIGURE 5.14
EEG cap.

the dermis layer so as to avoid pain and bleeding. Since the dry electrode is expected to circumvent the high impedance characteristics of the *SC*, there is no need for skin preparation and electrolytic gel application. A system using these dry electrodes provides convenient experiment setup and enables experiments which require long-term recording. Also, the number of wires needed to interface with a large number of electrodes is reduced, since the power, clocks and measured signals are daisy-chained from one board to another. Also, mobility is possible, because only a small amount of power is required for the whole signal-processing system. This system consumes only 3 mW of power, making battery power feasible. Small batteries can be employed to power the system for an extended period of time.

5.5.9 Wireless EEG System

In clinic, insomnia, that is sleeplessness, is a main sleep disorder. It leads to severe fatigue, anxiety, depression and lack of concentration. To treat insomnia, it is important to study patient's status of wake by detection of alpha wave in EEG of all night. Polysomnography (PSG) is used for recording EEG all night. There are two deficiencies noted in PSG while recording EEG: The patient whose EEG is recorded by PSG does not feel relaxed as longer electrode

FIGURE 5.15
Wireless EEG headsets.

wires always wrap his head and body, which causes patient to not to sleep on both ears so that he or she is often awake. In this situation, doctor may think it as a sleep disorder. The other limitation of PSG is power-line interference suppression. Power-line interference is very prominent in EEG, which can even lead to saturation of the amplifier of EEG which may lead to failure of EEG recording. The detection software cannot detect characteristic wave in EEG properly. So, to overcome these limitations, a wireless EEG sensor system has been designed. The system includes an EEG amplifier and a wireless transmitter and receiver. A personal computer is also used. The wireless EEG sensor system is good at detecting alpha wave and power-line interference suppression. And patient feels relaxed as it is small in size and has very short electrode wires. A wireless EEG platform and headset is also available which offers new perspectives towards a fully integrated, high-performance, wearable EEG monitoring system. This wireless EEG headset works reliably in a controlled environment where movement and interference are minimized (Figure 5.15). Dry electrodes provide signal quality comparable to gel electrodes. Low power wireless system enables the transition from lab to real life environment.

5.6 Brain–Computer Interface

Many BCIs are based on some of the features of EEG signals. To acquire these signals, a set of electrodes is used; most of the times these are attached to a brain cap. Several brain cap designs, as well many different electrodes, have been presented. So far, all the brain caps are uncomfortable to wear. Moreover, the electrodes, wet or dry, are also difficult to place. Despite the huge potential that a BCI offers, for disabled and healthy people, new brain caps are required to overcome these two main problems. The wearable cap is obtained using a flexible polymeric material, with new integrated contactless electrodes. The electrodes may be obtained using a new electroactive gel, which is used to read the EEG signals.

A BCI relies on the measurement of brain activity in order to provide solutions for communication and environmental control without movement. Although a BCI was initially intended for people with severe disabilities (e.g. spinal cord injury, brainstem stroke, etc.), it can also be used as an alternative communication channel by their healthy counterparts. Over the past two decades, several studies were performed towards the development of these BCI systems, which do not require muscle control. Although still in its infancy, BCI is no longer within the realms of science fiction, but an evolving area of research and applications. BCI may be used to augment human capabilities by enabling people to interact with a computer through a conscious and spontaneous modulation of their brainwaves after a short training period. They are indeed, brain-actuated systems that provide alternative channels for communication, entertainment and control. Nowadays, the typical BCI systems measure specific features of brain activity and translate them into device control signals, generally making use of EEG techniques. Essentially four modules of hardware/software compose the standard BCI. The signal acquisition module is generally constituted by a cap, electrodes and proper instrumentation hardware for neural data acquisition. The brain activity is then recorded, amplified and digitized. The recorded and pre-treated signals are further subjected to a series of processing techniques, which can be divided into two modules: feature extraction module and translation module. The first one employs several filters so as to extract significant measures from neural data. These features used include, for instance, slow cortical potentials (SCPs), sensorimotor rhythms (SMRs) and P300 evoked potentials. Some examples of possible filters to apply are: time (e.g. moving average), frequency (e.g. narrow-band power) and spatial (e.g. Laplacian filters) filters. The next stage, translation module, converts these signals features into device commands that carry out the user intents, by detecting a previously identified EEG pattern. Finally, the actuator module is the sub-system that performs the action corresponding to the output command, allowing for the operation of the desired device (e.g. wheelchair, neuroprosthesis, etc.).

Many factors determine the performance of a BCI system. These factors include the brain signals measured, the signal-processing methods responsible for extracting signal features, the algorithms that translate these features into device commands, the output devices that execute these commands, the neurofeedback provided and the characteristics of the user as well. The achievement of a superlative combination of these factors is the key to establish the ideal BCI system. The ideal solution would be a contactless measurement of brain activity, allowing the design of a wearable system which would provide a more comfortable and practical device for the user. In fact, the majority of the existent BCI technologies are based on EEG, which despite being the most straightforward method for measuring bio-potentials carries strong disadvantages. These are related to the need for an electrolytic paste to promote the impedance reduction as well as with the signal-to-noise ratio (SNR). Likewise, the electrodes with contact carry strong disadvantages, since they are difficult to place, resulting in time-consuming and complex attachment procedures. Concerning the wearability, the existing brain caps are difficult and uncomfortable to wear, not allowing its daily and normal use. In addition, portable and discreet solution for brain activity detection must be found out from the perspective of the user. Taking this into account, a new solution for the measurement of brain biopotential, which consists of a wearable device system with integrated sensors – a Wearable Brain Cap, has been proposed.

5.6.1 Wearable Brain Cap

The proposed wearable brain cap allows the use of a new generation of electrodes that besides being more flexible, allows to be integrated in a wearable device, and does not require any electrical contact between its surface and the subject scalp. Consequently, the disadvantages concerning the use of the standard EEG tests will be reverted with the use of contactless electrodes, since there will not be any need for an electrolyte gel neither for time-consuming and uncomfortable attachment procedures. Moreover, as opposed to other techniques, such as near-i spectroscopy or magnetoencephalography, with our sensors, it is possible to actually measure brain activity in terms of standard potentials, as it happens with EEG. This can be achieved by selecting one reference methodology: besides the contactless working electrodes, there will be a reference establishing contact with the subject skin. Usually the reference is placed on the ear lobes, since most of the times, they are not influenced by the electrical activity of the temporal lobe and by the muscle activity. On the other hand, from the user's perspective, it would be better to place the reference in a more discrete place, since the objective is to use this brain cap as normal daily garment. Therefore, another alternative would be to use the reference electrode on the back of the neck. With this approach, the electric field magnitude would most likely be higher, since the distance between the working electrodes and the reference increases, its detection is easily achieved.

5.6.2 Fabrication Technique

The novel fabrication technique used to design our device is based on the multilayer integration of different types of materials, allowing for the integration of several components, for example sensors, actuators, optical fibres, electrical wires and antennas. This integration is made concurrently to the deposition and resorting to printing techniques. The layers may be composed of different materials such as polymeric, metal or synthetic materials. For instance the fabric might consist of a polymeric material sandwiched in a synthetic layer, and the sensors and actuators, or even other components would be printed, for instance, in the polymeric layer. Thereby, the final result would be, to the naked eye, a normal fabric such as common t-shirt, or even a scuba diving suit. In this case, we are able to design a normal and wearable cap or even, if necessary, a normal swimming cap.

5.6.3 Contactless Sensor Design

The sensor proposed is a fibre-optic-based device whose operation relies on the electro-active principle. The main functional stages of sensor are depicted and consist of the following: optical signal generation, control and modulation and detection. Briefly, the first stage requires a light source in order to obtain an optical signal that will be further modulated and controlled (second stage) through a series of devices and steps. This second step is dependent on the electric field, and resides in an electro-active component. This can be used simply as a coating material like, for example a hydrogel or other piezoelectric material. Afterwards, the modulated light is guided to a photo-detector for further analysis. Electroactive hydrogel is used as the sensing/modulator component of sensor – polyacrylamide hydrogel (PAAM). This low-cost polymeric material allows the easy modification of its physical and chemical properties. When exposed to an external electric field, PAAM undergoes a bending process, altering its mass and volume properties. Likewise, the input light passing through this hydrogel will experience modifications, not only with respect to the refractive index, but also the amount of light that is transmitted back to the photo-detector. The hydrogel is constrained in a cavity at the end of the optical fibre, as near as possible to the subject scalp to increase the detection sensitivity. After the modifications of PAAM properties due to the electric field arising from the subject brain activity, the input light is modulated, that is change of the refractive index and amount of light transmitted.

5.6.3.1 Measurements and Analysis

In order to validate the wearable brain cap system proposed, some parameters through some experiences have been studied. Human scalp EEG experiments were carried out in order to have an idea of the typical electric field resulting from brain activity in the relaxed state. Likewise, we performed some tests to evaluate the electro-active behaviour of the PAAM hydrogel, in order to obtain the minimum electric field that could be detected by using it as a sensing element.

5.6.3.2 *Brain Activity Electric Field*

The component of E (electric field) in any direction can be expressed as the negative of the rate at which the electric potential (dV) changes with distance (ds) in that direction, that is $\mathbf{E} = \mathbf{dV/ds}$. Since the electric field can be related with the electrical potential (V), its determination can be achieved by using the EEG standard test.

5.6.3.3 *Electroactive Properties of the Hydrogel*

PAAM hydrogel is an electroactive polymer with great sensing capabilities to physical, chemical and biological environments and response to external stimulus in a controllable way. PAAM shows abrupt and vast volume changes as well as bending phenomena when submitted to an external electric field. To produce PAAM hydrogel, acrylamide 99%, N,N'-methylenebisacrylamide (BIS) 98% as cross-linker, N,N,N',N'-tertramethylethylenediamine 99% (TEMED), ammonium persulfate 98% (APS) and aniline purum 99% were used. All chemicals were purchased from Aldrich and used as received without any further purification. Deionized water was used for all the dilutions, the polymerization reactions, as well as for the gel swelling. PAAM is synthesized by the standard free radical polymerization method using 1 mL acrylamide (30%), 60 mL APS (25%), 20 mL Temed with no cross-linker under vacuum. After complete polymerization, the resulting gel was diluted in 4 mL of deionized water and 5 mL acrylamide (30%), 250 mL BIS (2%), 10 mL APS (25%) and 4 mL Temed were added to form the precursor solution. After its preparation, the gel was placed between two metallic plates and an external voltage was applied. The gel has a thickness of 1 cm and the voltage applied was 2 V.

5.7 Soldier's Status Monitoring Software Using Visual Basic

The bullet wound signals and the soldier's vital signals were transmitted to the soldier monitoring station at the remote end, where an observer will monitor the soldier's status from the specially designed application software. The software contains the soldier's physical, biological data and the current bullet wound status. Communication between the soldier and the remote location was carried out using GSM technology. The various sensors integrated in the teleintimation garment and the smart shirt has been connected to a central processing control unit in the garment. The processing unit is equipped with a computer chip containing firmware with a series of soldier's status decision support (SSDS) algorithms. The SSDS algorithms process the sensory information so as to produce meaningful information for combat medics and commanders. Specifically, as a first step, a inert, alive or unknown (IAU) status is estimated by these algorithms and transmitted to the field medic or elsewhere as part of a larger SSM system.

5.7.1 Soldier's Status Monitoring Software

The remote end server is loaded with the soldier's-monitoring application developed using Visual Basic software. At the server end, SSM provides the bullet information in numbers starting from 0 to 79 rows and columns. This will give the graphical view of the exact bullet information. The graphical representation is divided into 10 blocks in rows and 10 blocks in column. Each block will represent 8 lines of row and column intersection. Thus, the 80 lines from row and 80 lines from column are graphically separated. The screen shots of the developed application using VB 6.0 is shown in the following. Figure 5.16 shows the front view of the SSM software. After pressing the enter button, it will display the soldier's physical data as shown in Figure 5.17. The bullet wound intimation has been represented in the graphical matrix format, the matrix format will initially be in black colour which indicates that there is no bullet penetration as shown in Figure 5.18.

Whenever a soldier got bullet wounded, the vital signs from the smart shirt have been transmitted to the SSM through the telemonitoring system by means of GSM technology. The display of vital signs in the SSM is as shown in Figure 5.19. The bullet wound in the teleintimation garment, in the

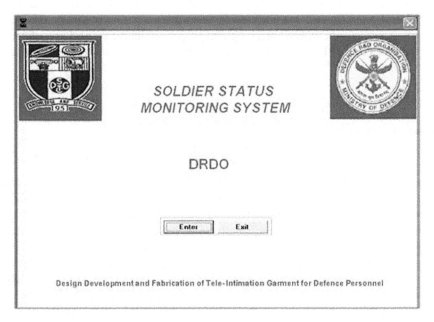

FIGURE 5.16
Snapshot of the front view of SSM software.

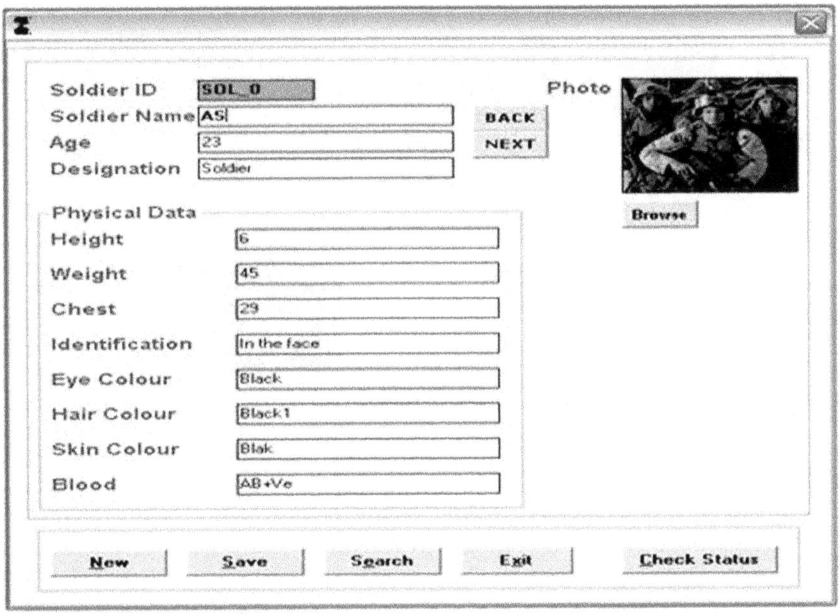

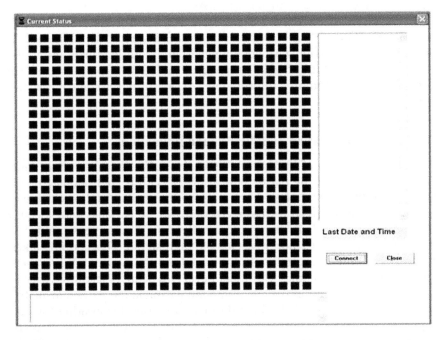

FIGURE 5.17
Soldier's database of SSM software.

FIGURE 5.18
Number of bullet wound intimation in matrix format.

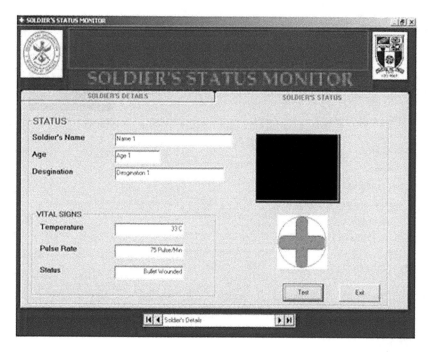

FIGURE 5.19
Display of vital signs from the smart shirt in the SSM.

corresponding location, is highlighted in the SSM in red colour. Figure 5.20 shows that teleintimation garment got penetrated with 16 bullets and in the text bar the row, column details on the date and time of penetration have been represented. From this information, IAU status of the combat casualty unit can be predicted from the SSM software.

5.7.2 Location Evaluation of Bullet Wound Intimation

To count the number of bullet wounds in the soldier's body, voltage level at the receiver end of the plastic optical fiber (POF) is continuously monitored. If any POF is interrupted between the transmitter and receiver end, logic high signal is given to the microcontroller. A variable with a count increment will be made in the microcontroller, and the counter variable has been incremented if any port gives a logic high signal. This count is taken as number of bullet penetration in the teleintimation garment. When there is no bullet wound, the microcontroller ports will be in logic low. Whenever there is a bullet wound in the teleintimation garment, the particular matrix coordinates will be affected. This information is used for locating the bullet penetration. When a fibre is cut at the location (1,1), the location could be displayed. But at the same time when a bullet hits at the location (2,2), the new bullet wound location could be displayed. In addition to this, two more

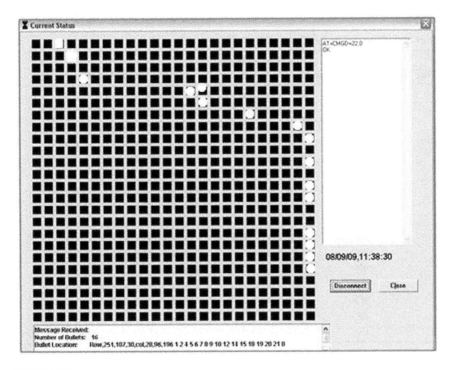

FIGURE 5.20
Bullet wounds in the teleintimation garment were highlighted in the SSM.

locations (1,2) and (2,1) are also displayed even though there is no bullet penetration. This is because the location (1,1) has got already wounded; hence, whenever another bullet hits at a different row or column, the previously wounded row and column are taken into account. This gives the false reading of the bullet location. To overcome this issue, the bullet wound location in terms of rows and columns are displayed individually. A POF is removed at the row end which in turn counts the bullet and shows its location.

The POF woven into the teleintimation garment sends and receives the optical signals to/from the microcontroller circuit, which will tell the presence or absence of the bullet in a particular location. The number of POF lines required for the horizontal and vertical section is designed for an ideal measurement. It has been found that for the efficient number of bullet counts and bullet wound location, it is necessary to have 80 POF lines with 0.5 cm distance each in vertical section. Also 80 POF lines with 0.5 cm in the horizontal section are required. It is formed in the array format so as to detect the bullet wound location. The data received from the PIC microcontroller is in the form of hexadecimal values and this data varies from 00 to FF according to the bullet wound location and the presence of bullet. Table 5.1 shows the bullet location with respect to the hexadecimal value.

TABLE 5.1

Bullet Location and Bullet Count Technique

S. No.	Hexadecimal Value	Equivalent Binary Value	Bullet Location
1.	00	0000 0000	No bullet
2.	01	0000 0001	1 bullet at 0th location
3.	02	0000 0010	1 bullet at 1st location
4.	03	0000 0011	2 bullets at 0th and 1st location
5.	04	0000 0100	1 bullet at 2nd location
6.	05	0000 0101	2 bullets at 2nd and 0th location
7.	06	0000 0110	2 bullets at 2nd and 1st location
8.	07	0000 0111	3 bullets at 2nd, 1st and 0th location
9.	08	0000 1000	1 bullet at 3rd location
10.	09	0000 1001	2 bullets at 3rd and 0th location
11.	0A	0000 1010	2 bullets at 3rd and 1st location
12.	0B	0000 1011	3 bullets at 3rd, 1st and 0th location
13.	0C	0000 1100	2 bullets at 3rd and 2nd location
14.	0D	0000 1101	3 bullets at 3rd, 2nd and 0th location
15.	0E	0000 1110	3 bullets at 3rd, 2nd and 1st location
16.	0F	0000 1111	4 bullets at 3rd, 2nd, 1st and 0th location

5.8 Soldier's Status Monitoring Software Using LabVIEW

The SSM software has been created using LabVIEW software to monitor the place and number of bullet wounds. The remote monitoring system is defined as a computer-controlled laboratory that can be accessed and controlled externally over some communication medium. The monitoring system is being processed and is running on a LabVIEW platform with the ability to be monitored and controlled over the internet within a web browser. The teleintimation garment has been connected to the LabVIEW system using GSM technology.

5.8.1 LabVIEW

LabVIEW is virtual instrumentation software from National Instruments. LabVIEW programmes are called virtual instruments (Vis), because their appearance and operation imitate physical instruments, such as oscilloscopes and multimeters. LabVIEW contains a comprehensive set of tools for acquiring, analysing, displaying and storing data, as well as tools to troubleshoot the programming code. LabVIEW Vis contains three components, namely (1) front panel (2) block diagram and (3) icon and connector pane. Front panel contains controls and indicators. Controls are knobs, push buttons, dials and other input devices. Indicators are graphs, LEDs and other displays. The block diagram contains code to control the front panel objects. In some ways, the block diagram resembles a flowchart. An icon is a graphical representation of a VI which can contain text, images or a combination of both used to identify it as a subVI on the block diagram of the VI. The connector pane is a set of terminals that correspond to the controls and indicators of that VI, similar to the parameter list of a function call in text-based programming languages. LabVIEW has been used to communicate with the teleintimation garment by using data acquisition card. LabVIEW has built-in features for connecting any application to the web using the LabVIEW Web Server and software standards such as TCP/IP networking and ActiveX.

5.8.2 Implementation of LabVIEW-Based SSM Software

The implementation of LabVIEW-based software on teleintimation garment has been done with the help of integrating the teleintimation garment with microcontroller through an interface to LabVIEW-based SSM software. The interface is specific that it contains components to control and measure parameters of the teleintimation garment. LabVIEW uses TCP/IP to communicate with ethernet controller which in turn uses serial peripheral interface (SPI) and interrupts to communicate with microcontroller. The client invokes LabVIEW program corresponding to the teleintimation garment through internet as a web page. The LabVIEW run time engine displays the front panel with controls and indicators and displays in matrix format as well as in text, the number and the place of the bullet wound. The signals from the teleintimation garment have been given to the data acquisition card using GSM technology. The ethernet controller used in this setup is ENC28J60 from Microchip, Inc. The main advantage is that it is a 28 pin package, which simplifies the design and reduces the overall board space when compared to conventional ethernet controllers containing more than 80 pins. This ethernet controller also employs the industry-standard SPI serial interface, which only requires four lines to interface to the host microcontroller.

5.8.3 Integration of Teleintimation Garment and SSM Software

The microcontroller used for this design is an 8 bit PIC18F-series microcontroller for applications including instrumentation and monitoring, data acquisition, power conditioning, environmental monitoring, telecom and consumer audio/video applications. It provides SPI interface through which the ethernet controller is interfaced. It also eliminates the requirement of a standard PC by its computing power just sufficient for these types of applications, making it a cost-effective solution. The interface contains components like ADCs, DACs, counters, relays, sensors to measure voltage, current, speed and power electronic components to control parameters of the experimental setup. The interface components are not specific and differ as per requirements of individual experimental setup. Transmission control protocol/internet protocol (TCP/IP) stack has to be implemented in the microcontroller to connect it to the local server. The microcontroller has to be programmed simultaneously to transmit and receive data to the local server while it is driving the experimental setup. The microcontroller supports MPLAB compiler tools and HI-TECH PICC C compiler for program development. The microcontroller can be programmed to work in power-down mode when there is no request for the experiments.

5.8.4 Soldier's Status Monitoring Software

The important aspects that should be taken into account by the software in developing SSM software are restricting access to authorized users and handling multiple simultaneous users. To prevent unauthorized access to the facility, user identification has to be implemented with LabVIEW. Multiple users can be easily handled by LabVIEW remote panels. A LabVIEW program can be enabled for remote control through a common web browser without any additional programming. The user interface for the application shows up in the web browser and is fully accessible by the remote user. The acquisition is still occurring on the host computer, but the remote user has total control and identical application functionality. To reduce confusion, only one client can control the application at a time, but the client can pass control easily among the various clients at run-time. At any time during this process, the operator of the host machine can assume control of the application back from the client currently in control. As only data is transmitted back and forth, the bandwidth requirement is minimal. Once connected to the SSM software, the soldier's status will automatically be in a monitoring state.

5.8.5 Server Operation

Visual feedback is an excellent way to monitor the status of the soldier. Visual feedback can be added to a remote laboratory in many different ways. One can choose to embed the live images into the actual VI front panel, place the picture on the web page created by LabVIEW or use a separate web page to display the picture. It consists of option to choose the communication port from the teleintimation garment to the remote end software. Also it consists of matrix format indication of the location of the bullet wound: normally, the matrix coordinates will be in black colour, once the soldier gets wounded by the bullet, the matrix coordinates will change to white colour. Apart from this, SSM indicates the location and number of bullet wounds in text message. The screen shots of the front view of the LabVIEW-based SSM software are shown in Figure 5.21.

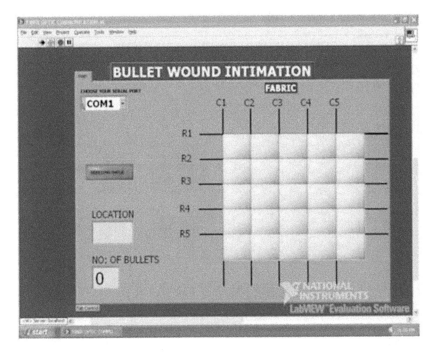

FIGURE 5.21
Front view of bullet wound intimation – LabVIEW-based SSM software.

5.8.6 Location Evaluation Using LabVIEW-Based SSM Software

To locate the number of bullet wounds and places of the bullet wound in the teleintimation garment, IR transmitter and receiver is used for transmitting and receiving the signals. When the signal got interrupted, meaning, physically the POF got because of the bullet penetration, the signals have been sent to the LabVIEW-based SSM through the microcontroller. The testing of teleintimation fabric integrated with LabVIEW-based SSM software kit is shown in Figure 5.22. If any POF is interrupted between the transmitter and receiver end, logic high signal is given to the microcontroller. A variable with

IR transreceiver circuits

Teleintimation fabric

FIGURE 5.22
Testing of teleintimation fabric using LabVIEW-based SSM software kit.

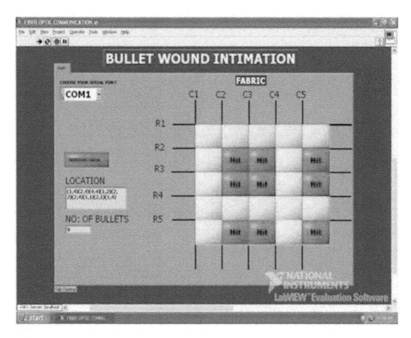

FIGURE 5.23
Bullet wound intimation in matrix and text format – SSM software.

a count increment will be made in the microcontroller, and the counter variable has been incremented if any port gives a logic high signal. This count is taken as the number of bullet penetrations in the teleintimation garment. The POF woven into the teleintimation garment sends and receives the optical signals to the microcontroller circuit, which will throw light on the presence or absence of the bullet in a particular location. Figure 5.23 shows the screen shots of the LabVIEW-based SSM software showing bullet penetration in matrix as well as in text format.

The implementation of LabVIEW-based SSM software is essential, because the huge investments made in replicating valuable resources can be diverted for other purposes. The advantage of having such software is that they provide a network of resources which can be utilized efficiently.

5.9 Communication System between Garment and SSM Software

The communication between the wearable electronic products, smart shirt and the teleintimation garment to the remote end has been processed using GSM technology for transmitting the vital signs and bullet wound intimation signals. The vital signs were measured and transmitted to the remote station by using telemonitoring system. In the telemonitoring system, the measured vital signs, temperature, respiration rate and pulse rate were sent to microcontroller unit and the microcontroller sends these signals to the modem (mobile) through RS232 cable. The signal received from the remote end is connected to the server with the receiver circuitry, which employs RS232 connection method. The remote station SSM can receive these data by using mobile phone and know the status of the wearer.

In the teleintimation garment, the method of addressing a transmitter is to address a location on the user's body, for example 'a wound on the left arm'. GSM module is used to send the information to remote station if the controller finds any break in optical cable. The various communication methods were tested to send the vital signals to the remote end server. The RF transmission technique is limited to short distance communication, whereas the mobile communication technique uses long distance signal transmission irrespective of the distance. From the tests, it was found that the mobile communication technique provides the optimum way of signal transmission. The cut in the optical signal from the POF is transmitted by GSM module. The received signal is sent to the server PC through RS232 cable. The row and column information is received by the hyper-terminal window of the server. The data-receiving methods are programmed using VB 6.0 in such a way that the row and column information was received in a sequential manner.

5.10 GPRS-Based Soldier-Tracking System

The system proposes a low-cost object-tracking system using GPS and GPRS. The current short message service (SMS) technology is costlier when compared to GPRS technology. GPRS also enjoys the advantages of faster and continuous data transmission. This system uses a Telit GM862-GPS module to track the location details like latitude, longitude, speed, altitude and accuracy, in a specific data format known as National Marine Engineering Association (NMEA) protocol. From this location data, these systems extract the details in a format that can be processed using 16F877 PIC microcontroller. The processed information is then sent to GPRS module which employs RS232 logic, whereas PIC-16F877 uses TTL logic. Hence, to interface these two, MAX-232 is employed to convert the logic levels. The system then sends the processed information to a web server containing database created using Microsoft Access, for storing the location details so as to allow the user to view the present and past positions of a target object on Google map through Internet. A web application has been developed using ASP.NET, MYSQL, Google Earth Geographical Information Service to track wounded soldiers and also to track military vehicles.

5.10.1 Schematic Diagram of Soldier-Tracking System

The system consists of two parts: the tracking device and the database server. The schematic of the soldier-tracking system is as shown in Figure 5.24. The tracking device is attached to the solider and it gets the position from GPS satellite in real-time. Then the signals send the position information to the International Mobile Equipment Identity (IMEI) number as its own identity to the server. The data is checked for validity and the valid data is saved into the database. When a user wants to track the soldier, the user can log into the service provider's website and get the live/recorded position of the soldier on the Google Map.

The tracking device is compatible with 850/900/1800/1900 MHz frequencies of cellular networks. This device is capable of working in any GSM network around the world and it consists of 3–5 MB volatile and non-volatile memory in order to store the data received from satellites to be stored. The size of the mobile device is small, since the device operates without any external controller. High sensitive GPS receiver capability and built-in SIM card holder, make the system compact and power efficient. Also the system supports complete standard AT command set plus custom AT command set in order to make the data transfer through the protocol via a GSM network. After turning on the device, the network is automatically initialized. Then it gets the GPS data in NMEA 0183 format and adds it with its own unique IMEI number. It then tries to connect to GPRS. If it fails due to GPRS

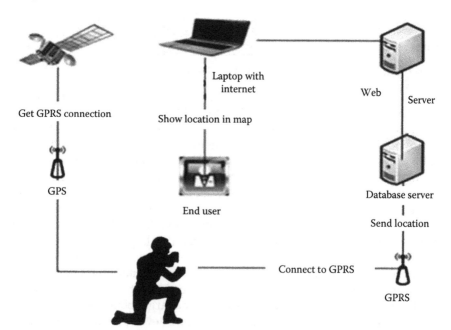

FIGURE 5.24
Schematic of the soldier-tracking system.

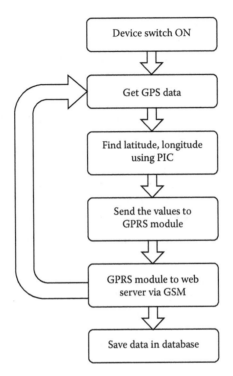

FIGURE 5.25
Flow of operations in the hardware.

unavailability, then it logs the data in the non-volatile memory and waits for a certain fixed period of time. After that, it tries to connect to the GPRS again. After establishing the GPRS connection, it connects to the service provider's server using the HTTP protocol. After successful connection, the GPS data with IMEI number is sent to the server as a string. After a certain time period, it checks the availability of GPRS and connects to the HTTP server and then current location of the device is sent. In this way, the device communicates with the server and sends the location. The purpose of the microcontroller is to process the information received from satellites via NMEA protocol and converts it into a format that can be used by the Google Maps. This is to be sent via GSM network to the web server already created. These operations are shown in the form of a flowchart in Figure 5.25.

5.10.2 Methodology for Soldier-Tracking System

The interfacing of the GPS/GPRS module and the PIC microcontroller is done by using MAX-232 ICs, which converts RS232 logic to the TTL logic and vice versa. This is because PIC microcontroller and almost all digital devices use TTL/CMOS logic, whereas GPS/GPRS module considered here uses RS232 logic. A 2 × 16 LCD display has been used to display latitude and longitude of the device and an indication that the GPS device is receiving the data from satellite and another indication showing the message is sent via GPRS. The SIM card and a 2G/EDGE/3G GSM network with high-speed circuit-switching capability has been used for tracking application. The program is developed in the microcontroller with AT commands to access the required online server and this server can be located anywhere in the world. The .NET server has been used, since it can provide multiple accesses, so that multiple users can access the data and receive the required information and process it further. To view the current position of the soldier, a web-based application has been developed. Using this web application, user can view the live position of the soldier and also the past position by selecting a specific date and time interval. An example of a packet of information sent via NMEA protocol and received using GPS device is shown here.

$ G P G G A : 1 6 0 2 2 3 . 9 9 9 , 2 3 4 5 . 3 5 2 2 N , 0 9 0 2 2 . 0 2 88E,2.4,33.9,3,51.30,1.62,0.87,090408,06.

After accepting the IMEI and NMEA data from the device by the web server, a method is used for tokenizing all the particular data as shown in Figure 5.26. This is done after verifying the IMEI number of the device and the NMEA formatted data have been converted to the decimal format. After converting NMEA formatted data, the decimal latitude 2345.3522N

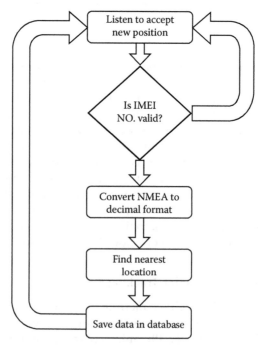

FIGURE 5.26
Operation flow on the server side.

is converted into 23.755895 and the longitude 09022.0288E has been converted into 90.367205. The spherical law of cosines is used to find out the
device's location. This formula is used generally for computing great-circle
distances between two pairs of coordinates on a sphere. Spherical law of
cosines is given in Equation 5.1:

$$d = R * a\cos\left(\cos(\text{lat1}) \cdot \cos(\text{lat2}) \cdot \cos(\text{lng2} - \text{lng1}) + \sin(\text{lat1}) \cdot \sin(\text{lat2})\right) \quad (5.1)$$

Here, d is the distance between two coordinates (lat1, lng2) and (lat2, lng2).
Live tracking of the soldier is the major part of this web application. This
enables the user to view the live position on the map.

Google map satellite version is used to locate the position. After logging
in, user will automatically be redirected to tracking web page; in this page,
Asynchronous JavaScript and XML (AJAX) function is used to track the new
position from the server. This is done at fixed intervals in order to update it
on the map without reloading the whole page repeatedly. Data is retrieved
using the XML Http request object that is available to scripting languages
running in modern browsers, or, alternatively, through the use of Remote
Scripting in browsers that do not support XML Http Request.

PIC-16F877 microcontroller controls all the operations taking place in
the soldier-tracking system. Figure 5.27 shows the different blocks used in
the soldier-tracking system. It is programmed in such a way that it gets the
required data from GPS/GPRS module and then extracts required information like latitude and longitude and is then given back to the GPS/GPRS

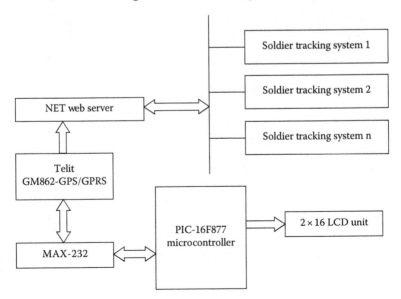

FIGURE 5.27
Block diagram of the soldier-tracking system.

module; then this GPS/GPRS module sends the extracted information to the web server in which these information are stored in a database. From this web server, end-user can easily view the location details of the selected soldier and also the required locations can be mapped on to the Google maps. The hardware unit of the system is as shown in Figure 5.28.

The PIC-16F877 microcontroller is interfaced to 16 pin LCD connector which is then connected to an LCD display for displaying the current location of the unit containing the GPS module. Initially, after GPS location data is received by the Telit GM862-GPS module, information like latitude, longitude, speed and altitude is extracted using PIC-16F877. Pin number 26 and pin number 27 of PIC-16F877, which are RX and TX, are used for the purpose of receiving required information from Telit GM862-GPS module. This information is the data packet of GPGGA string of the NMEA protocol used for GPS communication, received by the GPS part of the module. This is sent via TX pin to the PIC-16F877, which is programmed to extract information such as latitude and longitude. This data is sent to the GPRS part of the Telit GM862-GPS module which is then to be sent to the web server for storing the data. This interfacing of PIC microcontroller and LCD displays these extracted values of latitude and longitude of the current position of the soldier.

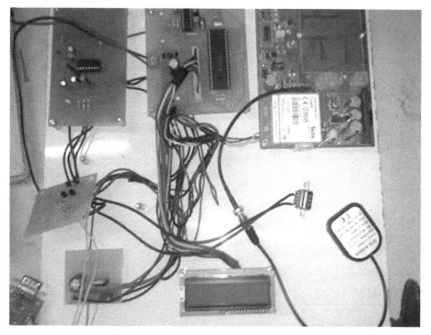

FIGURE 5.28
GPS/GPRS-based soldier-tracking system.

5.10.3 Hyper-Terminal Testing of AT Commands

The AT commands which are to be provided by the PIC are manually provided using the hyper-terminal in a PC for testing purposes. The following shows the experimental results of AT commands tested using hyper-terminal. In this research work, GPS-based service developed and provided by United States Department of Defense is used for the tracking purposes. It provides accurate, three-dimensional information of the location as well as precision velocities and timing services. A web application for plotting the location of the soldier has been developed using .NET, Visual Studio, MYSQL and Google Earth. Google Earth has Google maps embedded within it and it uses a .kml file for location mapping. The application gets the values of latitude and longitude from the database created already in the web server and then creates a corresponding .kml file with the obtained latitude and longitude values. This .kml file is then used by Google Earth, and the location is plotted. The web application created includes a first page, by the name default. aspx, which provides features for the users to login, as shown in Figure 5.29. There is a 'login' button, which when clicked checks the authenticity of the username and password. If the username and/or password does not match with those values in the database, it displays an error message. If the values

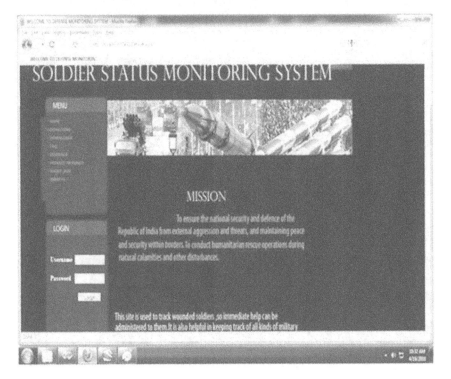

FIGURE 5.29
Front page of soldier-monitoring system.

of username and password match exactly with those of the values in the database, then the user will be redirected to the main page.

The main page has a drop-down list, a 'go' button and a grid display as the main controls. The soldier whose position has to be located can be selected from the drop-down list. The 'go' button when clicked displays the details of the location of the selected soldier. This is done by using SQL commands for filtering the information about the selected soldier from the database. There is a 'load' button which when clicked loads the entire database, showing the location details of all the soldiers. The preferred location can now be plotted on Google maps using the 'plot' button. An example of plotting of the location of soldiers in Google Maps is shown in Figures 5.30 and 5.31. Also, the values of the latitude and longitude of the current location are displayed in the LCD unit, as shown in Figure 5.32; similarly, the latitude and longitude of the current location will be displayed in the webpage, as shown in Figure 5.33.

In the current scenario, safety measures to the soldiers are one of the important factors. In addition to the protection of the soldiers during combat

FIGURE 5.30
Location of soldiers in Google maps.

FIGURE 5.31
Location of soldiers in Google maps.

FIGURE 5.32
Longitude and latitude of the current location.

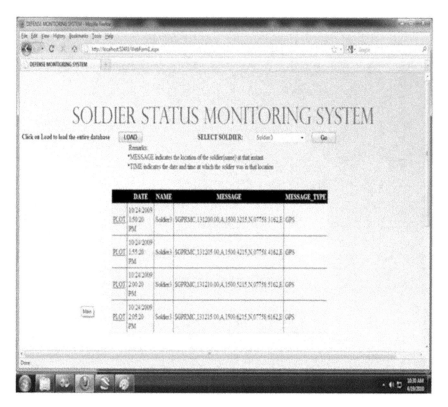

FIGURE 5.33
Tracked details in the web page of SSM system.

situation, monitoring them using GSM and gathering their health status and bullet wound information is important. The developed teleintimation garment gives the bullet wound and location of a soldier at the remote end server from the SSM system. The signal from the garment is transmitted to the remote end with RF and GSM communication technique. SSM software gives the physiological details and IAU status of the soldier. Apart from this, a low-cost tracking system using GPS/GPRS of a GSM network, suitable for a wide range of applications like soldier tracking in defence fields, has been developed. The combination of the GPS and GPRS provides continuous and real-time tracking. The cost is much lower compared to SMS-based tracking systems currently available in the market. It is expected that the full implementation of the proposed system would ultimately replace the traditional and costly SMS-based tracking systems. Also, having a single module for both GPS and GPRS purposes and using a PIC microcontroller instead of an external controller has made the overall size of the unit smaller and hence makes it compact.

5.11 Future Research Scope

Research is required in the field of wireless EEG systems to tackle the issues like robustness to motion artefacts to increase autonomy so that the system could be used anywhere under any condition. The electrodes used in headset make contact with the skin, hence reducing the wear ability and comfort. So, wearability must be improved by developing the electrodes which do not require contact with the skin. Further research may be carried out to reduce the power consumption of the system to improve its overall performance. Due to limitation of available technology, the height of MEMS-based dry electrodes is not sufficient to penetrate human hairs to contact even epidermis. Much of work is required to fabricate longer electrodes. Rapid growth of wireless infrastructure in following years will allow a range of new medical applications that will significantly improve the quality of health care. Wider acceptance of physiological monitoring hardware will allow development of devices based on natural human–computer interfaces. MEMS made possible the development of networks of intelligent wireless sensors for military and space applications through the increase of processing power, miniaturization, wireless communication and decreased power consumption. Defense Advanced Research Projects Agency (DARPA) and Army Research Laboratory (ARL), with their key partners – UCLA Electrical Engineering Department and Rockwell Science Center, are developing wireless integrated network sensors (WINS). Department of Commerce, through National Institute for Standards, sponsors Smart Spaces. This is NIST's approach to pervasive computing that is impossible without wireless sensors. DOE, and its Office of Industrial Technology, sponsors Oak Ridge National Laboratory to work on the project named 'Intelligent Wireless Sensors for Industrial Manufacturing'. Among the most important projects are Smart Dust (Berkeley), Sensor Web (JPL), SCADDS (UCS/ISI) and other projects supported by DARPA's Sensor Information Technology programme (SensIT). The same technology may be used for an intelligent monitor that is able to detect or predict emergency health medical situations. The most critical features of a wearable health monitor are long battery life, lightweight and small dimensions.

There are many companies selling patient-monitoring systems, like *Agilent Technologies, Protocol Systems*, etc. Their systems often include wireless connection between the portable monitor and the CS, processing of transferred data, issuing of warnings, archiving, etc. These and other commercial and experimental medical systems currently use only wireless data acquisition devices and wireless data presentation devices, such as palmtop PCs, pagers and cellular phones. A wearable device monitoring multiple physiological signals (polysomnograph) usually includes multiple wires connecting sensors and the monitoring device. Our goal was to go one step further and to bring in the wireless technology to the whole signal path, including the wireless connection to the sensors. Moreover, we are introducing intelligent sensors that will take part both in signal processing organized in hierarchical network, and in transmission of data. In this way, we are able to provide hassle-free environment for patients and other users, without excessive wires that limit movement and attract undesired attention during continuous monitoring. The whole system is integrated into hierarchically organized wireless personal area network (PAN). This type of organization is inevitable for promotion of implantable sensors, which are finding their place in the treatment of diabetics and other similar patients.

Recent development of low-power DSP technology allows implementation of intelligent personal monitoring devices. With further development of bio-sensor technology, personal health monitors will become standard part of personal mobile devices. We propose the implementation of hierarchical monitoring in a telemedical environment with power-efficient signal-processing algorithms and adaptive system configuration and operation. Limited capabilities of different components of our system dictate hierarchical organization. Proposed system organization allows flexible design space for optimum trade-off between processing power, device power consumption and battery life and storage capacity, sufficient for most medical applications. A new generation of wearable and contactless electrodes for brain potential measurement was developed based on an innovative fabrication technique. Sensor design presents the uniqueness of requiring no electrical contact between their surface and the subject skin, avoiding time-consuming and uncomfortable attachment procedures. This contactless sensor is a fibre-based device, whose basis of operation relies on the electro-active principle. The use of an electroactive hydrogel as a candidate for the sensing/modulator component – PAAM. This polymeric material showed a 1 cm deformation, when submitted to an external voltage of 2 V. Flexible brain cap is an attractive solution for BCI applications, since it overcomes some of the limitations of the existent brain cap systems, mainly the wearable issue, as well as the disadvantages regarding the dry or wet electrodes.

5.12 Summary

In this chapter, the software developed at the user end to monitor the wearer is discussed conversed besides the various communication methods used to interface the wearable electronic products to the remote end are also discussed. Whenever a soldier is wounded by a bullet, the condition of the vital signs from the smart shirt will be transmitted to the SSM software through the telemonitoring system by GSM technology. In the telemonitoring system, the measured vital signs, temperature, respiration rate and pulse rate are sent to microcontroller unit and the microcontroller sends these signals to the modem (mobile) through RS232 cable. The signal received from the remote end is connected to the server with the receiver circuitry, which employs RS232 connection method. The remote station SSM can receive these data by using a mobile phone and know the wearer's status. In the teleintimation garment, GSM module is used to send the information to the remote station if the controller finds any break in the optical cable. The various communication methods were tested to send the vital signals to the remote end server. The RF transmission technique is limited to short distance communication, whereas the mobile communication technique uses long distance signal transmission irrespective of the distance. From the tests, it was found that the mobile communication technique provides the optimum way of signal transmission. The cut in the optical signal from the POF is transmitted by GSM module. The received signal is sent to the server PC through RS232 cable. The row and column information is received in the hyper-terminal window of the server. The bullet wound in the teleintimation garment, in the corresponding location, is highlighted in the software developed by using Visual Basic and LabVIEW. The data-receiving methods are programmed using VB 6.0 and LabVIEW, in such a way that the row and column information was received in a sequential manner. From this information, combat casualty and IAU status can be predicted from the SSM software.

A low-cost tracking system using GPS/GPRS of a GSM network, suitable for a wide range of applications like soldier tracking in defence fields, has been developed. The combination of the GPS and GPRS provides continuous and real-time tracking. The tracking system consists of a single module for both GPS and GPRS purposes and using a PIC microcontroller instead of an external controller has made the overall size of the unit smaller and hence compact. In the current scenario, providing safety measures to the soldiers is one of the important factors. In addition to the protection of the soldiers during combat situation, monitoring them using GSM and gathering their health status and bullet wound information is important. The developed SSM software gives the information on bullet wound and location of a soldier at the remote end server from the SSM system.

Bibliography

Afonso, V.X., W.J. Tompkins, T.Q. Nguyen and S. Luo, ECG beat detection using filter banks, *IEEE Trans. Biomed. Eng.*, 46, 192–202 (1999).

Agre, J. and L. Clare, An integrated architecture for cooperative sensing networks, *IEEE Comput.*, 33, 106–108 (2000).

Ajiboye, A.B., J.D. Simeral, J.P. Donoghue, L.R. Hochberg and R.F. Kirsch, Prediction of imagined single-joint movements in a person with high-level tetraplegia, *IEEE Trans. Biomed. Eng.*, 59(10), 2755–2765 (2012).

Akin, T. and K. Najafi, A telemetrically powered and controlled implantable neural recording circuit with CMOS interface circuitry, In *Proceedings of the IEEE Seventh Mediterranean Electrotechnical Conference*, Antalya, Turkey, pp. 545–548 (1994).

Amft, O., M. Kusserow and G. Troster, Bite weight prediction from acoustic recognition of chewing, *IEEE Trans. Biomed. Eng.*, 56, 1663–1672 (2008).

Amft, O. and G. Troster, Recognition of dietary activity events using on-body sensors, *Artif. Intell. Med.*, 42, 121–136 (2008).

Anliker, U., J.A. Ward, P. Lukowicz, G. Troster, F. Dolveck, M. Baer, F. Keita et al., AMON: A wearable multiparameter medical monitoring and alert system, *IEEE Trans. Inform. Technol. Biomed.*, 8(4), 415–427 (2004).

Atallah, L., J. Zhang, B.P.L. Lo, D. Shrikrishna, J.L. Kelly, A. Jackson, M.I. Polkey, G.Z. Yang and N.S. Hopkinson, Validation of an ear worn sensor for activity monitoring in COPD, *Am. J. Respir. Crit. Care Med.*, 181, 1211–1217 (2010).

Axelrod, D.A., D. Millman and D.D. Abecassis, US Health Care Reform and Transplantation. Part I: Overview and impact on access and reimbursement in the private sector, *Am. J. Transplant.*, 10, 2197–2202 (2010).

Aziz, O., L. Atallah, B. Lo, M. Elhelw, L. Wang, G.Z. Yang and A. Darzi, A pervasive body sensor network for measuring postoperative recovery at home, *Surg. Innov.*, 14, 83–90 (2007).

Baba, A. and M.J. Burke, Measurement of the electrical properties of ungelled ECG electrodes, *Int. J. Biol. Biomed. Eng.*, 2(3), 89–97 (2008).

Babak, A., S. Taheri, T. Robert, T. Knight and L. Rosemary, A dry electrode for EEG recording, *Electroencephal. Clin. Neurophysiol.*, 90(5), 376–383 (1994).

Bachlin, M., M. Plotnik, D. Roggen, I. Maidan, J.M. Hausdorff, N. Giladi and G. Troster, Wearable assistant for Parkinson's disease patients with the freezing of gait symptom, *IEEE Trans. Inform. Technol. Biomed.*, 14, 436–446 (2009).

Baek, J.-Y., J.-H. An, J.-M. Choi, K.-S. Park, and S.-H. Lee, Flexible polymeric dry — electrodes for the long-term monitoring of ECG, *Sens. Actuat. A: Phys.*, 143(2), 423–429 (2008).

Bahl, P. and V. Padmanabhan, RADAR: An in-building RF based user location and tracking system, In *Proceedings of the IEEE Infocom*, Tel Aviv, Israel, pp. 775–784 (2000).

Bai, J., A portable ECG and blood pressure telemonitoring system, *IEEE Eng. Med. Biol.*, 18, 63–70 (1999).

Barro, S., Intelligent telemonitoring of critical care patients, *IEEE Eng. Med. Biol.*, 18, 80–88 (1999).

Bashashati, A., M. Fatourechi, R.K. Ward and G.E. Birch, A survey of signal processing algorithms in brain-computer interfaces based on electrical brain signals, *J. Neural Eng.*, 4, 32–57 (2007).

Benedetti, M.G., A. Di Gioia, L. Conti, L. Berti, L.D. Esposti, G. Tarrini, N. Melchionda and S. Giannini, Physical activity monitoring in obese people in the real life environment, *J. Neuroeng. Rehabil.*, 6, 47 (2009).

Berman, D.S., A. Rozanski and S.B. Knoebel, The detection of silent ischemia: Cautions and precautions, *Circulation*, 75(1), 101–105 (1987).

Bianchi, F., S.J. Redmond, M.R. Narayanan, S. Cerutti and N.H. Lovell, Barometric pressure and triaxial accelerometry-based falls event detection, *IEEE Trans. Neural Syst. Rehabil. Eng.*, 18, 619–627 (2010).

Bonato, P., Wearable sensors/systems and their impact on biomedical engineering, *IEEE Eng. Med. Biol. Mag.*, 22, 18–20 (2003).

Bonato, P., Wearable sensors and systems. From enabling technology to clinical applications, *IEEE Eng. Med. Biol. Mag.*, 29, 25–36 (2010).

Bourland, J.D., L.A. Geddes, G. Sewell, R. Baker and J. Kruer, Active cables for use with dry electrodes for electrocardiography, *J. Electro Cardiol.*, 11(1), 71–74 (1978).

Bowman, M.C., R.S. Johansson and J.R. Flanagan, Eye-hand coordination in a sequential target contact task, *Exp. Brain Res.*, 195, 273–283 (2009).

Bradberry, T.J., R.J. Gentili and J.L. Contreras-Vidal, Reconstructing three-dimensional hand movements from noninvasive electroencephalographic signals, *J. Neurosci.*, 30, 3432–3437 (2010).

Caudill, T.S., R. Lofgren, C.D. Jennings and M. Karpf, Commentary: Health care reform and primary care: Training physicians for tomorrow's challenges, *Acad. Med.*, 86, 158–160 (2011).

Chan, L., L.G. Hart and D.C. Goodman, Geographic access to health care for rural medicare beneficiaries, *J. Rural Health*, 22, 140–146 (2006).

Chi, Y.M., T.-P. Jung and G. Cauwenberghs, Dry-contact and noncontact biopotential electrodes: Methodological review, *IEEE Rev. Biomed. Eng.*, 3, 106–119 (2010).

Choon Po, S.N., G. Dagang, M.D.B.M. Hapipi, N.M. Naing, W.J. Shen and A. Ongkodjojo, Overview of MEMS wear II-incorporating MEMS technology into smart shirt for geriatric care, *J. Phys. Conf. Ser.*, 34, 1079–1085 (2006).

Clippingdale, A.J., R.J. Clark and C. Watkins, Ultrahigh impedance capacitive coupled heart imaging array, *Rev. Sci. Instrum.*, 65(1), 269–270 (1994).

Cole, C.R., E.H. Blackstone, F.J. Pashkow, C.E. Snader and M.S. Lauer, Heart-rate recovery immediately after exercise as a predictor of mortality, *New Engl. J. Med.*, 341(18), 1351–1357 (1999).

Corbett, E.A., E.J. Perreault and K.P. Kording, Decoding with limited neural data: A mixture of time-warped trajectory models for directional reaches, *J. Neural Eng.*, 9, 1–10 (2012).

Dario, P., R. Bardelli, D. De Rossi, L.R. Wang and P.C. Pinotti, Touch-sensitive polymer skin uses piezoelectric properties to recognize orientation of objects, *Sensor Rev.*, 2, 194–198 (1982).

Delin, K.A. and S.P. Jackson, Sensor web for in situ exploration of gaseous biosignatures, In *Proceedings of the 2000 IEEE Aerospace Conference*, Big Sky, MT, pp. 465–472 (March 1999).

Dipietro, L., A.M. Sabatini and P. Dario, Artificial neural network model of the mapping between electromyographic activation and trajectory patterns in free-arm movements, *Med. Biol. Eng. Comput.*, 41, 124–132 (2003).

Duchamp, D., K.F. Steven and Q.M. Gerald, Software technology for wireless mobile computing, *IEEE Network Mag.*, 12(18), 218–225 (1991).

Dunne, L.E., S. Brady, B. Smyth and D. Diamond, Initial development and testing of a novel foam-based pressure sensor for wearable sensing, *J. Neuro Eng. Rehabil.*, 2(4), 158–164 (2005).

Eggins, B., Skin-electrode impedance and its effect on recording cardiac potentials, *Analyst*, 118, 439–442 (1993).

Engelse, W.A.H and C. Zeelenberg, A single scan algorithm for QRS-detection and feature extraction, *Comput. Cardiol.*, 6, 37–42 (1979).

Farfan, F.D., J.C. Politti and C.J. Felice, Evaluation of EMG processing techniques using information theory, *Biomed. Eng. Online*, 9(72), 1–18 (2010).

Fleischer, C. and G. Hommel, A human-exoskeleton interface utilizing electromyography, *IEEE Trans. Robot.*, 24(4), 872–882 (2008).

Freiherr, G, Wireless technologies find niche in patient care, *Medical Device & Diagnostic Industry*, 8, 83–93 (1998).

Friesen, G., T. Jannett, M. Jadallan, S. Yates, S. Quint and H. Nagle, A comparison of the noise sensitivity of nine QRS detection algorithms, *IEEE Trans. Biomed. Eng.*, 37, 85–98 (1990).

Fukaya, K. and M. Uchida, Protection against impact with the ground using wearable airbags, *Ind. Health*, 46, 59–65 (2008).

Gallivan, J.P., C.S. Chapman, D.K. Wood, J.L. Milne, D. Ansari, J.C. Culham and M.A. Goodale, One to four, and nothing more: Nonconscious parallel individuation of objects during action planning, *Psychol. Sci.*, 22, 803–811 (2011).

Gebrial, W., R.J. Prance, C.J. Harland and T.D. Clark, Noninvasive imaging using an array of electric potential sensors, *Rev. Sci. Instrum.*, 77(6), 708–712 (2006).

Geddes, L. and M. Valentinuzzi, Temporal changes in electrode impedance while recording the electrocardiogram with "dry" electrodes, *Ann. Biomed. Eng.*, 1, 356–367 (1973).

Giansanti, D., G. Maccioni and S. Morelli, An experience of health technology assessment in new models of care for subjects with Parkinson's disease by means of a new wearable device, *Telemed. J. e-Health*, 14, 467–472 (2008a).

Giansanti, D., G. Ricci and G. Maccioni, Toward the design of a wearable system for the remote monitoring of epileptic crisis, *Telemed. J. e-Health*, 14, 1130–1135 (2008b).

Giorgino, T., P. Tormene, F. Lorussi, D. De Rossi and S. Quaglini, Sensor evaluation for wearable strain gauges in neurological rehabilitation, *IEEE Trans. Neural Syst. Rehabil. Eng.*, 17, 409–415 (2009).

Goonewardene, S.S., K. Baloch and I. Sargeant, Road traffic collisions-case fatality rate, crash injury rate, and number of motor vehicles: Time trends between a developed and developing country, *Am. Surg.*, 76, 977–981 (2010).

Gorlick, M., Electric suspenders: A fabric power bus and data network for wearable digital devices, In *Proceedings of the Third International Symposium on Wearable Computers*, San Francisco, CA, pp. 114–121 (1999).

Griffith, M.E., W.M. Portnoy, L.J. Stotts and J.L. Day, Improved capacitive electrocardiogram electrodes for burn applications, *Med. Biol. Eng. Comput.*, 17(5), 641–646 (1979).

Grozea, C., C.D. Voinescu and S. Fazli, Bristle-sensors low-cost flexible passive dry EEG electrodes for neurofeedback and BCI applications, *J. Neural Eng.*, 8, 2 (2011).

Grundlehner, B., J. Penders, M. Op de Beeck and F. Yazicioglu, EEG monitoring at home: The technical challenges, *J. Med. Biol. Eng.*, 7, 519–528 (2013).

Guidali, M., A. Duschau-Wicke, S. Broggi, V. Klamroth-Marganska, T. Nef and R. Riener, Arobotic system to train activities of daily living in a virtual environment, *Med. Biol. Eng. Comput.*, 49, 1213–1223 (2011).

Gulley, S.P., E.K. Rasch and L. Chan, Ongoing coverage for ongoing care: Access, utilization, and out-of-pocket spending among uninsured working-aged adults with chronic health care needs, *Am. J. Public Health*, 101, 368–375 (2011).

Habetha, J., The MyHeart project–fighting cardiovascular diseases by prevention and early diagnosis, In *Conference Proceedings of the IEEE Engineering in Medicine and Biology Society*, New York, pp. 6746–6749 (2006).

Hammon, P.S., S. Makeig, H. Poizner, E. Todorov and V.R.D. Sa, Predicting reaching targets from human EEG, *IEEE Signal Process. Mag.*, 69, 69–77 (2008).

Harland, C.J., T.D. Clark and R.J. Prance, Electric potential probes – New directions in the remote sensing of the human body, *Meas. Sci. Technol.*, 13(2), 163–170 (2002).

Hassibi, A., R. Navid, R.W. Dutton and T.H. Lee, Comprehensive study of noise processes in electrode electrolyte interfaces, *J. Appl. Phys.*, 96(2), 1074–1082 (2004).

He, P., G. Wilson and C. Russell, Removal of ocular artifacts from electroencephalogram by adaptive filtering, *Med. Biol. Eng. Comput.*, 42, 407–412 (2004).

Heidenreich, P., C. Ruggerio and B. Massie, Effect of a home monitoring system on hospitalization and resource use for patients with heart failure, *Am. Heart J.*, 138(4), 633–640 (1999).

Hill, J. and D. Culler, Mica: A wireless platform for deeply embedded networks, *IEEE Micro*, 22(6), 12–24 (2002).

Huang, H., F. Zhang, L.J. Hargrove, Z. Dou, D.R. Rogers and K.B. Englehart, Continuous locomotion-mode identification for prosthetic legs based on neuro-muscular-mechanical fusion, *IEEE Trans. Biomed. Eng.*, 58(10), 2867–2875 (2011).

Huigen, E., A. Peper and C.A. Grimbergen, Investigation into the origin of the noise of surface electrodes, *Med. Biol. Eng. Comput.*, 40(3), 332–338 (2002).

Jayaraman, S., Designing a textile curriculum for the '90s: A rewarding challenge, *J. Text. Inst.*, 81(2), 185–194 (1990).

Johansson, R.S., G. Westling, A. Backstrom and J.R. Flanagan, Eye hand coordination in object manipulation, *J. Neurosci.*, 21, 6917–6932 (2001).

Jovanov, E., P. Gelabert, B. Wheelock, R. Adhami and P. Smith, Real time portable heart monitoring using low power DSP, In *International Conference on Signal Processing Applications and Technology (ICSPAT)*, Dallas, TX, pp. 19–24 (2000a).

Jovanov, E., A. Milenkovic, C. Otto and P.C. de Groen, A wireless body area network of intelligent motion sensors for computer assisted physical rehabilitation, *J. Neuro Eng. Rehabil.*, 2(6), 45–67 (2005).

Jovanov, E., J. Price, D. Raskovic, K. Kavi, T. Martin and R. Adhami, Wireless personal area networks in telemedical environment, In *Proceedings of the 2000 IEEE EMBS International Conference on Information Technology Applications in Biomedicine*, Arlington, VA, pp. 8–10 (2000b).

Jovanov, E., D. Raskovic, J. Price, J. Chapman, A. Moore and A. Krishnamurthy, Patient monitoring using personal area networks of wireless intelligent sensors, *Biomed. Sci. Instrum.*, 37, 373–378 (2001).

Jovanov, E., D. Starcevic, V. Radivojevic, A. Samardzic and V. Simeunovic, Perceptualization of biomedical data, *IEEE Eng. Med. Biol. Mag.*, 18(1), 50–55 (1999a).

Jovanov, E., D. Starcevic, A. Samardzic, A. Marsh and Z. Obrenovic, EEG analysis in a telemedical virtual world, *Future Generat. Comput. Syst.*, 15, 255–263 (1999b).

Kańtoch, E., J. Jaworek and P. Augustyniak, Design of a wearable sensor network for home monitoring system, In *Proceedings of the Federated Conference on Computer Science and Information Systems*, Poland, ISBN 978-83-60810-22-4, pp. 401–403 (2011).

Kohler, B.U., C. Hennig and R. Orglmeister, The principles of software QRS detection, *IEEE Eng. Med. Biol. Mag.*, 21(1), 42–57 (2002).

Krajbich, I., C. Armel and A. Rangel, Visual fixations and the computation and comparison of value in simple choice, *Nat. Neurosci.*, 13, 1292–1298 (2010).

Lee, J.B. and V. Subramanian, Organic transistors on fiber: A first step toward electronic textiles, In *IEDM Technical Digest*, Washington, DC, pp. 199–202 (2003).

Lee, J.M., F. Pearce, C. Morrissette, A.D. Hibbs and R. Matthews, Evaluating a capacitively coupled, noncontact electrode for ECG monitoring, *Sens. Mag.*, 12(4), 85–90 (2005).

Lin, C.T., Y.C. Chen, T.Y. Huang, T.T. Chiu, L.W. Ko, S.F. Liang, H.Y. Hsieh, S.H. Hsu and J.R. Duann, Development of wireless brain computer interface with embedded multitask scheduling and its application on real-time driver's drowsiness detection and warning, *IEEE Trans. Biomed. Eng.*, 55(5), 1582–1591 (2008a).

Lin, C.T., L.W. Ko, J.C. Chiou, J.R. Duann, R.S. Huang, T.W. Chiu, S.F. Liang and T.P. Jung, Noninvasive neural prostheses using mobile & wireless EEG, *Proc. IEEE*, 96(7), 1167–1183 (2008b).

Lin, C.T., R.C. Wu, T.P. Jung, S.F. Liang and T.Y. Huang, Estimating driving performance based on EEG spectrum analysis, *J. Appl. Signal Process.*, 19, 3165–3174 (2005).

Logar, V., A. Belic, B. Koritnik, S. Brezan, J. Zidar, R. Karba and D. Matko, Using ANNs to predict a subject's response based on EEG traces, *Neural Networks*, 21, 881–887 (2008).

Lotte, F., M. Congedo, A. Lécuyer, F. Lamarche and B. Arnaldi, A review of classification algorithms for EEG-based brain-computer interfaces, *J. Neural Eng.*, 4, R1–R13 (2007).

Luu, S. and T. Chau, Decoding subjective preference from single-trial near-infrared spectroscopy signals, *J. Neural. Eng.*, 6, 16–23 (2009).

Makikallio, T.H., H.V. Huikuri, A. Makikallio, L.B. Sourander, R.D. Mitrani, A. Castellanos and R.J. Myerburg, Prediction of sudden cardiac death by fractal analysis of heart rate variability in elderly subjects, *J. Am. College Cardiol.*, 37(5), 1395–1402 (2001).

Mann, S., Wearable computing: A first step toward personal imaging, *Computer*, 30(2), 25–32 (1997).

Manson, A.J., P. Brown, J.D. O'Sullivan, P. Asselman, D. Buckwell and A.J. Lees, An ambulatory dyskinesia monitor, *J. Neurol. Neurosurg. Psychiatry*, 68, 196–201 (2000).

Marchal-Crespo, L., R. Zimmermann, O. Lambercy, J. Edelmann, M.-C. Fluet, M. Wolf, R. Gassert and R. Riener, Motor execution detection based on autonomic nervous system responses, *Physiol. Meas.*, 34, 35–51 (2013).

Martin, T., E. Jovanov and D. Raskovic, Issues in wearable computing for medical monitoring applications: A case study of a wearable ECG monitoring device, In *International Symposium on Wearable Computers ISWC*, Atlanta, GA, pp. 43–50 (2000).

Martinez-Mendez, R., M. Sekine and T. Tamura, Detection of anticipatory postural adjustments prior to gait initiation using inertial wearable sensors, *J. Neuroeng. Rehabil.*, 8, 17 (2011).

Matthews, R., N. McDonald, P. Hervieux, P. Turner and M. Steindorf, A wearable physiological sensor suite for unobtrusive monitoring of physiological and cognitive state, *Proc. IEEE Annu. Int. Conf. Engineering Medicine Biology Soc.*, 17(3), 5276–5281 (2007).

Merilahti, J., J. Parkka, K. Antila, P. Paavilainen, E. Mattila, E.J. Malm, A. Saarinen and I. Korhonen, Compliance and technical feasibility of long-term health monitoring with wearable and ambient technologies, *J. Telemed. Telecare*, 15, 302–309 (2009).

Metting Van Rijn, A., A. Peper and C. Grimbergen, High-quality recording of bioelectric events, *Med. Biol. Eng. Comput.*, 28, 389–397 (1990).

Morse, W., Medical electronics, *IEEE Spectrum*, 34(1), 99–102 (1997).

Mundt, C.W., K.N. Montgomery, U.E. Udoh, V.N. Barker, G.C. Thonier, A.M. Tellier, R.D. Ricks, R.B. Darling, Y.D. Cagle and N.A. Cabrol, Multi parameter wearable physiologic monitoring system for space and terrestrial applications, *IEEE Trans. Inform. Technol. Biomed.*, 9, 382–391 (2005).

Naranjo, J.R., A. Brovelli, R. Longo, R. Budai, R. Kristeva and P.P. Battaglini, EEG dynamics of the frontoparietal network during reaching preparation in humans, *NeuroImage*, 34(4), 1673–1682 (2007).

Naranjo-Hernandez, D., L.M. Roa, J. Reina-Tosina and M.A. Estudillo-Valderrama, SoM: A smart sensor for human activity monitoring and assisted healthy ageing, *IEEE Trans. Biomed. Eng.*, 59(11), 3177–3184 (2012).

Novak, D., M. Mihelj and M. Munih, A survey of methods for data fusion and system adaptation using autonomic nervous system responses in physiological computing, *Interact. Comput.*, 24, 154–172 (2012).

Novak, D., X. Omlin, R. Leins-Hess and R. Riener, Predicting targets of human reaching motions using different sensing technologies, *IEEE Trans. Biomed. Eng.*, 60(9), 2645–2652 (2013).

Nyan, M.N., F.E. Tay and E. Murugasu, A wearable system for pre-impact fall detection, *J. Biomech.*, 41, 3475–3481 (2008).

Oehler, M., V. Ling, K. Melhorn and M. Schilling, A multichannel portable ECG system with capacitive sensors, *Physiol. Meas.*, 29(7), 783–789 (2008).

Otto, C., A. Milenkovic, C. Sanders and E. Jovanov, Final results from a pilot study with an implantable loop recorder to determine the etiology of syncope in patients with negative noninvasive and invasive testing, *Am. J. Cardiol.*, 82, 117–119 (1998).

Pan, J. and Tompkins, W.J., A real-time QRS detection algorithm, *IEEE Trans. Biomed. Eng.*, 32(3), 230–236 (1985).

Paradiso, R., G. Loriga and N. Taccini, Wearable system for vital signs monitoring, *Stud. Health Technol. Inform.*, 108, 253–259 (2004).

Parkka, J., M. Ermes, P. Korpipaa, J. Mantyjarvi, J. Peltola and I. Korhonen, Activity classification using realistic data from wearable sensors, *IEEE Trans. Inform. Technol. Biomed.*, 10(1), 119–128 (2006).

Pfurtscheller, G., G. Muller-Putz, J. Pfurtscheller and R. Rupp, EEG based asynchronous BCI controls functional electrical stimulation in a tetraplegic patient, *EURASIP J. Appl. Signal Process.*, 19, 3152–3155 (2005).

Portet, F., A.I. Hernandez and G. Carrault, Evaluation of real-time QRS detection algorithms in variable contexts, *Med. Biol. Eng. Comput.*, 43(3), 379–385 (2005).

Prance, R.J., S.T. Beardsmore-Rust, P. Watson, C.J. Harland and H. Prance, Remote detection of human electrophysiological signals using electric potential sensors, *Appl. Phys. Lett.*, 93(3), 33–40 (2008).

Prance, R.J., A. Debray, T.D. Clark, H. Prance, M. Nock, C.J. Harland and A.J. Clipping dale, An ultra-low-noise electrical-potential probe for human-body scanning, *Meas. Sci. Technol.*, 11(3), 291–297 (2000).

Riley, R., S. Thakkar, J. Czarnaski and B. Schott, Power-aware acoustic beam forming, In *Proceedings of the Military Sensing Symposium Specialty Group on Battlefield Acoustic and Seismic Sensing, Magnetic and Electric Field Sensors*, New York (2002).

Rizzo, A., P. Requejo, C.J. Winstein, B. Lange, G. Ragusa, A. Merians, J. Patton, P. Banerjee and M. Aisen, Virtual reality applications for addressing the needs of those aging with disability, *Stud. Health Technol. Inform.*, 163, 510–516 (2011).

Salarian, A., F.B. Horak, C. Zampieri, P. Carlson-Kuhta, J.G. Nutt and K. Aminian, A sensitive and reliable measure of mobility, *IEEE Trans. Neural Syst. Rehabil. Eng.*, 18, 303–310 (2010).

Sazonov, E.S., S.A. Schuckers, P. Lopez-Meyer, O. Makeyev, E.L. Melanson, M.R. Neuman and J.O. Hill, Toward objective monitoring of digestive behavior in free-living population, *Obesity (Silver Spring)*, 17, 1971–1975 (2009).

Schmorrow, D.D., Why virtual?, In: J. Cohn & A. Bolton (Eds.), *Special issue on Optimizing Virtual Training systems in Theoretical issues of Ergonomics Science*, CRC Press, Boca Raton, FL, Vol. 3, pp. 257–258 (2009).

Sciacqua, A., M. Valentini, A. Gualtieri, F. Perticone, A. Faini, G. Zacharioudakis, I. Karatzanis, F. Chiarugi, C. Assimakopoulou and P. Meriggi, Validation of a flexible and innovative platform for the home monitoring of heart failure patients: Preliminary results, *Comput. Cardiol.*, 36, 97–100 (2009).

Searle, A. and L. Kirkup, A direct comparison of wet, dry and insulating bioelectric recording electrodes, *Physiol. Meas.*, 21(2), 271–276 (2000).

Sherrill, D.M., M.L. Moy, J.J. Reilly and P. Bonato, Using hierarchical clustering methods to classify motor activities of COPD patients from wearable sensor data, *J. Neuroeng. Rehabil.*, 2, 16–21 (2005).

Shi, G., C.S. Chan, W.J. Li, K.S. Leung, Y. Zou and Y. Jin, Mobile human airbag system for fall protection using MEMS sensors and embedded SVM classifier, *IEEE Sensors J.*, 9, 95–503 (2009).

Spach, M., R. Barr, J. Havstad and E. Long, Skin-electrode impedance and its effect on recording cardiac potentials, *Circulation*, 34, 649–656 (1966).

Spinelli, E. and M. Haberman, Insulating electrodes: A review on biopotential front ends for dielectric skin–electrode interfaces, *Physiol. Meas.*, 31(10), 183–189 (2010).

Subramanian, V., Clinical and research applications of ambulatory holter ST-segment and heart rate monitoring, *Am. J. Cardiol.*, 58(4), 11–20 (1986).

Sun, M., M. Mickle, W. Liang, Q. Liu and R.J. Sclabassi, Data communication between brain implants and computer, *IEEE Trans. Neural. Syst. Rehabil. Eng.*, 11(2), 189–192 (2003).

Sung, M., C. Marci and A. Pentland, Wearable feedback systems for rehabilitation, *J. Neuroeng. Rehabil.*, 2, 17 (2005).

Tamura, T., T. Yoshimura, M. Sekine, M. Uchida and O. Tanaka, A wearable airbag to prevent fall injuries, *IEEE Trans. Inform. Technol. Biomed.*, 13, 910–914 (2009).

Teng, X.F., Y.T. Zhang, C.Y. Poon and P. Bonato, Wearable medical systems for p-health, *IEEE Rev. Biomed. Eng.*, 1, 62–74 (2008).

Vivekananthan, D.P., E.H. Blackstone, C.E. Pothier and M.S. Lauer, Heart rate recovery after exercise is a predictor of mortality, independent of the angiographic severity of coronary disease, *J. Am. College Cardiol.*, 42(5), 831–838 (2003).

Want, R., iPhone: Smarter than the average phone, *IEEE Perv. Comput.*, 9, 6–9 (2010).

Want, R., A. Hopper, V. Falcao and J. Gibbons, The active badge location system, *ACM Trans. Inform. Syst.*, 10(1), 91–102 (1992).

Wood, A., J. Stankovic, G. Virone, L. Selavo, Z. He, Q. Cao, T. Doan, Y. Wu, L. Fang and R. Stoleru, Context-aware wireless sensor networks for assisted living and residential monitoring, *IEEE Network*, 22, 26–33 (2008).

Wu, W., A. Bui, M. Batalin, L. Au, J. Binney and W. Kaiser, MEDIC: Medical embedded device for individualized care, *Artif. Intell. Med.*, 42, 137–152 (2008).

Zong, W., G.B. Moody and D. Jiang, A robust open-source algorithm to detect onset and duration of QRS complexes, *Comput. Cardiol.*, 30, 737–740 (2003).

6

Design and Development of Flexible Solar Cell's Integration in Clothing

6.1 Introduction

This chapter deals with the development of flexible solar tents used for power generation and aims at supplying power to the load, which can be utilised for electronic equipment and wearable electronic products used by the soldiers in the war field. Soldiers carry daily supply of primary batteries, but limited power capacity and the continual need for the supply can limit the mobility, range and mission length required for effective field operations. This problem can be eliminated by using flexible solar tent in the defence field. The new technology aims to make military missions safer and more energy-efficient. Such capabilities would ultimately make travel safer for soldiers in countries during wartime, since extra battery packs carried by troops are sometimes left behind, providing unnecessary hints to enemy forces. The integration of photovoltaics into clothes imposes some novel challenges and restrictions which are uncommon to standard photovoltaics. The most important tasks in developing such integrated flexible solar tent are: (1) flexible cells, (2) determination of the energy yield under realistic operating conditions, (3) charge controller and system design and (4) proper integration of the solar cells and electrical connections in garments.

6.2 Electronic Display Technologies

Designers have been quick to jump on the high-tech fabric bandwagon, adopting electronic display technologies to create colourful, novelty clothing items. For example, the Italian-made fabric Luminex®, which contains coloured light-emitting diodes (LEDs), has been used to make a glow-in-the dark bridal gown, sparkly cocktail dresses and costumes for opera singers. Luminex is made by binding LED fibres to the ends of ordinary fabric, which then form the seams of tailor-made clothing. The fibres are powered by tiny, rechargeable batteries that are turned on by the wearer via a hidden switch. Flicking the switch causes the fibres to glow in one of the five different colours, giving Luminex garments an overall appearance of shininess when the lights are dimmed. France Telecom has gone one step further by developing a flexible, battery-powered optical fibre screen that can be woven into clothing. Each plastic fibre-optic thread is illuminated by tiny LEDs that are fixed along the edge of the display panel and controlled by a microchip. The threads are set up so that certain portions are lit when the LEDs are switched on, while other sections remain dark. These light and dark patches essentially act as pixels for the display screen. A prototype version integrated into a jacket displayed crude but readable symbols. More sophisticated versions may support advertising slogans, safety notices, or simply a range of different geometric patterns can be switched on and off. The integrating of woven fabric with electronics is finding favour in the world of interior design as well. Maggie Orth, cofounder and CEO of a Massachusetts Institute of Technology start-up, International Fashion Machines, is currently producing one-of-a-kind, electro-textile wall panels. Instead of self-illuminating optical fibres, she is working with a fabric known as Electric Plaid™ that exploits reflective colouring. The novel fabric contains interwoven stainless steel yarns, painted with thermochromic inks, which are connected to drive electronics. The flexible wall hangings can then be programmed to change colour in response to heat from the conducting wires (Figure 6.1). Elsewhere, garment manufacturers are focusing on functional benefits rather than aesthetics. The simplest of these so-called smart clothing items are made by adding the required circuitry, power sources, electronic devices and sensors to standard fabric garments. Batteries can be sewn into pockets, wires fed through seams and wireless antennae attached to collars and cuffs.

Researchers are also working to create a wearable version of a giant textile 'sensornet' designed to detect noise. The fabric, developed with support from the U.S. military, is fitted with an acoustic beam former capable of picking up and pinpointing the location of an approaching vehicle. Electrical

FIGURE 6.1
Optoelectronic fabrics may find a market in the world of interior design owing to their originality and aesthetic appeal. (Courtesy of Maggie Orth, International Fashion Machines, Cambridge, MA.)

FIGURE 6.2
Cloth sensor net developed at Virginia Tech could help military personnel detect and locate approaching enemy vehicles.

connections are made by weaving wires into the heavy-duty cloth, and discrete microphones are attached at suitable points (Figure 6.2), though these could also be replaced by piezoelectric film sensors in the future. The film's sensing properties will be different from a discrete microphone, because the sound will hit a larger surface area.

Kirstein regards the development of interwoven electronic textiles as a significant advance in the field of wearable computing, though accepts that the materials' complexity will keep intelligent garments off the market for a few more years. The researchers are also currently trying to integrate as much functionality into their fabrics as possible. Textile antennae developed at the Zürich labs will let the cloth computers communicate with

(a)   (b)

FIGURE 6.3
(a) Researchers at ETH Zürich are using woven fabric with embedded Cu fibres to produce context-aware clothing. (b) Their prototype 'intelligent' jacket uses the conductive fibres to transmit data between different sensors and to measure body movements. (From Kirstein, T. et al., Textiles for signal transmission in wearables. In *Proc. ACM of First Workshop on Electronic Textiles (MAMSET 2002)*, San Jose, CA, Vol. 3, pp. 47–51, 2002.)

each other or the outside world (Figure 6.3). The next step forward will be the creation of conductive thread-like elongation sensors to monitor body movement. The team is hoping to have such fibres embedded in a prototype context-aware garment within the next couple of years. Devising a novel way to power the clothing is a further challenge. Batteries may have reduced in size, but wire connection to a pocket-held power source still goes against the grain of ready-to-wear computing. Prototypes at the moment use simple rechargeable batteries, but, of course, if we really want to sell these products, we have to think about alternative energy generation.

6.3 Capacitive Sensing Fabrics and Garments

Advances in textile technology, computer engineering and materials science are promoting a new breed of functional fabrics. Fashion designers are adding wires, circuits and optical fibres to traditional textiles, creating garments that glow in the dark or keep the wearer warm. Meanwhile, electronics engineers are sewing conductive threads and sensors into body suits that map users' whereabouts and respond to environmental stimuli. Researchers agree that the development of genuinely interactive electronic textiles is technically possible, and that challenges in scaling up the handmade garments will eventually be overcome. Now they must determine how best to use the technology. The term 'smart dresser' could soon acquire a new meaning. An unlikely alliance between textile manufacturers, materials scientists and computer engineers has resulted in some truly clever clothing.

Metal wires are used in weaving within fabric (Figure 6.4) and also twisting around insulating fibres. Sinusoidally weaving through fabric often used to measure changes in resistance or inductance and it can be plated as well.

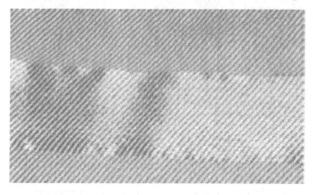

FIGURE 6.4
Woven fabrics with metal wire.

Utilizing Jacquard binary techniques, CuS bonded to polyester with touch-sensitive electrodes on non-conductive fabric was developed and also piezo-electric electrode interconnecting tracks of metal fibres (Figure 6.5).

Advantages

- Inherently antibacterial
- Lower degradation of conductance
- Stable conductance

Disadvantages

- Ageing effect
- Cannot be washed
- Difficult to calibrate
- Affected by humidity

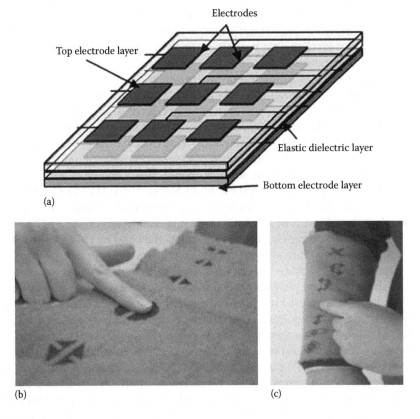

(a)

(b)   (c)

FIGURE 6.5
(a) Electrode layered fabric; (b) integration of electrode layered fabric in clothing; (c) garment developed using electrode layered fabrics.

6.4 Solar Cell Embedded on Clothing

Integration of flexible solar cells into clothing can provide power for portable electronic devices. Photovoltaics is the most advanced way of providing electricity far from any mains supply, although it suffers from the limits of ambient light intensity. But as the energy demand of portable devices is now low, clothing-integrated solar cells are able to power most mobile electronics. Introducing clothing-integrated photovoltaics, their scope and limitations, the status of flexible solar cells, charge controller and system design, as well as prototype solutions will provide for various applications. *Integrating fabric solar cells, for example, is a very promising idea, because you have a large surface area on the clothing, so you could use that for generating energy.* In the wireless world, whatever may be the technical obstacles, researchers involved in the development of interactive electronic clothing appear to be confident that context-aware coats and sensory shirts can be made but it is only a matter of time this is realized. The level of security required for electronic textile garments will vary according to their applications. Military battle dress, medical monitoring suits and fashion garments, for example, could each be installed with a different level of data protection software. Universal communication and encryption standards will also be required to ensure different products work together in an efficient and secure manner.

Photovoltaic power generation has grown rapidly worldwide (Barnett and Resch 2005), and is starting to contribute a noticeable amount of electricity production to public grids, especially in Japan (Tomita 2005) and Germany (Barnett and Resch 2005). Electric energy from solar cells is still too expensive to compete with established power plants, but Photovoltaic Island systems (Preiser 2003) located far from any grid connection have been economically successful for many years (Hegedus and Okubo 2005). The size and output power of such island systems varies over a considerable range from several kilowatts to less than 1 W. Common layout is comprised of storage batteries and power conditioning electronics, as well as the solar modules themselves. Specific types of island system, namely, carry-on photovoltaics integrated with clothing, will be added advantages. Jackets, coats, backpacks, accessories, even T-shirts and caps provide a much larger area for integrated photovoltaics (IPV) than the ever-shrinking portable devices themselves. Today, consumers frequently use the ubiquitous entertainment, voice and data communication, health monitoring, emergency and surveillance functions, all of which rely on wireless protocols and services. Consequently, portable electronic devices like mobile phones, MP3 players, personal digital assistants (PDAs), cameras, global positioning systems (GPS) or notebook computers need a wireless, mobile and sustainable energy supply in order to overcome the constant problem of batteries running out of power when most urgently needed. As a result of their steadily decreasing power demand, many portable devices can harvest enough energy from clothing-integrated solar modules (Schubert and Werner

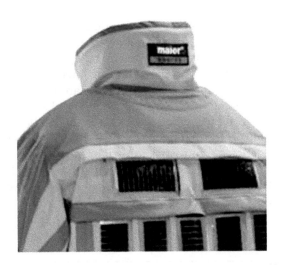

FIGURE 6.6
Prototype of a winter outdoor jacket with IPV. (Courtesy of Maier Sports, Köngen, Germany.)

2005) with a maximum installed power of 1–5 W. Figure 6.6 shows a recent prototype of a winter outdoor jacket with integrated solar modules that deliver a maximum output power of 2.5 W. This clothing-integrated photovoltaic system is designed to power an MP3 player, and after 3 h under full sun provides more than 40 h of music without any additional illumination.

This jacket was developed by Maier Sports (Chwang et al. 2003), with support from the Institute for Physikalische Elektronik and other partners in the German SOLARTEX project. The jacket was first presented in Munich at the International Trade Fair for Sports Equipment and Fashion (ISPO) in January 2006. It comprises nine amorphous Si solar modules from Akzo Nobel (Contreras et al. 2005) with a maximum power output of 2.5 W under full sun. Since 2000, design studies on solar cells integrated into clothing have been regularly presented at fairs and exhibitions on *smart textiles* or 'smart clothes', for example the Avantex fairs in Frankfurt, Germany, or the Nixdorf Innovation Forum. Hartmann et al. (2000) have been presented a vision of high-tech fashion including the use of photovoltaic power. A first systematic approach to the layout of complete IPV systems was published by Werner (2003). Although consumers and the clothing industry seem to be very interested in clothing-integrated photovoltaics, the advent of real products in the market has been hindered and delayed by the limited availability and performance of flexible solar cells.

From a customer point of view, an IPV system should be easy to use, comfortable and reliable; offer a universal socket for the countless different charging adapters and devices and, of course, deliver plenty of energy at an affordable price. If parts of the system need to be visible, they should be attractive and integrate well with the particular design of the garments. Connecting wires, charge controllers and batteries ought to be invisible, lightweight and maintenance-free. As an additional requirement, clothing with integrated electronics and photovoltaics should be as washable as every other textile. The most decisive and restricting demand, however, is the conformal flexibility of solar cells used for clothing integration. Existing cells on plastic or metal foils with protective laminates can only bend in one direction rather than exhibiting full conformal flexibility like a woven textile. The various flexible solar cell technologies, with a more detailed focus on amorphous Si (a-Si) and protocrystalline Si (pc-Si), have been reviewed, where protocrystalline denotes a film structure right at the edge of crystallinity. Flexible, single-crystalline Si(c-Si) cells are presented, and IPV prototypes are described. The area demand for IPV is examined, as well as issues of optimizing system design and clothing integration.

6.5 Flexible Solar Cell Technologies

The interest in flexible solar cells is steadily increasing, since high altitude platforms, satellites for telecommunications and deep-space missions would benefit from rollable or foldable solar generators. Cars, aircraft and various electric appliances could also cover part of their power demand from ambient illumination of their free-form cases. The integration of photovoltaics with textiles is not only interesting for powering portable devices, which we address here, but also opens a wealth of opportunities for the integration of electronic features with architectural fabrics. Current thin-film solar cells consist of a layer stack that is continuous in two dimensions and very thin in the third. Because of their planar substrates, these cells bend but do not crinkle. Solar cells have been formed on Cu wires for fabrics made of photovoltaic fibres (Rojahn et al. 2001). Without a continuous planar substrate, however, the fibres arbitrarily move against each other, which gives rise to many problems like moving interconnect, shadowing and cancellation of the electric output of single fibres. Considering all the unsolved problems of manufacture and interconnection of photovoltaics fibres, woven solar modules are to be technically feasible in the foreseeable future research works. Finding a compromise between the minimum total thickness and, hence, maximum conformal flexibility of IPV modules on the one hand, and washability, mechanical resilience and durability on the other, is an important task that has not yet been solved for most of the flexible cell technologies described here. A lot of hope is focused on organic and dye-sensitized solar cells that could, in principle, be printed on polymer foils. In a major breakthrough in organic solar cell technology, Brabec (2004) have demonstrated bulk heterojunction cells with a conversion efficiency of 5%. The main challenge for all these organic and dye-sensitized cell concepts, however, is in their encapsulation and long-term stability. Organic and dye-sensitized cells work reasonably well in tightly sealed glass-glass packages

(Niggemann et al. 2004). For flexible cells, organic–inorganic multilayers can prevent moisture and oxygen permeation into the sensitive electrode or dye materials (Chwang et al. 2003). For these solar cells, such barrier films must limit the permeation of water to less than 10–8 g/m²/day, which is two orders of magnitude less than required for organic light-emitting displays, and this has not yet been achieved. Research goals also include enhancing the photochemical stability of the dyes and the absorption of visible light (Lagref et al. 2003). Further progress in these fields is needed before organic or dye-sensitized cells will be suitable for IPV applications.

Copper indium gallium selenide (CIGS) and copper indium sulfide (CIS) cells demonstrate their best performance on glass substrates, with a record efficiency of 19.5% (Contreras et al. 2005). Flexible CIGS cells reach efficiencies of up to 17.4% on steel foils (Kessler et al. 2005). A sophisticated barrier layer stack protects the CIGS heterojunction against impurity diffusion from the metal substrate (Herrmann et al. 2005). Since polyimide (PI) substrates limit the deposition temperature to $T_d < 450°C$, the efficiency drops to 8 ~ 11% (Herrmann et al. 2005). Figure 6.7 displays some examples of CIGS solar cells and modules on metal and PI foils. While all these data refer to co-evaporated CIGS, alternative deposition methods try to exploit the potential of low-cost, roll-to-roll deposition. Presently, the solar cell efficiency, as well as the maturity of these manufacturing methods clearly lags behind co-evaporation techniques. When considering CIGS for IPV, however, the most serious drawback is the poor low-light performance. The material composition and electronic properties of CIGS fluctuate on a micrometer scale, introducing local electric shunts into the heterojunctions that result in a high saturation current and deteriorate the low-light efficiency of CIGS solar cells. Gemmer and Schubert (2001) have compared the intensity dependence of realistic models of CIGS, a-Si and c-Si diodes, and evaluated their possible energy yield in various IPV scenarios. According to the results, CIGS can only compete with Si in outdoor applications. Virtuani et al. (2006) and coworkers have shown improved shunt control and low-light efficiency in CIGS cells. Whether this recent development will render flexible CIGS modules competitive for indoor use is still an open question. The suitability of CdTe cells for IPV has not yet been studied, but Tiwari et al. (2001) have achieved efficiencies of 8.6% for flexible CdTe cells. In comparison with the material systems discussed earlier, flexible thin-film Si is a good choice for IPV, especially because of its good low-light performance. Well-optimized c-Si cells with edge isolation exhibit a value of n = 1 in Shockley's ideal diode equation (n may also be termed the emission coefficient or diode quality factor). This results in an almost ideal intensity dependence with an open circuit voltage drop of V_{oc} = −70 mV/decade of the incident radiation, in contrast with V_{oc} = −130 mV for pc-Si. Consequently, c-Si would be the preferred material for IPV, if it could attain similar flexibility as ultrathin pc-Si cells. At a deposition temperature of 100°C, our pc-Si cells achieve ~5% on low-cost polymer foils. Flexible, single-crystalline c-Si transfer cells reach = 14.6%.

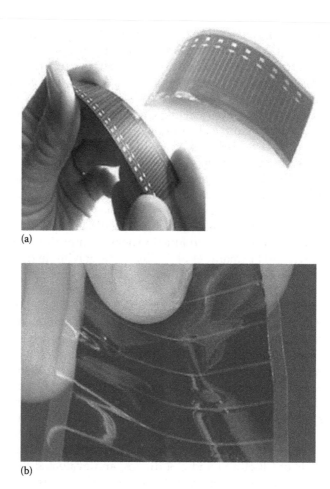

FIGURE 6.7
Examples of flexible (a) CIGS test cells and (b) an integrated module.

6.5.1 Flexible a-Si and pc-Si Cells

The industrial development of flexible a-Si modules mostly targets the power market and tries to realize the cost advantages of roll-to-roll production. Several players are active in this field with differences in the maturity of their technologies. Smaller companies are more interested in small-scale applications like IPV. The Swiss company VHF technologies is developing a-Si modules on 50 μm PI or polyethylene terephthalate (PET) foils in cooperation with the University of Neuchâtel, and delivers modules with a total thickness of only 150 μm. Bailat et al. (2005) have reported record efficiencies of 7% for single a-Si cells on nanostructured PET, and equals 8.3% for a double-stacked, so-called micro morph tandem cell. This comprises an a-Si cell on top of a microcrystalline Si cell. Iowa Thin Films Technologies took up former 3M technology (Jeffrey et al. 1985) in 1988. While the company is now focusing on military applications, it also produces a-Si modules with a specified 3.5% that we use for IPV demonstrators described later and is shown in Figure 6.8. Since research at our institute develops solar cells to comply with the textile feeling of regular clothes like shirts, jackets, etc.

Under full-sun illumination, cells on the PI foil reach 11% (Herrmann et al. 2005) and those on stainless steel achieve 17.4% (Contreras et al. 2005). Monolithic series connection of the cells yields a module efficiency of 6.8%. Deposition starts with the transparent superstrate and front contact where the incident light will enter the cell. During growth of the undoped pc-Si absorber layer, nano-crystals start to form in an amorphous matrix so that nanocrystalline volume fraction increases with film thickness. Further there is a need to reduce the substrate thickness, and also therefore the substrate temperature during deposition. It is only in a deposition temperature range of $T_d = 100°C–130°C$, that ultrathin PET or polyethylene naphthalate (PEN) foils with a thickness of d = 10–20 μm retain enough mechanical stability for solar cells and module processing. The pioneering work of Koch et al. (2001) has revealed that at $T_d \sim 100°C$, protocrystalline absorbers yield better solar cells than amorphous ones. Even at $T_d = 70°C$, an initial conversion efficiency 3.8% is feasible, while $T_d = 100°C$ yields 6% for various configurations of double-stacked cells. Light-induced degradation results in stabilized efficiencies close to 5%, for example 4.8% at $T_d = 110°C$. At even lower values, for example $T_d = 40°C$, the parameter window for pc-Si growth becomes very narrow (Koch et al. 1999, 2000); hence, a range of $T_d = 70°C–130°C$ proves feasible for solar cell optimization. All a-Si- or pc-Si-based solar cells are drift-controlled. That is, p- and n-type doped layers of only 10–20 nm thickness induce an internal electric field F across a nominally undoped, intrinsic i-layer. Photogenerated electron–hole pairs are spatially separated by F and contribute to the external photocurrent of these p–i–n cells. Typical i-layer thicknesses are in the range from 150 to 300 nm. With decreasing T_d, the density of electronic defects in the i-layer

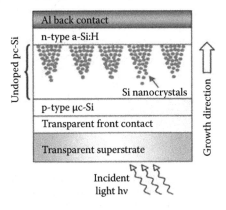

FIGURE 6.8
Schematic of a pc-Si solar cell.

rises by one to two orders of magnitude. Therefore, F collapses because of the charge trapped at these additional defects, and consequently drops. Moreover, i-layer deterioration not only limits the efficiency of low T_d cells, but also causes a remarkable loss of doping efficiency (Koch et al. 2001; Schubert and Werner 2005). Figure 6.8 shows the structure of pc-Si solar cells. The optimum absorber material is right at the edge of crystallinity, with dispersed but not yet interconnected nanocrystals in an amorphous matrix. For depositing high-quality pc-Si at $T_d < 120°C$, plasma excitation in the very-high-frequency (VHF) range is favourable; we routinely use 80 MHz. The transition from amorphous to nanocrystalline growth depends not only on the H_2 dilution of the process gases during plasma deposition, but also on the substrate temperature and film thickness. The thicker the films is to be greater the nanocrystalline volume fraction. Figure 6.9 demonstrates the clear coincidence of the absorber layer quality and solar cell performance for $T_d = 110°C$. When tuning the H_2 dilution ratio $R = (H_2)/(SiH_4)$, our deposition setup produces the best material at $R_{opt} = 17$. Figure 6.10a presents the current density versus voltage (J/V) characteristics and key performance parameters of solar cells grown on rigid glass substrates, as well as on 50 μm thick PET foils. Figures 6.5c and 6.10b show arrays of pc-Si test cells deposited on 23 μm thick PET foils at the upper limit of $T_d = 130°C$. These pc-Si cells, optimized for direct lamination with textiles, exhibit a promising >5%. Nevertheless, further challenges arise from the mismatch in thermal expansion coefficients of the polymer foils and inorganic solar cell materials, and from moisture permeation into the electronically

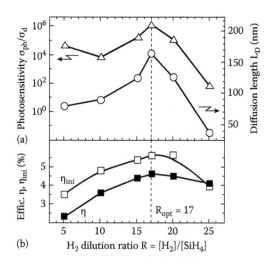

FIGURE 6.9

Optimum H_2 dilution for pc-Si growth. (a) Two figures of merit, the photo-to-dark conductivity ratio and the ambipolar diffusion length from steady-state photo carrier grating (SSPG) measurements, mark the best i-layer quality at an H_2 dilution ratio of $R_{opt} = 17$. (b) Highest i-layer quality at $R_{opt} = 17$ directly translates into the greatest solar cell efficiency. Each data point represents an average over 40 test cells of 3 mm × 3 mm.

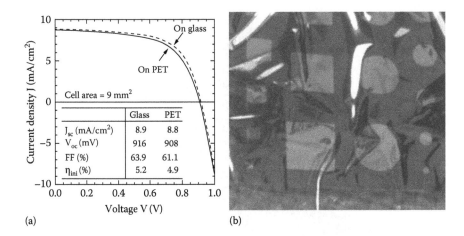

(a)   (b)

FIGURE 6.10
(a) At Td = 110°C, optimized pc-Si test cells reach very similar efficiencies on non-textured glass and plastic substrates; (b) showing the front view in the direction of the incident light.

active layers. Further studies need to explore these limits, especially on long-term stability and washability, for real photovoltaics laminates with textiles.

The J/V characteristics under full-sun-like illumination yield the cell's performance parameters short circuit current density J_{SC}, open circuit voltage V_{oc}, fill factor FF and initial efficiency. Flexible cells on PET slightly lag behind corresponding cells on glass. This may be the result of a small difference in substrate temperature that cannot be eliminated in our substrate holder setup pc-Si cells on 23 μm thick PET foil. Since the active cell thickness is less than 1 μm, the PET substrate controls mechanical stability and flexibility. The photographs indicate that the test cells faithfully follow textile wrinkles, but it must be kept in mind that additional encapsulation by protecting foils is mandatory for clothing integration.

6.5.2 Single-Crystalline Si Transfer Technique

While the development of low-temperature a-Si cells is aimed at the large-area integration of photovoltaic modules with clothing and other textiles, the single-crystalline Si transfer technique provides high efficiency cells with a somewhat restricted flexibility, but much smaller area demand. Transfer cells are therefore well suited for implementing eye-catching design highlights in clothing. Figure 6.11 shows a Bogner jacket that demonstrates the potential of small-area photovoltaic add-ons. The small and brittle c-Si cells exceed > 20%, but need some care for clothing integration. The single-crystalline c-Si layer transfer technique developed at the institute is able to replace the brittle and rigid cells with flexible c-Si cells, which have an efficiency ~ 15%. In contrast to pc-Si based cells with a total thickness of d ~ 1 μm, such crystalline cells need d = 15–30 μm for

FIGURE 6.11
BOGNER/MUSTANG denim jacket, including a small-area photovoltaic add-on designed for the Avantex fair in 2005. Brittle c-Si cells are fixed onto a rigid metallic backplane that ensures mechanical stability, with their front and back contacts series connected at the same time like roofing shingles. Two modules, each of 16 cells, power blinking or walking lights along the arms of the jacket where high-brightness white light-emitting diodes are incorporated. (Courtesy of MUSTANG Bekleidungswerke, Kunzelsau, Germany.)

sufficient absorption of ambient light. Although they consist of perfect single crystals, the c-Si films become flexible at d < 50 μm. A brief description of the steps essential for forming flexible c-Si cells starts with the electrochemical formation of a porous double layer over a 6 in. wafer surface. The left-hand scanning electron micrograph in Figure 6.12 shows the initial

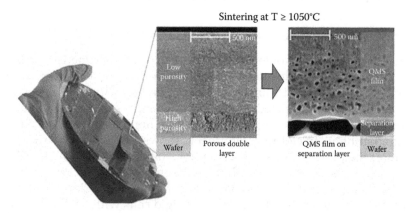

FIGURE 6.12
Flexible c-Si solar cells by layer transfer.

porous double-layer structure that extends over a 6 in. host wafer. This converts into a quasi-monocrystalline Si (QMS) layer on top of a separation layer during a high-temperature anneal. After standard epitaxy and front-side processing, the thin-film c-Si cells are transferred to a flexible polymer foil before completing backside passivation and contacts (Kessler et al. 2005).

A subsequent annealing step restructures the porous layer structure into a separation layer and a so-called QMS layer on top (Bergmann et al. 2002; Rinke et al. 1999). This step also closes the top QMS surface to provide a perfect template for high temperature Si epitaxy. All solar cell front-side processing proceeds at high temperature on the device-grade epitaxial layer. The separation layer still fixes the half-processed, thin-film solar cell to the original host wafer. By laminating a transparent superstrate like glass or appropriate plastic foils onto the front of the cell, the separation layer finally releases the thin-film cell from the host wafer. Completion of the cell backside must now precede at low temperature, since either the flexible foil or a transparent adhesive on the front side limits the process temperature to T < 200°C (Brendle et al. 2005; Rostan et al. 2005). For further details of our transfer process, competing approaches, and solar-cell processing, see Werner (2003) and Bergmann et al. (2002) and references therein. Bending experiments (Figure 6.13a) demonstrate that free-standing 25 μm thick c-Si films can be bent to a curvature radius of 2 mm. Figure 6.13b shows a monocrystalline, 150 mm diameter sheet of transfer Si laminated between protective foils. Solar cells manufactured into the transfer layers reach conversion efficiencies of up to 16.6% if transferred to glass plates (Feldrapp et al. 2003; Tayanaka et al. 1998), or up to 14.6% if transferred to flexible plastic foils. These record efficiencies require careful optimization and application of photolithographic patterning. However, production technologies mostly apply screen-printing for solar cell formation. Without lithographic patterning, Auer and Brendel (2001) have demonstrated a small module yielding 9.9%. Figure 6.13c demonstrates that our transfer layers are suitable for use in an industrial screen-printing process. We expect that transfer cells will reach > 18% within the next 2 years.

6.5.3 Photovoltaic Thin-Film Cells into Wearables

Consumer electronics are typically designed with a battery power source and are carried in light limited pockets or bags. The integration of electronic devices and wearables provides the opportunity to utilize surfaces exposed to the sun to generate energy to power the electronic devices. Photovoltaic flexible thin film converts solar energy into electrical energy. The photovoltaic flexible modules come in various sizes ranging from 2 × 4 in. to 8.5 × 11 in. sheets. They can be rolled into a 3 in. diameter without physical damage and continue to function if scratched or punctured. The durability and

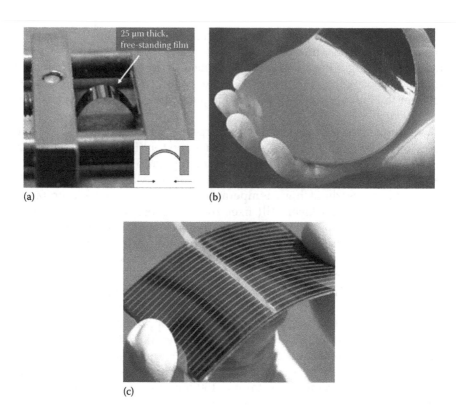

(a)

(b)

(c)

FIGURE 6.13

Flexible c-Si films and solar cells made using our layer transfer technique: (a) Layer transfer produces c-Si films that bend to a 2 mm radius of curvature and the crystalline structure favours crack formation and propagation along the crystal axes, the edges of such films and solar cells need to be well protected; (b) a flexible, 150 mm diameter, 25 μm thick c-Si sheet laminated between two, 200 μm thick protective foils; (c) industrial screen printing at the research facilities of Shell Solar60 forms transfer cells of 5 cm × 5 cm.

efficiency of these photovoltaic flexible thin-film modules have improved to a point where they are a viable option for incorporation into wearables (Hynek et al. 2005). Ultimately, these thin-film photovoltaic cells can reduce the amount of battery storage engineered into electronic devices and eliminate maintenance related to replacing batteries. Devices such as the Burton Shield iPOD Jacket, Memswear Fall Sensing Shirt and Shimadzu DataGlass 2/A are examples of smart garments where battery maintenance or battery weight could be reduced by using photovoltaic flexible thin film for charging. The successful integration of photovoltaic thin-film cells into wearable is tightly tied to the consumer concept of fashion. In the consumer market, smart clothing must remain visually attractive and complement or enhance the wearer's appearance; otherwise, it will not be commercially successful. In order to maximize energy collection, it is necessary to place the photovoltaic film in visibly prominent areas on the wearable. These solar cells are graphically strong and as a result need to be more visually integrated into the garment structure.

Digital textile printing enables the designer to incorporate unusual components into a design by printing fabric that matches the pattern of the component. Direct digital textile printing technologies typically employ the use of ink jet printing to allow the user to print designs directly from the computer to fabric. Digital textile printing technologies refer to a number of integrated software and hardware components that are typically used to create digitally printed fabrics. These include off-the-shelf software packages (such as Adobe Photoshop® and Illustrator®), industry specific-design and pattern-making software (PAD® patternmaking system, Pointcarré®), wide-format ink jet printers for direct textile printing (Encad 1500TX, Colorspan Fabrijet), RIP software used for processing the image files into information that is useful for the printer and other support hardware (Parsons and Campbell 2004). The wide variety of wearables to which photovoltaic thin-film cells can be applied requires a wide variety of fabrics to make these garments which eliminates the possibility of utilizing a large order of fabric. Digital textile printing makes it possible to do small-scale production on a variety of fabrics (Ross 2005). Digital textile printing has been used previously by Campbell and Parsons (2001) to intersect two design problems to create multiple points of exploration in the development of a uniquely patterned jacket. When the garment design process is mixed with an engineering design process, there is a greater need for pattern matching to incorporate electronic components and circuitry into the wearable. Digital textile printing provides the capability to create unique patterns that open doors to a wide range of design activities and diverse disciplines (Parsons and Campbell 2004).

6.5.4 Protocrystalline Silicon Cells

Although clothing industry is strongly interested in such IPV, further progress is delayed by the limited availability of high-efficiency flexible solar cells. One promising approach which follows on a laboratory scale is the transfer of thin, and hence flexible, monocrystalline silicon solar cells to appropriate plastic substrates; this option seems well capable of providing 15% efficient flexible cells with a good low-light performance. The faster route to substantial volumes of IPV applications, although at a lower performance level around 5% of AM 1.5 efficiency, will utilize protocrystalline silicon (pc-Si)-based cells; recently, optimization of such flexible pin and nip solar cells on plastic substrates down to 20 nm in thickness has been carried out. These substrates (e.g. PET) limit the deposition temperature to 110°C. Following the pioneering work, not only protocrystalline absorber layers are needed to be optimized at a deposition temperature of 110°C, but also the doped layers require careful tuning of hydrogen dilution and plasma excitation frequency.

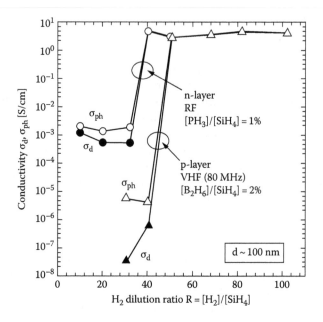

FIGURE 6.14
Optimization of p- and n-type layers at 110°C deposition temperature.

Figure 6.14 clearly demonstrates the benefit of hydrogen dilution and VHF excitation for achieving high p- and n-layer conductivities. All films and solar cells are grown from a 6 in. cluster tool deposition system. Optimization of the intrinsic absorber layer quality is monitored by the constant photocurrent method (CPM) as well as by the steady-state photo-carrier grating (SSPG) technique. At 110°C deposition temperature, an optimum hydrogen dilution ratio $R = (Hz)/(SiH_4) = 17$ lowers the mid-gap defect density of undoped pc-Si by two orders of magnitude, resulting in a CPM absorption $u(ZeV) = 3 \ cm^{-1}$, and a maximum bipolar diffusion length of 165 nm from SSPG. A detailed comparison of the performance and of the light-induced degradation of pc-Si solar cells grown at 110°C in both, nip as well as pin deposition sequences, reveals that the initial efficiency qo of optimized nip cells degrades much less ($Aq/qa = -8\%$ with $Aq = q - qo$) than that of pin cells ($Aiyqo = -16\%$). The progressive formation of nano-crystalline silicon helps in pc-Si growth (Koch et al. 2004). The nip deposition sequence produces a high crystalline volume fraction at the interface of the cell with low-light-induced degradation, whereas pin deposition results in an amorphous, and hence degrading pli interface region.

6.5.5 Energy Yield

When introducing the idea of clothing-integrated photovoltaics, the key question immediately arises: how much energy will be collected under real working conditions of a solar jacket, or a similar IPV garment. This complex problem by (1) modelling the energy yield of single solar cells, and (2) mobile measurements of complete current voltage characteristics under real-life operating conditions. Several aspects determine the performance of an IPV system. Different solar cell materials produce greatly varying output, depending on the illumination spectrum and intensity. Modelling results include the comparison of crystalline and amorphous silicon, as well as copper indium galtium diselenide (CIGS) cells under four different spectra, namely, AM1.5G. D65 (representing 'standard' cloudy sky), halogen lamp (Planck's radiation), as well as a typical fluorescent tube spectrum. Moreover, the illumination intensity of these prototype spectra varies over six orders of magnitude. Indeed, the study proves a-Si to perform well at very low illumination intensity (400 pW/cm^2), or under fluorescent tube light, but under medium to high intensity. Crystalline silicon output is higher under most illumination conditions. Under low-light conditions, CIGS always performs worse than its silicon counterparts. Real IPV applications need to make the best use of medium range intensities of 0.1–10 mW/cm^2, where a-Si suffers from its diode quality factors n > 1.5, whereas c-Si cells fully benefit from nul. Concluding on the material issue, for most IPV applications that experience some outdoor illumination, c-Si seems favourable, whereas a-Si governs pure indoor use. But in addition, IPV demands for flexible modules which are available from a-Si, but not yet from c-Si (Gemmer 2003). In parallel to the development of flexible solar cells, first prototypes of IPV garments, equipped with commercial amorphous silicon cells and realized in cooperation with our partners from clothing industry. Numerical investigations of the attainable energy yield demonstrate the prospects and limits of garment-integrated photovoltaics, proving that an ample choice of mobile electronic devices can be sustainably powered by IPV. The proprietary layout of charge controllers makes best use of the abundant low-light intensity situations: these applications are designed for such cases. Experimental verification of the attainable energy yield is currently in progress by using mobile *in* monitors in real-life IPV use.

6.6 Clothing-Integrated Photovoltaics

The integration of flexible solar cells into clothing or accessories enables ubiquitous power generation for mobile electronic equipment. This contribution reports on recent progress in important areas of IPV development, such as (1) direct deposition of flexible solar cells on plastic foils, (2) forecast of the energy yield under realistic operating conditions and (3) charge controller and system design. Modelling of the attainable energy yield proves the validity of the concept, and first prototypes demonstrate its feasibility. Mobile and wireless electronic devices are increasingly popular and rapidly developing (Schuber et al. 2006). Entertainment, voice and data communication, medical and emergency needs foster the further diversification of applications, devices and services. Whereas continuously evolving wireless protocols enable all sorts of services and communications, a sustainable wireless power supply is obviously rather difficult to accomplish. At present, photovoltaic power is the only alternative to changing batteries, or carrying additional charging units and finding a mains plug just in the nick of time. Here, first steps into a clothing-integrated photovoltaics for powering mobile electronic equipment. Since miniaturization proceeds and the dimensions of mobile electronic devices continuously shrink, directly attaching solar cells onto the devices themselves will not succeed due to their very limited surface areas, and due to the limited light intensity, especially under indoor or average outdoor illumination conditions. Clothing-integrated photovoltaics (IPV), however, can solve this problem by providing sufficiently large areas in the range of several 100 cm^2 up to about 2000 cm^2, directly on garments or accessories.

The integration of photovoltaics into clothes imposes some novel challenges and restrictions which are uncommon to standard photovoltaics. The most important tasks in developing such IPV focus on: (1) flexible cells, either by low-temperature deposition on plastics, or by the use of transfer techniques; (2) determination of the energy yield under realistic operating conditions; (3) charge controller and system design and (4) proper integration of the solar cells, electrical connections, etc. into the garments. On the one hand, high-performance, flexible monocrystalline silicon solar cells reach conversion efficiencies of up to 14.6% after being transferred to a plastic foil (Wemer et al. 2003). On the other hand, I cells incorporating so-called protocrystalline (i.e. right at the transition between the amorphous and crystalline phases) silicon absorber layers, directly deposited onto plastic substrates at temperatures below 390 K, enable flexible cells in the 5% efficiency range. The optimization of such protocrystalline silicon (pc-Si) cells, including the introduction of a novel undoped buffer layer, is discussed. In order to predict the performance of clothing-integrated solar cells under varying illumination and operating conditions, model their energy yield for a wide range of illumination intensities and spectra. Based on these data, the potential of crystalline silicon (c-Si), amorphous silicon (a-Si) and copper indium gallium diselenide (CIGS) for IPV is evident. For demonstrating the feasibility of such clothing-integrated photovoltaics, we built first prototypes from standard, rigid and brittle crystalline silicon solar cells (Wemer et al. 2003). For practical use, of course, we need to incorporate flexible solar cells. In this chapter, we have for the first time presented prototypes which use commercially available, flexible a-Si solar cells. Jeans jacket with 400 cm^2 of flexible solar cells are incorporated by one of our partners from clothing industry. As the first application, the solar cells are powering a brand logo made out of 40 LEDs with a total power demand of 10 mW. The target application is a multipurpose 'photovoltaic plug' for powering arbitrary devices like CD OF MP3 players, mobile phones, etc. That presents a different type of application, namely, a function for protection wear and professional clothing (Berge et al. 2004). In this case, a few high brightness LEDs are fully integrated into the jacket in order to enhance the visibility of, for example, street constructions workers after sundown. Since the idea of a ubiquitous power supply from the photovoltaic conversion of ambient light is very appealing, one needs to investigate how much energy is harvested during a cloudy or rainy day, or during indoor use of IPV. Because of a considerable lack of data in the literature, Gemmer (Gemmer and Schubert 2001; Tiwari et al. 2001) performed a detailed study in our laboratory to link empirical, long-term data on the incident solar radiation that reaches the earth's surface under different weather conditions, as well as various indoor illumination scenarios, with the spectral and intensity dependence of different types of solar cells. His modelling of these different conditions enabled, for the first time, a rough estimation of what IPV can accomplish and where its limits are reached. Based upon typical parameters of real (rigid) solar cells, Gemmer compared the performance of a-Si, c-Si and CIGS. Figure 6.14 shows the intensity dependence under a typical cloudy sky in northern Europe, represented by the D65 spectrum (Colorimetry 1986); because of its higher band gap, $E_g = 1.8\,eV$, a-Si is the best choice for very low illumination intensities and for spectra with a comparatively high blue share, like D65 or fluorescent light. For most real-life situations, c-Si performs best because of its high efficiency and favourably low diode ideality n ~ 1. Despite a high efficiency under full sun, the output voltage of CIGS cells collapses for light intensities below 1% because of their comparatively low shunt resistances. The outdoor intensity under these spectral conditions averages only $80\,W/m^2$, whereas a clear sky yields a long-term average of $560\,W/m^2$ in Stuttgart, Germany. Because of their high efficiency and low V_{oc}, c-Si cells perform best, while a-Si still operates at very low-light intensities. CIGS suffers from its low shunt resistance and ceases to function at $2\,W/m^2$. The modelling of conversion efficiencies uses the parameters of real, commercial solar cells, rather than laboratory-scale world record data (Figure 6.15).

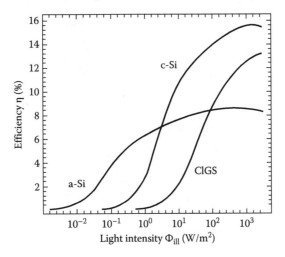

FIGURE 6.15
Photovoltaic conversion efficiency of c-Si, CIGS and a-Si cells as a function of illumination intensity under a D65 spectrum.

Tentative available areas on jackets range from 400 cm² for children to more than 1000 cm² for adults; the dashed line marks a typical value of 600 cm². These results indicate that IPV is feasible for powering a wide range of mobile electronic devices. Taking these intensity and spectral dependencies of the different technologies into account, Gemmer deduced annual energy yields for IPV under real weather conditions in Stuttgart, Germany, and calculated those for a set of prototype user profiles, for example a 'regular clerk', an 'outdoor construction worker' and a 'night-shift nurse'. Figure 6.16 summarizes the results for a clerk in terms of the demand for clothing-integrated solar-cell area that is capable of powering the indicated devices. This estimate uses a realistic mix of indoor and outdoor exposure and frequent use of electronic devices. For different user profiles, the area demand varies by a factor of 5, with the clerk in Figure 6.16 representing a mean value. The performance of different cell technologies is strongly dependent on the spectrum and intensity of the dominant light source. Outdoor conditions at light intensities of 10–1000 W/m² result in similar areas for CIGS, c-Si and a-Si, while indoor use down to 1 W/m² favours the Si cells, with a-Si being most efficient below 0.1 W/m². In order to make best use of photovoltaic energy, accumulators and power conditioning circuits must be carefully selected and optimized. Rechargeable alkaline manganese (RAM) cells are a good choice for IPV because of their very low self-discharge, robust charging protocol and extended temperature range for charge and discharge.

Figure 6.17 shows prototypes realized as part of the German SOLARTEX project. The Bogner/Mustang jacket in Figure 6.17a was designed to be a

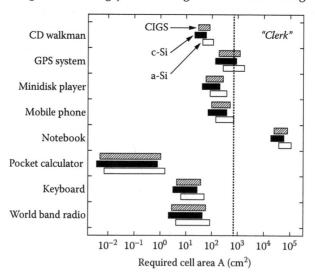

FIGURE 6.16
Solar cell area required for clothing-integrated photovoltaics to power the indicated devices sustainably, based on an annual average of the individual illumination conditions experienced by a prototype user 'clerk'.

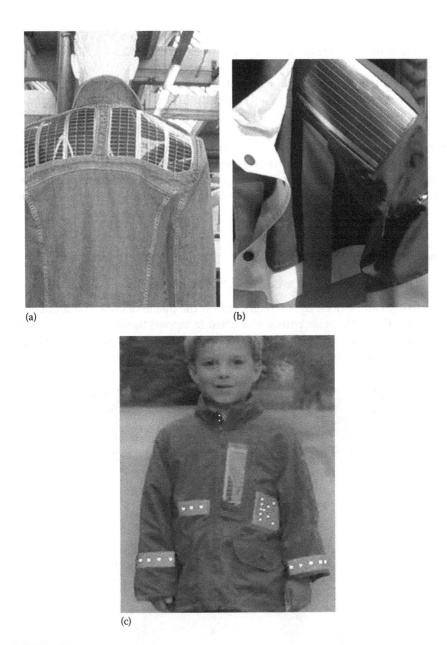

(a)

(b)

(c)

FIGURE 6.17
SOLARTEX prototypes for different applications: (a) BOGNER/MUSTANG jacket provides a universal socket and energy buffer for operating and charging USB-compliant devices. Iowa Thin Films modules deliver a maximum power of 2.5 W. Three hours of full sun enable 40 h of MP3 operation; (b) In the TEMPEX jacket, bright photovoltaic-powered warning lights supplement the built-in passive safety features, for example for street construction workers; (c) Using photovoltaic power harvested during the day, light-emitting diodes enhance the visibility of children at night in this KANZ coat. (Courtesy of Institut für Textil- und Verfahrenstechnik.)

universal, solar-powered source for operating MP3 players or other USB-compliant devices. Figure 6.17b shows an early example of IPV for driving warning LEDs in Tempex protection wear for street construction workers, while Figure 6.17c shows IPV warning lights in a Kanz coat that is designed to enhance child safety at dusk and night time. There is a range of truly inter-disciplinary tasks involved in the design and manufacture of IPV products. In the framework of Solartex, the Hohenstein Institute has developed the first working solutions for the placement and fixing of solar modules into clothing, and the Institut für Textile and Verfahrenstechnik in Germany is working on textile wiring and integration procedures fit for garment production. Results of this joint research will be published elsewhere in the near future.

6.7 Integration of Garment and Solar Panel

The fixing can be done by a straight cut on the inside of back and a pouch is made for fixing the solar panels with closing zippers as shown in Figure 6.18. The size of pouch is $15 \times 30 \times 2$ cm^3. The transparent material is stitched at the backside of the garment for getting the sunlight to the solar panel for the process

FIGURE 6.18
Solar panel pouch.

of converting light energy from sun into electrical energy. This energy is transformed to the circuit board with the help of wires. The circuit board is fixed inside of pouch at the right side of the garment underside which is enclosed by the zippers. Figure 6.19 shows the attachment of circuit board in the pocket, and Figure 6.20 shows the back view of the circuit board.

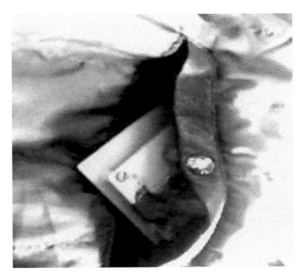

FIGURE 6.19
Fixing circuit board in the pocket.

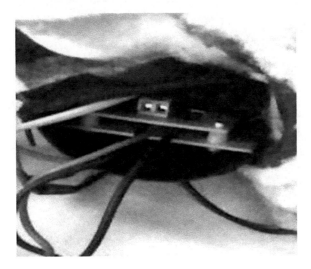

FIGURE 6.20
Back view of circuit board in the pocket.

One wire is connected to the LED and another one to the cell phone charger as shown in Figures 6.21 and 6.22. These wires are embedded inside the jacket. The cell phone charger is fixed inside the left side pocket of the jacket which is enclosed by the zippers. Finally, the LED connection is fixed on the right and sleeve and it is just above the wrist. Then the foam sheet of 1.2 cm width is attached around the solar panel pouch for supporting, which reduce the restrictions for the body movement as shown in Figure 6.23. Another small pouch at the left side front placket (just below the collar and

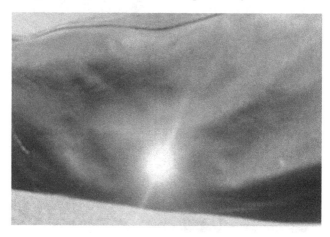

FIGURE 6.21
LED fixed on the right sleeve.

FIGURE 6.22
Head set and charger connection.

FIGURE 6.23
Foam sheet in the jacket.

near the zipper) is made for storing the headset and one metal snap fastener is stitched for closing this pouch.

The solar jacket is tested for its functionality using the conventional method of testing the solar panels and by testing the rechargeable batteries: (a) nickel metal hydrate battery (for charger) and (b) nickel cadmium battery (for LED). The solar panel's average output voltage is 7.4 V and the output current is 320 mA which is sufficient to power the LED and the mobile phones. The rechargeable nickel metal hydride is giving output of 6.4 V and nickel cadmium gives an average output of 4.5 V which is measured using multimeter when it is charged for 1 h duration. The successful integration, on a commercial basis, of solar cells into textiles opens up a whole range of fresh applications. Many of these applications are likely to exploit the flexibility and lightweight of textile fabrics. They can, for example, be placed over the curved surfaces of buildings. They can also be installed in spaces which may otherwise be inaccessible, as in automotive, marine and aerospace equipment. It may also be that the presence of solar cells in apparel will be particularly beneficial, for example, to maintain the temperatures of young children, the elderly and hospital patients, or to power-sensing elements for temperature and movement. In this research work, flexible solar cells were integrated with the defence tent to make solar tent. The flexible solar cloth adopted by this concept can be utilized for tent. The power generated can be used to meet excess electrical demands.

The power requirements of smart garments have a direct impact on the photovoltaic thin-film module surface area needed. The voltage requirement of the circuit in the garment also dictates how the photovoltaic thin-film module pattern will repeat. The tie was designed with the intent of charging a 3 V lithium ion battery which is typically found in cell phones. Three Iowa Thin Film MPT3.6-75 solar panels were applied to the front area of tie. These panels are shunted to 3.6 V and when connected in series produce approximately ~150 mA in full sun. The flexible solar panels were coated with a laminate that protects the solar material. When energized, the solar panels act similar to batteries. For connecting the panel to a circuit, a connection bus made of electrically conductive tape is adhered to the plastic panel. This makes it possible to solder connections. A multi-filar wire, having insulation on each filament, was soldered to the solar panel buses to connect them in parallel to a female plug that was removed from a Nokia cell phone charger. To integrate the solar panels into the design, an Iowa Thin Film MPT3.6-75 thin-film module was digitally scanned into Adobe Photoshop and visually colour-matched. The solar panel image was then adjusted for colour correctness and modified to become a perfect repeating design. Next, cotton sateen fabric was inkjet printed with the solar panel imagery. Minor adjustments were made to create more contrast between dark and light areas to compensate for slight colour changes that occur as the inks soak into the fabric structure during fabric printing. The ENCAD NovaJet 880 inkjet printer was loaded with fibre reactive dyes. The tie was constructed from a commercially available tie pattern. A small elastic pocket was sewn onto the back of the tie to provide a holding place for the phone and the charging wire. There was adequate slack provided in the wire to compensate for any elasticity in the fabric (Hynek et al. 2005).

The scale of the digitally printed image matched the solar panel, making it easy to line up the designs and visually integrate the panel into the surface of the tie. The cotton sateen was intentionally chosen for its lustrous quality, as a means to simulate the reflective nature of the solar panels. The tie does not stretch during use; so, there is no distortion of the digitally printed pattern that would create an aberration. The solar panels were a close chromatic match to the inkjet printed material (Figure 6.24). The difference between the sheen of the printed material and the reflectivity of the solar panels was the most revealing trait. In the future, changing or modifying the laminate on the solar panel would help to match the reflective properties. Selection of fabric with more lustre, or treatment of the fabric with a reflective coating, could help eliminate differences in surface appearance between the fabric and the solar panel (Figure 6.25). The solar panels were attached using 'liquid stitch' adhesive, which eliminated the concern of accounting for a fastener in the design of the digitally printed pattern. The dot of solder used to connect the circuitry to the solar panels was a close colour match to the connection bus adhered to the solar panels and was not visually distracting. Application of the solar panels to the tie made it difficult to tie the knot (Figure 6.26). In typical tie knots, a small loop is created through which the large end of the tie is passed. At this step in the process, the physical width of the solar panel forces the user to enlarge their tie loop to allow the solar panel to pass through. The solar panels are limited to bending around a

FIGURE 6.24
Solar panels integrated with digitally printed fabric. (From Hynek, S.J. et al., *Apparel Technol. Manage*, 2005.)

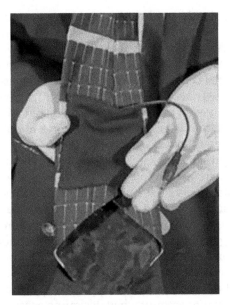

FIGURE 6.25
Back of tie shows pocket to hold the cell phone and 3.6 V plug-in to charge the cell phone. (From Hynek, S.J. et al., *Apparel Technol. Manage*, 2005.)

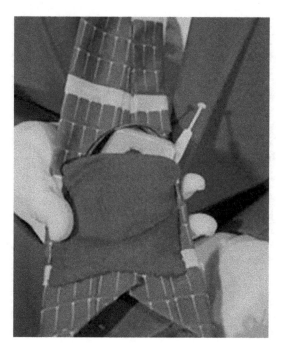

FIGURE 6.26
Phone in elastic pocket where the phone is charged. (From Hynek, S.J. et al., *Apparel Technol. Manage*, 2005.)

3 in. diameter and should not be creased. Usage of narrower solar panels was considered, but there would be sacrifices in power generation (Campbell and Parsons 2001). Applying solar panels to clip-on ties would eliminate this problem. The tie was tested by a user for performance in a typical business environment. The trial consisted of wear during 5 business days and storage in a closet. During the trials, there were no safety infringements with respect to electrical shock. A visual inspection of the insulated wire revealed no degradation of the insulation after the trials. The wires maintained their ability to conduct electricity without resistance.

Smart garments present new challenges to the textiles and apparel industry. There is a balance between integrating technology, so it adds value to a garment and maintaining visual appeal. Smart garment solutions appeal to niche consumers: small quantity manufacturing will be necessary. Digital textile printing can effectively meet these challenging demands by providing a visually appealing medium for integration of flexible photovoltaic thin-film cells or other electronic components. The digital print allowed the solar panels to be placed on the visually prominent face of the tie and create an unbroken pattern that gave the tie a formal wear appearance. The rigidity of the solar panels must be taken into account when designing accessories. This application of digital textile printing technology shows that it can play a significant role in the creation of visually appealing solar-enabled wearables.

6.7.1 Flexible Solar Cells

The flexible solar cells can be created on flexible surfaces such as plastic with low initial expenditure. Unlike conventional solar cells, the new, cheap material has no rigid silicon base. Instead, it is made of thousands of inexpensive silicon beads sandwiched between two thin layers of aluminium foil and sealed on both sides with plastic. Each bead functions has a tiny solar cell, absorbing sunlight and converting it into electricity. The flexible solar cells were sourced from different solar cell–manufacturing companies all over the world, and these collected samples were tested by means of load test and feasibility of integration with cloth material was verified. The panel PT 15-300, manufactured by M/s. Flex Solar Cells, United States, is approximately giving the expected test results. Based on calculations, 110 panels were purchased. The integration of the solar cells with the cloth was designed with stitching and fusing method.

6.7.2 Battery

When introducing the idea of flexible solar tent, the key question immediately arises: how much energy will be collected under real working conditions of a solar tent. This complex problem can be addressed by (1) modelling the energy yield of single solar cells, and (2) mobile measurements of complete current voltage characteristics under real-life operating conditions. Based on the simulations and energy calculations, the number of batteries required for solar tent was estimated and the interconnection of battery is shown in Figure 6.27. The batteries selected are sealed and maintenance-free. There is no corrosive gas generation during normal use and no need to check the specific gravity of the electrolyte or to add water during the service life. Also there is a low self-discharge, because of the use of lead calcium grids alloy and high purity materials. The solid copper terminals ensure highest current-carrying capability.

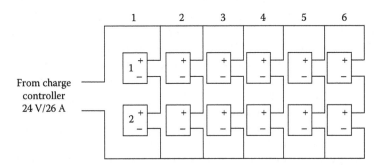

FIGURE 6.27
Interconnection of batteries.

6.7.3 Charge Controller

Charge controllers are designed to protect batteries from overcharge and overdischarge, which can be seriously damaging to the lead acid battery. Charge controllers use pulse width–modulated charging, for charging the batteries than the simple on/off logic of older and simpler units. This type of controller maximizes the amount of current coming out of the panels and going into the batteries. In this research work, the calculated solar cell output current is 11 A and the allowable current limit is 15%. Phocos Model charge controller operates in 12/24 V DC, with maximum charging current of 20 A selected for this work.

6.7.4 Inverter

A 2.2 kVA sine wave inverter operates in 48 V DC and 230 V AC output is selected. It generates a sinusoidal AC voltage with an exceptionally precise voltage and stabilized frequency. The inverter is protected against overload and short circuit. The advantage in giving the AC load is supplied load that can be utilized for charging the battery bank also. The load test is carried out and the voltage and current between the inverter is calculated.

6.8 Integration of Tent and Solar Cells

The integration of the solar cells and the cloth was carried out using two methods: (1) stitching of flexible solar cells and (2) fusing of flexible solar cells.

6.8.1 Stitching of Flexible Solar Cells

The integration of the flexible solar cells with the tent cloth has been made by stitching the solar panels with the surface of the tent cloth. The stitching line should fall in the outer surface of the solar panel so that it does not affect the solar cells. The area of the stitching line on the solar cells is marked with the lines in Figure 6.28. In this method, the integration of the flexible solar panel on grey cotton cloth is done by means of sewing operation. Stitching with the sewing thread is very easy, but life of the sewing thread is low in the case of frequent use of the panel. So fusing, along with sewing, to withstand the stitch, was selected. Initially the solar panel was stitched with double-side gum foam material. The double-side gummed foam material is a woven fabric of around 100 GSM and gum coating is given on both sides of the fabric, which acts as fusing binders

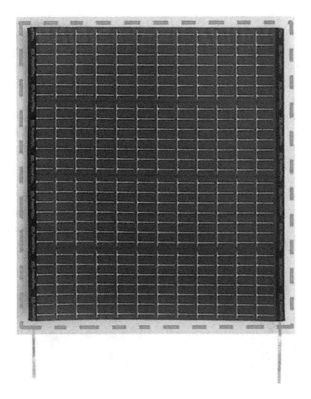

FIGURE 6.28
Stitching method of flexible solar cells.

during fusing. The foam was used in between the solar cells and grey cotton cloth. The binding materials in the foam will glue into the grey cotton cloth by the application of high pressure and temperature.

Flexible solar cells are stitched on the one-side gummed foam cloth with the gum side on the bottom. The stitching, cutting work has been carried out in M/s. Sri Nivas Fashion, Tirupur. After stitching, the solar cells have to be fused to the cotton cloth, which in turn should be attached to the tent. For the proper positioning of the solar panels on the grey cotton cloth, proper marking was done.

6.8.2 Fusing of Flexible Solar Cells

Fusing is the process of uniting the solar panel with surface of the tent by the application of high temperature and pressure with the help of the fusing machines. Normally, the fusing machines are used for the application of the sublimation printing on the fabric. In this research work, Single

Bed Pneumatic Classic/Digital Type Meters – Silicon Rubber Pad fusing machine with a 230 V power supply was used. To avoid the melting of the panels while applying high temperature, it was planned to stitch the panel with appropriate binders and then fusing was done. After stitching, the solar cells have to be fused to the cotton cloth which in turn should be attached to the tent. For the proper positioning of the solar panels on the grey cotton cloth proper, marking is done which is as shown in Figures 6.29 and 6.30. The fusing of the flexible solar cells is done on the cotton cloth which is stitched with one-sided gum foam by applying high temperature and pressure. Due to high temperature, the gum in the foam will evaporate and make the fabric to stick firmly on the cotton cloth. If the temperature goes beyond 150°C, there is a possibility that solar panels might melt. Hence, great care was taken to maintain the temperature within this limit. Fusing of flexible solar cells is carried out in M/s. Lion Textiles India, Tirupur, and it is shown in Figures 6.31 and 6.32.

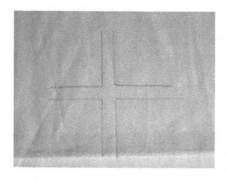 

FIGURE 6.29
Flexible solar cells fusing area marking. (From Ashok Kumar, L., *Int. J. Emerg. Trends Electric. Electron.*, 5(1), 108, 2013.)

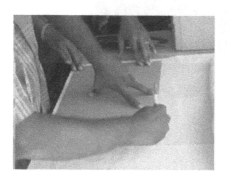 

FIGURE 6.30
Flexible solar cells fusing area after marking. (From Ashok Kumar, L., *Int. J. Emerg. Trends Electric. Electron.*, 5(1), 108, 2013.)

FIGURE 6.31
Panel fixed in the marked areas. (From Ashok Kumar, L., *Int. J. Emerg. Trends Electric. Electron.*, 5(1), 108, 2013.)

FIGURE 6.32
Pressure applied using fusing machine to the FSC. (From Ashok Kumar, L., *Int. J. Emerg. Trends Electric. Electron.*, 5(1), 108, 2013.)

6.9 Finishing and Integration of Flexible Solar Array

After fusing the cloth of the flexible solar cells each array has to be integrated properly for getting the full voltage. Totally six arrays were integrated using Velcro and buckles to get the completed tent. This type of finishing is shown in Figures 6.33 and 6.34. Figure 6.34 shows the completed solar tent structure.

FIGURE 6.33
Stitching the edge of the integrated FSC cloth. (From Ashok Kumar, L., *Int. J. Emerg. Trends Electric. Electron.*, 5(1), 108, 2013.)

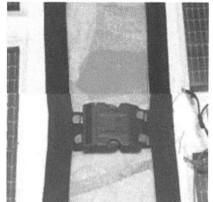

FIGURE 6.34
Flexible solar cells after finishing. (From Ashok Kumar, L., *Int. J. Emerg. Trends Electric. Electron.*, 5(1), 108, 2013.)

6.10 Testing of Flexible Solar Cell

The selection of the suitable flexible solar cells was based on the load test conducted on the samples collected from the different flexible solar cell–manufacturing companies. Load test is one of the methods used for testing the flexible solar cell (Figure 6.35). The voltmeter is connected to the solar

FIGURE 6.35
Completed solar tent. (From Ashok Kumar, L., *Int. J. Emerg. Trends Electric. Electron.*, 5(1), 108, 2013.)

panel and ammeter; rheostat is connected in series with the solar cell to vary the load in the circuit. Depending upon the variations in the load, there will be simultaneous change in the value of the current/voltage. By taking the average value from the recorded value, the capacity of the solar cell is finalized. The readings were taken on July 27, 2007, at 11 am. The arrangement of the circuit for the load test is as shown in Figure 6.36. The VI characteristics of model VHF-FSC, PT 15-300 and PM-TF-1 are shown in Figures 6.37 through 6.39. In the load test, the flexible solar cell's panel output current, weight, area and temperature-withstanding capacity are the four main parameters tested. Not all the panels satisfied the required parameters. The panel PT 15-300 is approximately giving the expected test results; hence, it was decided to employ the same for the power generation with the solar tent.

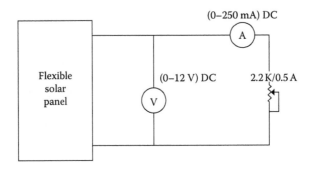

FIGURE 6.36
Load test on flexible solar cell.

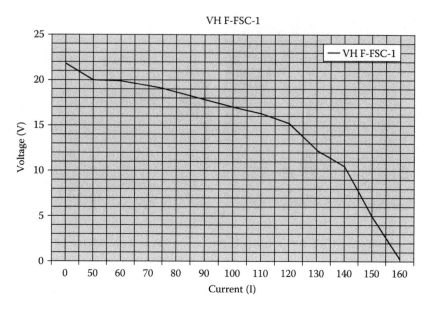

FIGURE 6.37
V-I characteristics of VHF-FSC-1.

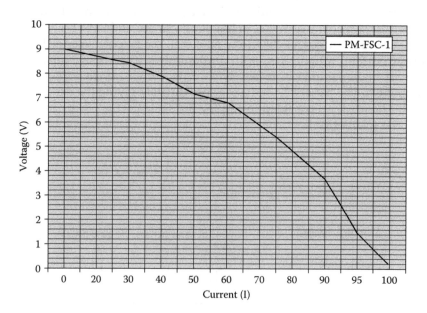

FIGURE 6.38
V-I characteristics of PM-FSC-1.

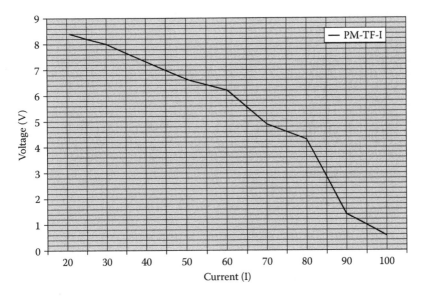

FIGURE 6.39
V-I characteristics of PM-TF-1.

6.11 Testing of Charge Controller

The testing of the 20 A charge controller has been carried out as follows. The testing arrangement of the charge controller circuit is shown in Figure 6.40. If the load current is maximum than the input current of array output, the battery will be supplying the rest of the current. In this test, a load of 10 W CFL has been connected and this can consume load of 0.42 A and the total output from array is 1 A, 24 V DC. The load is calculated for the following cases and the results are listed in the following text. The result of the test is shown in the following text in Table 6.1

Case 1: When the input current is removed, (solar power) the battery will be supplying the current to load.

Case 2: When battery is removed, the solar array will be supplying the power to load.

Case 3: When both are connected, 0.5 A will be supplied to the load and 0.5 A will be charging the battery.

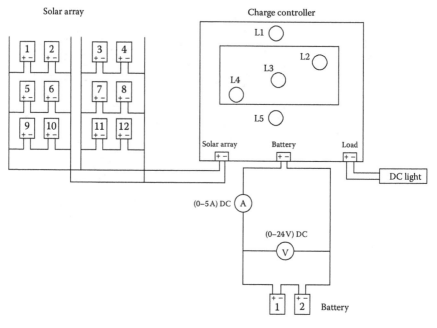

FIGURE 6.40
Connection diagram of charge controller.

TABLE 6.1

Output Loads in Charge Controller

Description	Specification	Load
Open circuit test	30 V	No load
Short circuit test	24 V, 1 A	10 W

6.12 Testing of Inverter

The 2 kVA inverter has been tested with and without supplying AC voltage. The voltage and current reading in the voltmeter and ammeter is varying, depending upon the load given to the circuit. With the increase in the load, the current increases with constant voltage. For testing the inverter without AC supply, DC load is given to the inverter circuit directly from the battery bank. The voltage and current reading in the voltmeter and ammeter varies depending upon the load given to the circuit. Table 6.2 shows the test results of the inverter. By analysing the values from the Table 6.2, it is concluded that the inverter can withstand a current rate from 15 to 20 A.

TABLE 6.2

Test Results of the Inverter

S. No.	Load (W)	With AC Supply		Without AC Supply	
		Voltage (V)	Current (A)	Voltage (V)	Current (A)
1	200	240	0.7	225	1.5
2	400	235	1.7	225	2.5
3	600	230	2.6	225	3.4
4	800	230	3.7	220	4.2
5	1000	230	4.8	220	5.1
6	1200	230	5	220	6
7	1400	230	6.1	220	7
8	1600	230	7	220	8

6.13 Testing of Cloth-Integrated Flexible Solar Cells

The testing of the solar cells which have been integrated into the cloth by means of stitching and fusing is done for its functionality. The main requirement is that it has to withstand in water. To confirm this property, flexible solar cell integrated with the cloth was kept in water for 6 h. After this period, the double-sided gummed foam material is detached from the cloth. In the integration, the double-sided gummed foam would not be suitable for flexible solar cells integrated with cloth. The other option in this fusing method is the use of one-sided gummed foam for stitching and fusing with the cloth. The solar cells are stitched with polyester sewing thread in the one-sided gummed foam which is 100 GSM in weight. The stitching of the panel is done such that the gum side of the foam is utilized for the fusing purpose. The stitched foam is then fused to the grey cotton cloth with the panel facing the top surface. The same testing procedure is carried out with this sample also. The solar cells are not damaged and it can withstand high temperature, and the panel is fixed firmly on to the cotton cloth.

6.14 Integration and Testing

The hut shape defence tent integrated with 110 panels of Power Film PT 15-300 flexible solar cells is designed for 1 kW load. Figure 6.41 shows the complete block diagram with charge controller, inverter and battery setup.

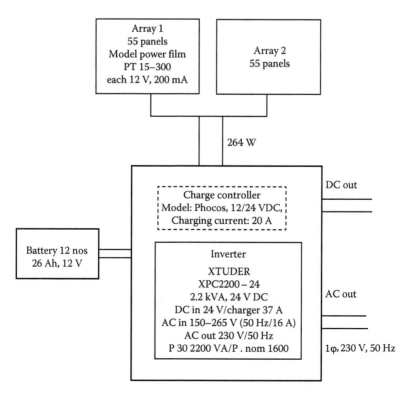

FIGURE 6.41
Block diagram of the solar tent system.

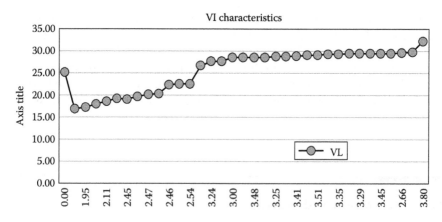

FIGURE 6.42
Characteristics of 110 panels of solar array for the load of 10 Ω.

TABLE 6.3

Material Used for Solar-Powered Defence Tent

S. No.	Description	Specification
1.	Hut shape defence tent	5.50 m length, 5.5 m width, 3.85 m centre height and 1.85 m wall height
2.	Flexible solar cells	PowerFilm PT 15-300, 12 V, 200 mA, 2.4 W, Dimension: 27 × 32.5 cm, Weight: 94.5 g, panel area: 0.087 m²
3.	Charge controller	Model: Phocos, 12/24 V DC, charging current: 20 A, Max. load current: 20 A
4.	Inverter	Sine wave inverter, 2.2 kVA, input: 48 V DC, output: 230 V AC, ±3%, frequency: 50 Hz

Testing was carried out on a typical sunny days having maximum solar radiation of 805 W/m² measured using digital pyranometer during March 29 and 30, 2010. For the single flexible solar panel, open circuit voltage is 17.5 V and the short circuit current is 200 mA, the VI characteristics of the flexible solar array consisting of 110 panels are shown in Figure 6.42. Table 6.3 shows the materials used for solar-powered defence tent.

6.15 Solar Jacket

The purpose of solar jacket is to utilize the solar energy and to create wearable electronic garments for functional purpose like military and for the persons who are in need of electrical energy in the remote areas. Here the solar cells are embedded in various parts of the garment and energy is generated and converted into electrical energy and stored in suitable devices for charging electronic gadgets. A solar-powered jacket concentrates on the integration of electronic functions, components and systems into textile products. These jackets will allow wearers to carry, connect and charge their portable digital devices like light mobile charger in one convenient and fully integrated package. Strategically mounted panels are pre-wired, providing solar power to individual devices for power storage (charging) or immediate consumption. Combined with integrated charge control and optional battery/charger systems, it provides the conveniences of back-up and always on, on-demand small-scale solar electrical power. The jacket has one photovoltaic panel that fits onto its back. The batteries are wired to all the pockets, which can have almost any mobile devices plugged into them like light mobile charger, etc. Two-inch-by-three-inch (30 × 15 × 1.5 cm³) silicon-based polycrystalline solar panels are embedded in the jacket capable of converting 6 V of sunlight. A 5 V connection charges the iPod and a 6 V connection charges the mobile for about 4 h, but the battery is completely loaded in 8 h of sun exposure. The solar modules placed in the coat charge the NiMH battery, which can be used to charge a device directly or store the power until needed. The battery weighs around 100 g. It requires 10–15 h to load and, when full, can charge a mobile phone or iPod in less than 4 h. The panel is water resistant and removable; it can be used separately from the jacket when not needed. Electronic components like solar cells, resistors and LEDs are integrated directly into the textile and wired together into working circuits using thin wires.

6.16 Future Research Scope

For the moment, much of the buzz surrounding wearable technology has to do with the upcoming release of Google Glass. While Google Glass may be the future, all types of wearable technology exist that you can buy today. From wearable cameras and sleep monitors to augmented reality glasses and gaming gear, the technology exists now. Wearable technology devices collect information from biofeedback sensors attached to your clothing or mobile devices and can be applied to matters of individual health and safety, or just entertainment. Much like assistive technology, the application of wearable technology has only just begun to be realized for its possibilities. The field of wearable technology is broad, and the list of applications is limited only by developers' imaginations and undeveloped markets. Much wearable technology as it exists today resides as a $700 million industry in the form of sports and activity trackers, and that figure is conservatively expected to at least double in the next year. Looking at what is already available in the area of wearable technology is an interesting exercise in both seeing the possibilities of tomorrow as well as taking in the history of technology.

6.17 Summary

In this chapter, demonstration of clothing integrated flexible solar cells, equipped with commercial amorphous silicon, was discussed. The PT 15-300 flexible solar cells integrated with defence tents which can be utilized for the power generation were successfully developed with the help of fusing and stitching method and also it can be used for makeshift office in the defence field to power the wearable electronic products. With the help of this tent, a load up to 1 kW can be powered and in future, this load can be increased, depending upon the requirements. Clothing-integrated photovoltaics are an interesting option for powering mobile electronic devices. Among various technologies for flexible solar modules, Si thin films currently offer the most promising combination of flexibility and performance, especially under typical low-light conditions. Ultrathin pc-Si cells are ready for direct lamination with textiles, while high-efficiency c-Si transfer cells are suitable for add-on accessories or pocket integration. Low-loss charge controllers, textile wiring and specialized knowhow on integration with garment production will pave the road to the first IPV products. High performance and high end use in the military mission in combat environment, the integration of flexible solar cells into garment is essential. The energy generation and storage level of solar technology will give a new dimension to the photovoltaic research in the future.

Bibliography

Ashok Kumar, L., Design and development of illuminated clothing with PMMA for versatile applications, *Indian J. Fibre Text. Res.*, 31(4), 577–579 (2006).

Ashok Kumar, L., Emerging trends in flexible wearable electronics for versatile applications, *Indian Text. J.*, 2(4), 62–66 (2010).

Ashok Kumar, L., Stand-alone solar power generation system with constant current discharge, *Int. J. Emerg. Trends Electric. Electron.*, 5(1), 108–115 (2013), ISSN: 2320-9569.

Ashok Kumar, L. and T. Ramachandran, Emerging trends in flexible wearable electronics for versatile applications, *Indian Text. J.*, 7, 62–66 (2010), ISSN: 0019-6436.

Ashok Kumar, L. and T. Ramachandran, Design and development of telemonitoring system using mobile network for medical applications, *Natl. J. Technol.*, 7(1), 1–7 (2011).

Ashok Kumar, L., A.N. Saikrishnan and A. Venkatachalam, Design and development of illuminated clothing with PMMA for versatile applications, *Indian J. Fibre Text. Res.*, 31(4), 577–579 (2006).

Ashok Kumar, L. and A. Venkatachalam, Electrotextiles: Concepts and challenges, *Natl. Conf. Funct. Text. Apparels*, 2(1), 75–84 (2007).

Auer, R. and R. Brendel, Textured monocrystalline thin-film Si cells from the porous silicon (PSI) process, *Progress in Photovoltaics*, 9(3), 217–221 (2001).

Axisa, F., P.M. Schmitt, C. Gehin, G. Delhomme, E. McAdams and A. Dittmar, Flexible technologies and smart clothes for citizen medicine, home healthcare, and disease prevention, *IEEE Trans. Inf. Technol. Biomed.*, 9(3), 325–336 (2005).

Baard, M., E-fabrics still too stiff to wear, *Wired News*, 5, 2–4 (December 5, 2002), www.wired.com/news/technology/0,1282,56708,00.html.

Bailat, J., V. Terrazzoni-Daudrix, J. Guillet, F. Freitas, X. Niquille, A. Shah, C. Ballif et al., *Proceedings of the 20th EU-PVSEC*, Barcelona, Spain, p. 1529 (2005).

Ball, P., Shoes and sheets get wired, *Nature Science Update*, 151, 856–857 (December 6, 2002a), www.nature.com/nsu/nsu_pf/021202/021202–11.html.

Ball, P., TV on a T-shirt, *Nature Science Update*, 84, 537–541 (May 22, 2002b), www.nature.com/nsu/020520/020520-4.html.

Barnett, A. and R. Resch, *Technical Digest of the 15th International Photovoltaic Science and Engineering Conference*, Shanghai Scientific & Technical Publishers, Shanghai, China, p. 20 (2005).

Berge, C., R.B. Bergmann, T.J. Rinke and J. H. Werner, Monocrystalline Si thin film solar cells by layer transfer, In *Proceedings of the 17th European Photovoltaic Solar Energy Conference*, WIP-Renewable Energies, Munich, Germany, p. 1039 (2002).

Berge, C., T.A. Wagner, W. Brendle, C. Craff-Castillo, M.B. Schubert and J.H. Werner, Flexible monocrystalline Si films for thin film devices from transfer processes, *Mater. Res. Soc. Symp. Proc.*, 769, 162–167 (2003).

Berge, C., M. Zhu, W. Brendle, M.B. Schubert and J.H. Werner, 150 mm layer transfer for monocrystalline silicon solar cells, In *Technical Digest of the 14th International Photovoltaic Science and Engineering Conference PVSEC-14*, Bangkok, Thailand, pp. 31–32 (2004).

Bergmann, R.B., C. Berge, T.J. Rinke, J. Schmidt and J.H. Werner, Advances in mono-crystalline Si thin film solar cells by layer transfer, *Sol. Energy Mater. Sol. Cells*, 74, 213–218 (2002).

Bergmann, R.B. and J.H. Werner, The future of crystalline silicon films on foreign substrates, *Thin Solid Films*, 162, 403–404 (2002).

Berry, P., G. Butch and M. Dwane, Highly conductive printable inks for flexible and low-cost circuitry, In *Proceedings of the 37th International Symposium on Microelectronics, IMAPS*, California (2004).

Bonderover, E. and S. Wagner, Woven inverter circuit for e-textile, *Appl. IEEE Electron Device Lett.*, 25(5), 295–297 (2004).

Brabec, C.J., Organic photovoltaics: Technology and market, *Solar Energy Mater. Solar Cells*, 83, 273–292 (2004).

Brendel, R., R. Auer and H. Artmann, Textured monocrystalline thin-film Si cells from the porous silicon (psi) process, *Mater. Res. Soc. Symp. Proc.*, 9, 217–221 (2001).

Brendle, W., V. Nguyen, K. Brenner, P.J. Rostan and A. Grohe, Low temperature back contact for high efficiency silicon solar cells, In Palz, W. et al. (eds.), *Proceedings of the 20th European Photovoltaic Solar Energy Conference*, WIP-Renewable Energies, Munich, Germany, p. 745 (2005).

BST Berger Safety Textiles, German Textile Innovations group, Tapes and ropes, Maulburg, Germany. www.berger-safety-textiles.com. Accessed 3 August, 2012.

Campbell, J.R. and J. Parsons, Intersecting two design problems to create multiple points of exploration, In *Exploring the Visual Future: Art Design, Science and Technology*, Selected Readings of the International Visual Literacy Association, Blacksburg, VA, pp. 51–56 (2001).

Carter, E.C., Y. Ohno, M.R. Pointer, A.R. Robertson, R. Seve, J.D. Schanda and K. Witt, *Colorimetry*, 2nd edn., Commission Internationale de l'Eclairage, Vienna, Austria (1986).

Chwang, A.B., M.A. Rothman, S.Y. Mao, R.H. Hewitt, M.S. Weaver, J.A. Silvernail1, K. Rajan et al., Thin film encapsulated flexible organic electroluminescent displays, *Appl. Phys. Lett.*, 83, 413 (2003).

Contreras, M.A., K. Ramanathan, J. AbuShama, F. Hasoon, D.L. Young, B. Egaas and R. Noufi, Short communication: Accelerated publication: Diode characteristics in state-of-the-art ZnO/CdS/Cu $(In_{1-x}Ga_x)$ Se_2 solar cells, *Prog. Photovolt.: Res. Appl.*, 13, 209 (2005).

Dunne, L.E., S. Brady, B. Smyth and D. Diamond, Initial development and testing of a novel foam-based pressure sensor for wearable sensing, *J. Neuro Eng. Rehabil.*, 2(4), 7–13 (2005), accessed through www.jneuroengrehab.com.

Dunne, L.E., A. Toney, S.P. Ashdown and B. Thomas, Subtle integration of technology: A case study of the business suit, In *Proceedings of the 1st International Forum on Applied Wearable Computing (IFAWC)*, Bremen, Germany, pp. 35–44 (2004).

Eisenberg, A., For the smart dresser, electric threads that cosset you, *New York Times*, 13, 7 (February 6, 2003), www.nytimes.com/2003/02/06/technology/circuits/06next.html?ex=1045545602&ei=1&en=419c725550669347.

Elton, S.F., Ten new developments for high-tech fabrics & garments invented or adapted by the research & technical group of the defence clothing and textile agency, In *Proceedings of the Avantex, International Symposium for High-Tech Apparel Textiles and Fashion Engineering with Innovation-Forum*, Frankfurt-am-Main, Germany, pp. 27–29 (2000).

Feldrapp, K., R. Horbelt, R. Auer and R. Brendel, Thin-film (25.5 μm) solar cells from layer transfer using porous silicon with 32.7 mA/cm^2 short-circuit current density, *Prog. Photovolt.: Res. Appl.*, 11, 105 (2003).

Gemmer, C. and M.B. Schubert, Solar cell performance under different illumination conditions, *Mater. Res. Soc. Symp. Proc.*, 664, A25.9 (2001).

Gemmer, J., Distribution of local entropy in the Hilbert space of bi-partite quantum systems: origin of Jaynes' principle, *The European Physical Journal B - Condensed Matter and Complex Systems*, 31(2), 249–257 (2003).

Gould, P., Textiles gains intelligence, *Mater. Today*, 6(10), 38–47 (2003).

Guha, S., *Technical Digest of the 15th International Photovoltaic Science and Engineering Conference*, Shanghai Scientific & Technical Publishers, Shanghai, China, p. 35 (2005).

Hartmann, W.D., K. Steilmann and A. Ullsperger, *High-Tech Fashion*, Heimdall, Witten, Germany (2000).

Hatcher, M., Fiber-optic dress goes down the aisle, Optics.org, 17, 8–10 (October 8, 2002a), www.optics.org/articles/news/8/10/11/1.

Hatcher, M., France Telecom debuts fiber screen, Optics.org, 23, 21–23 (July 2, 2002b), www.optics.org/articles/news/8/7/1/1.

Hegedus, S. and N. Okubo, *Conference Record of the 31st IEEE Photovoltaic Specialists Conference*, IEEE, New York, p. 1651 (2005).

Herrmann, D., F. Kessler, F. Lammer and M. Powalla, *Nineteenth European Photovoltaic Solar Energy Conference*, *Proceedings of the 20th European Photovoltaic Solar Energy Conference*, WIP-Renewable Energies, Munich, Germany, p. 1875, F15.1 (2005).

Herrmanna, D., F. Kesslera, U. Klemma, R. Kniesea, T.M. Friedlmeiera, S. Spieringa, W. Wittea and M. Powalla, Flexible, monolithically integrated CuSe$_2$ thin-film solar modules, *Mater. Res. Soc. Symp. Proc.*, Boston, MA, pp. 865–867 (2005).

HighSolar, Plumber, Photovoltaics, Ulm, Germany. www.highsolar.de. Accessed 10 June, 2012.

Hogan, J., Fashion firm denies plan to track customers, *New Scientist*, 11, 178–181 (2003).

Hohenstein Institutes, Dr. Jürgen Mecheels, Testing Service, Bönnigheim, Germany. www.hohenstein.de/en. Accessed 25 May, 2012.

Hynek, S.J., J.R. Campbell and K.M. Bryden, Application of digital textile printing technology to integrate photovoltaic thin film cells into wearables, *J. Text. Apparel Technol. Manage.*, 4(3), 17–23 (2005).

Institut für Textil- und Verfahrenstechnik (ITV), Prof. Dr.-Ing. Götz T. Gresser's, Textile Innovations, Denkendorf, Germany. www.itv-denkendorf.de. Accessed 10 June, 2012.

Ishikawa, Y. and M.B. Schubert, *Proceedings of the 20th EUPVSEC*, Barcelona, Spain, p. 1525 (2005).

Ito, M., C. Koch, V. Svrcek, M.B. Schubert and J.H. Werner, Silicon thin film solar cells deposited under 80°C, *Thin Solid Films*, 383(1–2), 129–131 (2001).

Jeffrey, F., Iowa Thin Films, Boone, IA. www.iowathinfilm.com. Accessed 25 September, 2013.

Jeffrey, F.R., G.D. Vernstrom, F.E. Aspen and R.L. Jacobson, Fabrication of amorphous silicon devices on plastic substrates, *Mater. Res. Soc. Symp. Proc.*, 49, 41 (1985).

Jongerden, G.J., Monolithically series integrated flexible PV modules manufactured on commodity polymer substrates, In *Proceedings of the Third World Conference on Photovoltaic Energy Conversion*, Osaka, Japan, IEEE, New York, Vol. 2, pp. 2109–2111 (2003).

Kallmayer, C., T. Linz, R. Aschenbrenner and H. Reichl, System integration technologies for smart textiles, *MST News*, 2, 42–43 (2005).

Kampmann, A., J. Rechid, A. Raitzig, S. Wulff, M. Mihhailova, R. Thyen and K. Kalberlaha, Electrodeposition of CIGS on metal substrates, *Mater. Res. Soc. Symp. Proc.*, 763, B8.5 (2003).

KANZ, Children's clothing, Neufra, Germany. www.kanz.com. Accessed 25 May, 2012.

Kemble, W., Shell Solar Research Facilities, München, Germany. www.shellsolar.com. Accessed 17 June, 2012.

Kessler, F., D. Herrmann and M. Powalla, Approaches to flexible CIGS thin-film solar cells, *Thin Solid Films*, 480–481, 491 (2005).

Kirstein, T., D. Cottet, J. Grzyb, and G. Troester, Textiles for signal transmission in wearables. In *Proc. ACM of First Workshop on Electronic Textiles (MAMSET 2002)*, San Jose, CA, Vol. 3, pp. 47–51 (2002).

Koch, C., M. Ito and M. Schubert, Low-temperature deposition of amorphous silicon solar cells, *Sol. Energy Mater. Solar Cells*, 68, 227 (2001).

Koch, C., M. Ito and M.B. Schubert, Low temperature deposition of amorphous silicon solar cells, *Solar Energy Mater. Solar Cells*, 43, 7909–7911 (2004).

Koch, E., T.J. Rinke, R.B. Bergmann, and J.H. Werner, Efficient thin film solar cells by transfer of monocrystalline Si layers, In Scheer, H., McNelis, B., Palz, W., Ossenbrink, H.A., and Helm, P. (eds.), *Proc. 16th European Photovoltais Solar Energy Conference*, James & James Science Publishers Ltd., London, U.K., pp. 1128 (2000).

Koch, H., M. Ito, M. Schubert and J.H. Werner, Low temperature deposition of amorphous silicon based solar cells, *Mater. Res. Soc. Symp. Proc.*, 557, 749 (1999).

Lagref, J.J., Md.K. Nazeeruddin and M. Grätzel, Molecular engineering on semiconductor surfaces: Design, synthesis and application of new efficient amphiphilic ruthenium photosensitizers for nanocrystalline TiO_2 solar cells, *Synth. Met.*, 138, 333 (2003).

Leckenwalter, R., Innovations in new dimensions of functional clothing, In *Proceedings of the Avantex International Symposium for High-Tech Apparel Textiles and Fashion Engineering with Innovation-Forum*, Frankfurt am Main, Germany, pp. 3–16 (2000).

Linz, T. and C. Kallmayer, Embroidering electrical interconnects with conductive yarn for the integration of flexible electronic modules into fabric, In *IEEE ISWC*, Osaka, Japan (2005a).

Linz, T. and C. Kallmayer, New interconnection technologies for the integration of electronics on textile substrates, In *Ambience 2005*, Tampere, Finland (2005b).

Lippolda, G., H. Neumanna and A. Schindler, Ion beam assisted deposition of Cu(In,Ga)Se₂ films for thin film solar cells, *Mater. Res. Soc. Symp. Proc.*, 668, H3.9 (2001).

Mann, S., Smart clothing: The shift to wearable computing, *Commun. ACM*, 39(80), 23–24 (1996).

Mei, T., Y. Ge, Y. Chen, L. Ni, W. Liao, Y. Xu and W. J. Li, Design and fabrication of an integrated three-dimensional tactile sensor for space robotic applications, In *IEEE MEMS*, Orlando, FL, pp. 112–117 (1999).

Meoli, D. and T. May-Plumlee, Interactive electronic textile development: A review of technologies, *J. Text. Apparel Technol. Manage.*, 2(2) (2002).

Moore, S.K., Just one word-plastics, organic semiconductors could put cheap circuits everywhere and make flexible displays a reality, *IEEE Spectrum*, 13, 55–59 (2002).

MUSTANG Bekleidungswerke, MEHR LESEN, Denim Collection, Künzelsau, Germany. www.mustang.de. Accessed 23 May, 2012.

Niggemann, M., M. Glatthaar, A. Gombert, A. Hinsch and V. Wittwer, Diffraction gratings and buried nano-electrodes—Architectures for organic solar cells, *Thin Solid Films*, 451–452, 619–623 (2004).

Otto, C., A. Milenkovic, C. Sanders and E. Jovanov, Final results from a pilot study with an implantable loop recorder to determine the etiology of syncope in patients with negative noninvasive and invasive testing, *Am. J. Cardiol.*, 82, 117–119 (1998).

Park, S. and S. Jayaraman, Adaptive and responsive textile structures, In Tao, X. (ed.), *Smart Fibers, Fabrics and Clothing: Fundamentals and Applications*, Woodhead, Cambridge, U.K., pp. 226–245 (2001).

Parsons, J. and J.R. Campbell, Digital apparel design process: Placing a new technology into a framework for the creative design process, *Cloth. Text. Res. J.*, *Special Issue on Design*, 22, 88–98 (2004).

Post, R. and M. Orth, Smart fabric, wearable clothing, In *Proceedings of the First International Symposium on Wearable Computers*, IEEE, Cambridge, MA, pp. 167–168 (1997).

Power film solar Inc., Ames, IA, www.powerfilmsolar.com. Accessed 11 March, 2012.

Preiser, K., Photovoltaic systems, In Luque, A. and Hegedus, S. (eds.), *Handbook of Photovoltaic Science and Engineering*, John Wiley & Sons, Chichester, U.K., p. 753 (2003).

Rantanen, J. and M. Hännikäinen, Data transfer for smart clothing: Requirements and potential technologies, In Tao, X. (ed.), *Wearable Electronics and Photonics*, Woodhead Publishing in Textiles, Cambridge, U.K., p. 198 (2005).

Raskovic, D., T. Martin and E. Jovanov, Medical monitoring applications for wearable computing, *Comput. J.*, 47(4), 495 (2004).

Rau, U. and J.H. Werner, A multi-diode model for spatially inhomogeneous solar cells, *Thin Solid Films*, 487, 14 (2005).

Rinke, T.J., R.B. Bergmann and J.H. Werner, Quasi-monocrystalline silicon for thin-film devices, *Appl. Phys. A*, 68, 705 (1999).

Rojahn, M., M. Rakhlin and M.B. Schubert, Photovoltaics on wire, In *MRS Proceedings*, Material Research Society, Cambridge University Press, U.K., Vol. 664, A2.1 (2001).

Rostan, P.J., U. Rau, V.X. Nguyen, T. Kirchartz, M.B. Schubert and J.H. Werner, *Technical Digest of the 15th International Photovoltaic Science and Engineering Conference*, Shanghai Scientific & Technical Publishers, Shanghai, China, pp. 214–215 (2005).

Russell, D.A., S.F. Elton, J. Squire, R. Staples and N. Wilson, First experience with shape memory material in functional clothing, In *Proceedings of the Avantex, International Symposium for High-Tech Apparel Textiles and Fashion Engineering with Innovation-Forum*, Frankfurt-am-Main, Germany (2000).

Schroppa, R.E.I., C.H.M. Van Der Werf, M. Zeman, M.C.M. Van de Sanden, C.I.M.A. Spee, E. Middelman, L.V. de Jonge-Meschaninova et al., Novel amorphous silicon solar cell using a manufacturing procedure with a temporary superstrate, *Mater. Res. Soc. Symp. Proc.*, 557, 713 (1999).

Schubert, D., Dargusch, R., Raitano, J., and Chan, S., Cerium and yttrium oxide nanoparticles are neuro-protective, *Biochemical and Biophysical Research Communications*, 342(1), 86–91 (2006).

Schubert, M.B., Y. Ishikawa, J.W. Kramer, C.E.M. Gemmer and J.H. Werner, Clothing integrated photovoltaics, *Mater. Today*, 9(6), 42–47 (2006).

Schubert, M.B. and J.H. Werner, Flexible solar cells for clothing, In *Conference Record of the 31st IEEE Photovoltaic Specialists Conference*, IEEE, New York, p. 1488 (2005).

Schubert, M.B. and J.H. Werner, Flexible solar cells for clothing, *Mater. Today*, 9(6), 254–258 (2006).

Tanda, M., K. Tabuchi and M. Uno, Fabrication of high efficiency a-Si thin-film solar cell on flexible PDMS substrate by very high frequency plasma chemical vapor deposition, In *Conference Record of the 31st IEEE Photovoltaic Specialists Conference*, IEEE, New York, p. 1560 (2005).

Tayanaka, H., K. Yamauchi and T. Matsushita, Thin-film crystalline silicon solar cells obtained by separation of a porous silicon sacrificial layer, In *Proceedings of the Second World Conference on Photovoltaic Solar Energy Conversion*, European Commission, Ispra, Italy, p. 1272 (1998).

Tempex GmbH, Functional clothing, Heidenheim, Germany. www.tempex.com. Accessed 17 June, 2012.

Tiwari, A.N., A. Romeo, D. Baetzner and H. Zogg, Flexible CdTe solar cells on polymer films, *Prog. Photovolt.: Res. Appl.*, 9(3), 211–215 (2001).

Tobera, O., J. Wienkea, M. Winklera, J. Penndorfa and J. Grieschea, Current status and future prospects of CISCuT based solar cells and modules, *Mater. Res. Soc. Symp. Proc.*, 763, B8.16 (2003).

Tomita, T., *Conference Record of the 31st IEEE Photovoltaic Specialists Conference*, IEEE, New York, p. 7 (2005).

Troster, G., T. Kirstein and P. Lukowicz, Wearable computing: Packaging in textiles and clothes wearable computing lab, In *14th European Microelectronics and Packaging Conference & Exhibition*, Friedrichshafen, Germany, pp. 23–25 (2003).

Virtuani, A., E. Lotter, M. Powalla and U. Rau, Influence of Cu content on electronic transport and shunting behavior of Cu (In, Ga) Se$_2$ solar cells, *J. Appl. Phys.*, 99, 014906 (2006).

Werner, J.H., Institut für Physikalische Elektronik, Universität Stuttgart. www.ipe. unistuttgart. Accessed 10 June, 2012.

Wemer, J.H., F.A. Wagner, C. Gemmer, C. Berge, W. Brendle and M.B. Schubert, Recent progress on transfer-Si solar cells at ipe Stuttgart, In *Proceedings of the Third World Conference on Photovoltaic Energy Conversion*, Arisumi Printing, Tokyo, Japan, 1272-1, p. 275 (2003).

Werner, J.H., *Proceedings of the Third World Conference Photovoltaic Energy Conversion*, Osaka, Japan, IEEE, New York, Vol. 2, p. 1272 (2003).

Yang, J., A. Banerjee and S. Guha, Triple-junction amorphous silicon alloy solar cell with 14.6% initial and 13.0% stable conversion efficiencies, *Appl. Phys. Lett.*, 70, 2975 (1997).

Zeman, M., G. Tao, M. Trijssenaar, J. Willemen, W. Metselaar and R. Schropp. *Mater. Res. Soc. Symp. Proc.* 377, 639–642 (1995).

7

Garment-Integrated Wearable Electronic Products

7.1 Introduction

This chapter discusses the development of fabric-integrated wearable electronic products for the use of wearable mobihealth care system, smart clothes for ambulatory remote monitoring, electronic jerkin, heating gloves and pneumatic gloves. There is a brief discussion on product design and integration of electronic circuits, and the products discussed in this chapter are electronic jerkin, pneumatic gloves and heating gloves. Many researchers and scientists around the globe are trying to improve the design concept of wearable electronics products and their comfort level to wearers based on functional end uses. These kinds of efforts may bring wearable products in aesthetic and long service point of view. Heating gloves, in particular, will be very useful for defence personnel who are work under extreme climate conditions, and for physically challenged people, pneumatic gloves have been designed to enable them to lead a normal life.

7.2 Wearable Mobihealth Care System

Wearable mobihealth care systems (WS) aim at providing long-term continuous monitoring of vital signs for high-risk cardiovascular patients and its potential has been analysed (Zheng et al. 2007). They used a portable patient unit (PPU) and a wearable shirt (WS) to monitor electrocardiogram (ECG), respiration (*acquired with respiratory inductive plethysmography* [*RIP*]) and activity. Owing to the integration of fabric sensors and electrodes endowed with electro-physical properties into the WS, long-term continuous monitoring can be carried out without making patients uncomfortable or restricting their mobility. The PPU analyses physiological signals in real time and determines whether the patient is in danger or needs external help. The PPU will alert the patient, and an emergency call will be automatically made to a medical service centre (MSC) when life-threatening arrhythmias or falls are detected. With advanced gpsOne technology, the patient can be located and rescued instantly whether he/she is in indoors or outdoors in case of emergency. In recent years, there has been an increasing interest in wearable telemedicine monitoring system both in research and market areas. Wearable telemedicine, which can provide non-invasive and continuous monitoring of multiple physiological parameters, is expected to be the most important and feasible method under the new-generation medical care system.

Intelligent wearable electronics and telemedicine are very promising development areas, which will extend monitoring, increase patient's comfort and improve their living standard. Traditional personal medical monitoring systems, such as holter monitors, have been used to collect data for offline processing. Systems with multiple sensors typically feature several wires between the electrodes and the monitoring devices. These wires may limit a patient's activity and reduce the level of comfort. Therefore, there has been an increasing interest in health-care-monitoring devices with wearable technology. In recent years, there has been a proliferation of consumer wearable monitoring devices in sports and recreational field. There are sophisticated devices available today that provide real-time heart rate information and let users store and analyse their physiological data on their home PCs. BodyMedia has developed an armband that can measure galvanic skin response, skin temperature, activity and heat flux. However, in all cases, the physiological data are analysed on a home PC at a later time. Monitoring physiological parameters in a mobile environment has been widely studied by many research groups. However, majority of such existing devices are more or less not satisfying. Lifeguard has been successfully used in a number of application domains. But too many disturbing sensors, electrodes and wires make the device very difficult to use. AMON is a wearable (*wrist-worn*) medical monitoring and alert system. It includes multiple vital signs monitoring, online analysis and a cellular connection to a medical centre. But the system is unable to provide acceptable performance in the detection of ECG. Moreover, the monitoring of ECG is not continuous, since during a measurement, the device must touch the left hand, the right hand and the abdomen of the wearer.

Vital phones with built-in global positioning system (GPS) can record ECG with a mobile phone and transmit data to a service centre where the diagnosis can be made. But it does not provide online diagnosis and can only locate an outdoor patient with the GPS antenna having an unrestricted view of the sky. Several research projects consider processing the medical data on a local device. EPI-MEDICS is able to detect standard 12-lead ECG, make online diagnosis and raise alarms. But the monitoring is not continuous, since the patient turns on the device only when he/she feels uncomfortable. MOLEC provides real-time classification of ECG on a PDA and forwards alarms to the hospital in case of high-risk arrhythmias. But the previous two devices are both not really wearable devices.

Perhaps the most relevant project work is the WEALTHY aiming to design a garment with embedded ECG sensors, respiration sensors and activity sensor for continuous monitoring of vital signs. WEALTHY is based on a wearable interface implemented by integrating fabric sensors, signal processing and telecommunication on a textile platform. But the authors report neither experimental results in a real setting nor algorithms for automatic diagnosis and alarm. Effective GPS has not been integrated in the WEALTHY. The system aims to provide long-term continuous monitoring of vital signs for high-risk cardiovascular patients. It includes real-time collection, online diagnosis, automatic alarms, secure wireless links and effective locating mechanism. Taking all the previous devices into consideration, he most innovative ones are those that have the combination of wearable monitoring and real-time diagnosis; this system has the following several unique features:

- *Highly integrated wearable system*: Sensors and electrodes had been designed to be non-intrusive for health status monitoring.
- *User-friendly designs*: Since the hardware of the WS is designed such that it can be removed from the WS at will and the tracks embedded in the WS are made of special textile materials resistant to twist, the wash-ability and re-usage of the WS are made possible. Moreover, since the connection between the PPU and the WS is wireless, the hardware size of the WS is greatly reduced; therefore, patients can feel more comfortable.
- *Online diagnosis and automatic alarms*: The system provides online diagnosis and three degrees alarm on local portable device (PPU). This will reduce communication costs, because only anomalous physiological signals can be transmitted to the MSC.
- *Advanced locating mechanism*: Applying with GPS One technology, the patient can be located whether he/she is indoors or outdoors.

Wearable mobihealth care system is composed of three parts: the WS, the PPU and the MSC. Figure 7.1 shows a schematic overview of the system. The fabric sensors and electrodes endowed with electro-physical properties enable the making of the WS capable of continuously recording a full 12-lead ECG, respiration and activity. The physiological signals measured by the WS are transferred to the PPU via short-range wireless communication. The PPU provides real-time visualization, memorizing, analysis, diagnosis and three degrees of alarm. In case of emergency, an emergency call will be automatically initiated and it would be followed with immediate rescue action, according to the location information of the patient acquired through gpsOne. Moreover, the PPU also has three manually activated buttons for summoning help. The MSC is composed of a medical data server (MDS) and several monitoring terminals (MTs), where the doctors and medical

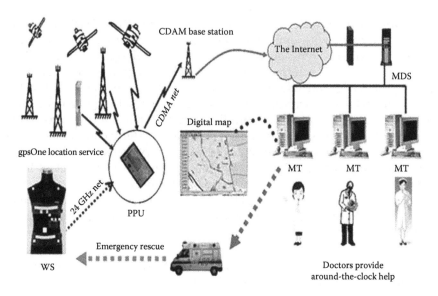

FIGURE 7.1
System overview of the wearable mobihealth care system. (From Zheng, J.W. et al., *Med Biol. Eng. Comput.*, 45, 877, 2007.)

assistants will provide round-the-clock medical service. The physiological data received from the PPU will be processed by specialized software running on the MT and more detailed diagnosis information would be provided.

7.2.1 Development of the Wearable Shirt

The WS mainly woven with cotton and Lycra® materials is comfortable like a common T-shirt. The position of the electrodes and sensors is fixed and the elasticity of the WS makes it fit around the body well. The WS consists of electrodes, sensors and a small-size printed circuit board (PCB). The PCB includes four blocks: power supply (a 3.7 V 14,500 size cylindrical Li-Ion cell with a capacity of 750 mAh), MCU (Cygnal C8051F330), analog circuit and short-range communication unit. To reduce the power consumption of the WS, the physiological signals are only detected and digitalized in the WS. The processing work will be implemented in the PPU. To facilitate washing, the PCB in the WS is designed to be removed from the shirt. The prototype of the WS is shown in Figure 7.2, where the position of the electrodes and sensors is highlighted. In the picture, eight ECG electrodes (RA, LA, LL, RL, V1–V6) are shown, respectively, while the two respiration sensors and the activity sensor are positioned on the thorax, the abdomen and the waist, respectively. Three types of sensors and the methods used to develop the WS are described in the following section.

Fabric electrodes used to measure ECG signals substituting for conventional gel electrodes are knitted with stainless steel threads and textile yarns. The advantage of fabric electrodes is that they are non-irritating character and

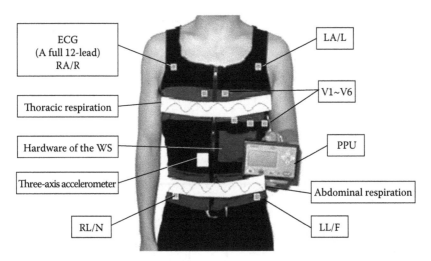

FIGURE 7.2
Prototype of WS and PPU. (From Zheng, J.W. et al., *Med Biol. Eng. Comput.*, 45, 877, 2007.)

it is possible to integrate them into a shirt. The major drawback is the inherent high skin-electrode impedance, which leads to a reduced signal-to-noise ratio (SNR). The AC coupled circ-common mode rejection. In order to reduce the contact impedance of the skin and improve the electrical signal quality in dynamic condition, the hydrogel membrane has been added to the surface of fabric electrode. The system is able to acquire a full 12-lead ECG by gathering signals through eight input channels simultaneously. The eight input channels (II, III, V1–V6) are sampled at 200 Hz with a 10-bit resolution analogue to digital (A/D) converter embedded in the MCU (C8051F330). Each input channel includes a differential preamplifier with a gain of eight, a high-pass filter with 0.05 Hz cut-off frequency, a fourth-order Butterworth low-pass filter with 100 Hz cut-off frequency and a main amplifier with a gain of 125. Figure 7.3 shows an example of acquiring ECG signals with fabric electrodes.

With regard to detecting respiration, an advanced respiratory monitoring method called RIP was designed and developed. By virtue of its design, RIP reduces the signal interference and distortion that is often associated with other monitoring technologies, enabling clinicians to obtain more accurate measurements of respiration. RIP measures changes in the cross-sectional area of the rib cage and abdomen over time and applies a series of proprietary algorithms to the data to calculate the amount of air either inhaled or exhaled during respiration. A new type patented RIP was designed to acquire high-quality signals and lower system power cost. To measure respiratory function, sensors are woven into the shirt around the patient's chest and abdomen. Two parallel, sinusoidal arrays of insulated wires embedded in elastic bands are woven into the shirt, surrounding the rib cage and abdominal areas of the torso. Usually, a colpitts oscillatory circuit is used to design the RIP by measuring the oscillatory frequency and the changes in cross-sectional area

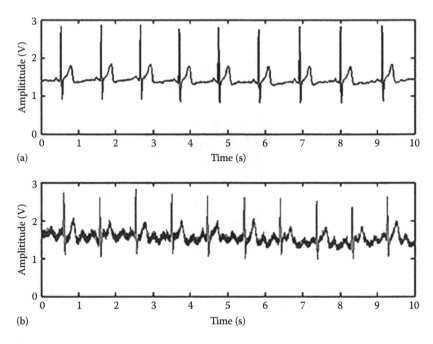

FIGURE 7.3
ECG signals obtained from fabric electrodes: (a) with hydrogel membrane and (b) without hydrogel membrane. (From Zheng, J.W. et al., *Med Biol. Eng. Comput.*, 45, 877, 2007.)

that occur during breathing that is indirectly measured. As the two body chambers expand and contract, the electrical sensors generate different magnetic fields that are converted into proportional voltage changes over time (i.e. *waveforms*). In this newly patented RIP, pulse amplitude modulation technology was used. Extremely high electrical current is passed through the wires in very short time; in this short time, the signal is sampled and held. A high SNR was acquired because of the high-input electrical current, but the system power cost was reduced because of the very short exciting time. The electrical current is passed through the wires of the abdomen and rib cage in turn in 100 ls and each wire is excited to 50 Hz. Thus, the sample rate of respiration is 50 Hz. Figure 7.4 shows an example of simultaneous acquisition of respiration and ECG signals from a walking person wearing the WS. It is very evident that high-quality signals could be acquired with the WS.

Accelerometer has been proposed as being suitable for monitoring human movements and has particular application in the monitoring of free-living subjects. It has been used to monitor a range of different movements including gait, sit-to-stand transfers, postural sway and falls. As for our system, the main purpose is to monitor abnormal events, such as falls or long periods without movement. Falls are very serious risk for the elderly, particularly for those living in the community. It has been proved that change in orientation from upright to lying that occurs immediately after an abrupt and great

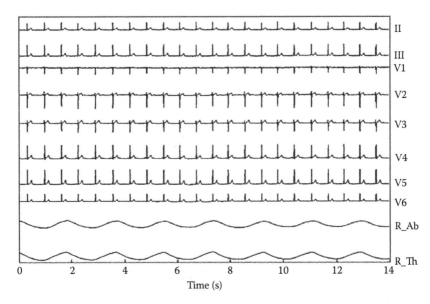

FIGURE 7.4
Simultaneous acquisitions of 8-channel ECG, abdominal respiration and thoracic respiration from a walking person wearing the WS. (From Zheng, J.W. et al., *Med Biol. Eng. Comput.*, 45, 877, 2007.)

negative acceleration (*due to impact*) is indicative of a fall. In order to assess daily physical activity, accelerometers must be able to measure accelerations up to ±1.5 g when they are attached at waist level (near the centre of mass of the subject). To meet these requirements, a tri-axial accelerometer, Freescale MMA7260, was used to implement human fall detection in our system. The MMA7260 features signal conditioning, low pass filter and temperature compensation. Figure 7.5 shows the signals from the MMA7260 sampled at 45 Hz when a subject performed a test.

7.2.2 Development of the PPU

The PPU includes six blocks: power supply (*two 3.7 V 14,500 size cylindrical Li-Ion cells with a capacity of 750 mAh*), MCU (*Cygnal C8051F120*), man–machine interface (MMI), memory, short-range communication unit and CDMA/gpsOne communication unit. The PPU can display two-channel ECG or respiration signals with a 160·64 dot-matrix LCD. Other information, such as heart rate, respiratory rate and battery status, is displayed on the right side of the LCD. To facilitate operation, the PPU provides five functional buttons, three emergency-aid buttons, an on–off button and a key-lock switch. The three emergency-aid buttons make the patient obtain rapid external help by establishing voice link with the MSC, local first-aid centre or a friend. The key-lock switch can avoid operating in a wrong manner.

In order to provide real-time health care for the patient without continuously transmitting medical data to the MSC, the physiological signals will

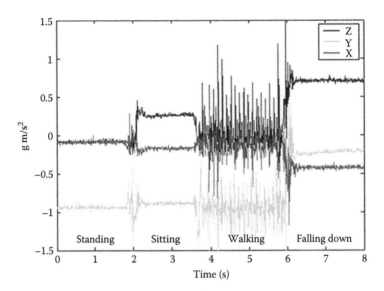

FIGURE 7.5
Tri-axial signals acquired from a waist-mounted accelerometer with Y axis in reverse gravitational direction during a test. (From Zheng, J.W. et al., *Med Biol. Eng. Comput.*, 45, 877, 2007.)

be processed and diagnosed in the PPU. The application in the PPU receives the measurements from the WS and determines whether an abnormal event had occurred and raises an alarm through real-time analysis and diagnosis. The PPU has three degrees of alarm. A minor one indicates that a not so important abnormality has been detected and some suggestions will be given to the patient. That includes ST segment (STS) changes and QRS amplitude (QRSA) changes. A medium one indicates life-threatening arrhythmias have been detected. To avoid raising false alarms, the PPU will receive physiological signals for a period of time (currently 10 s) and make a diagnosis. If the life-threatening situation is still detected, the PPU will ask the patient for the confirmation of an automatic emergency call. The patient can disable the alarm in case of a false alarm. If the patient does not react within a certain time (currently 10 s), an emergency call will be automatically initiated. A major one indicates a fall has been detected and the patient cannot stand up by himself or herself. An emergency call will be automatically initiated.

As for the respiration signals acquired with RIP, its main use would be to acquire a very accurate non-invasive measurement of ventilation and non-invasive detection of sleep apnoea syndrome in the future. In the current prototype, the tidal volume has been able to be accurately calculated, but the study of sleep apnoea monitoring is still in the development phase. To realize real-time arrhythmias classification in the PPU, some simple and accurate processing algorithms must be performed. These algorithms include removal of interferences, QRS complex, P wave and T wave detection, arrhythmias classification and fall detection. In order to acquire high-quality ECG for arrhythmias classification use, some digital filters must be applied to remove noise and interference. A subtraction procedure was applied to remove power-line interference. The procedure is simple and has good performance in coping with changes in amplitude of the power-line interference. A high-pass filter with QRS complex elimination was applied to remove baseline drift. The procedure applies a fast linear-phase low-pass filter to extract the baseline drift and then subtract it from the original ECG signals. The heart beat classification is the most important part in the software design of the PPU. In the first prototype, the minor alarm occurs in the situation: STS > 0.2 mV, STS < –0.1 mV, QRSA > 2.5 mV, or QRSA < 0.5 mV. Then apply 'So and Chan' QRS detection method to detect QRS complex, P wave and T wave. It selects the first derivative approach, which had proven the good performance even though the ECG signal was corrupted by high-level composite noise. The ST segment can be simply acquired by adding about 100 ms to the R point. As for the medium degree of alarm, focus on two life-threatening arrhythmias: ventricular tachycardia (VT) and ventricular fibrillation (VF). To detect these arrhythmias, the open source software was developed by E.P. Limited, which includes a C implementation of heart beat classifier. This classifier is able to classify the heartbeat as normal, premature ventricular contraction (PVC) or unknown. Since the open source software cannot recognize VT or VF, the PVC and unknown as VT or VF using the method of sample percentage in the dynamic range (SPDR) and complexity measure (CM). MIT/BIH arrhythmia database is used to test the heart beat classification algorithm. The accuracy percentages for VT and VF are 82.5% and 92.3%, respectively. The result proved the classifier can provide a very good accuracy for arrhythmias detection and classification. In the design of the major alarm, patient's falls must be detected in real time with high accuracy. Each of the three axis's signal sampled at 45 Hz was passed through a median filter of 13 samples in order to remove noise. Then a high-pass finite impulse response (FIR) filter of order 35, with a cut-off frequency at 0.25 Hz, was applied to each axis's signal. Finally, a non-overlapping moving window with a length of 0.8 s (36 samples) was applied to each axis signal.

7.2.3 Development of Wireless Communication

The main task of the short-range wireless communication is to transfer physiological signals measured from the WS to the PPU. The two main open standards currently used for medium data rate transfer are Bluetooth and Zigbee, which have an advantage of good connectivity between equipment of different manufacturers. But they neither have optimum price nor power consumption. Nordic nRF24L01 was used, which features low cost, low power and high speed, to build short-range communication unit. To avoid abrupt disconnection of the short-range communication, both the PPU and the WS would alert the wearer with sound and vibration when the PPU is taken at a great distance from the WS (*more than 10 m*). Utilize CDMA data link as air interface. To ensure the reliability of the data transfer, use TCP/IP proto-col to perform error detection and flow control. The gpsOne technology is an advanced positioning solution that utilizes assisted GPS (AGPS) alone or in combination with wireless network measurements to create reliable and accurate location information. The system uses a CDMA data module, Any DATA DTGS-800 is used, to establish CDMA communication and acquire the location information. The module supports not only CDMA 2000 1 air interface, but also GPS One position location capability. It uses QUALCOMM MSM6050 chipset, which provides three modes of operation in gpsOne func-tion (MS based, MS assisted and MS assisted/hybrid). Applying with GPS One technology, it can immediately locate the patient in case of emergency whether he/she is indoors or outdoors. In this way, the patient is able to walk out of room to any place without worrying whether he/she is in familiar sur-roundings or at a location that he/she does not know. The CDMA module was designed to support four types of communication channels.

1. *SMS*: It is used for command messages such as set emergency phone numbers.
2. *Voice call*: It is used to establish a voice link in case of emergency.
3. *Packet data service*: It is used to build a data connection with the MSC based on TCP/IP protocol.
4. *GpsOne function*: It is used to get the location information of wearer through gpsOne function.

The communication protocol has been designed in such a way that short-range wireless communication, the SMS and the packet data service rely on the same structure: header, raw data and tail. The header includes the flag explaining the type of the content, the length of the content and sequence of the current type of data. The raw data include medical data, the location information or the command information. The tail is used to check errors.

Wearable mobihealth care system with real-time monitoring, diagnosis, alarm, telecommunication and advanced GPS was described. The system is able to provide real-time analysis and diagnosis on a portable device and raise three degrees of alarm. In abnormal situations, an emergency call will be automatically initiated. In particular, the system provides three manual emer-gency aid buttons aimed at acquiring quick external help. The related studies point out that if immediate treatment is received, most of the cardiovascular patients would survive. By implementing online diagnosis and alarms locally on the PPU, we can supervise the patient in real time without being continu-ously connected to the MSC. This would reduce the workload of medical staff and communication costs. The WS makes the wearer feel increasingly comfort-able and convenient by integrating fabric electrodes and sensors into the shirt. The small-size PPU can be easily worn without connecting disturbing sen-sors and electrodes due to application of short-range wireless communication.

A working prototype of the system has been developed, and a medical experiment on 15 healthy volunteers had been taken. There are several advantages using the WS: a full 12-lead ECG, respiration and activity can be acquired; electrodes and sensors are always in the right place; a mix-up of electrode leads is impossible and there are no skin irritations even on long-term use. Regarding the ECG signals measured with the WS, although fabric electrodes can cause baseline drift owing to the high skin-electrode impedance and the variation in pressure of the electrode to the skin, high-quality signals with remarkable stabilities have been acquired most of the time in our daily life. However, the WS provides poor ECG signals as the participant is lying down. This may be related to the lack of adherence of fabric electrodes to the skin during body movement. To reduce the motion artefacts, we attempt to change the shape of the WS to fit the body of the wearer without reducing the comfort level of the wearer. Additionally, some algorithms, such as restraining the motion artefact with the help of activity information, will be implemented in the next prototype of the system. Nevertheless, the respiration signals measured with RIP are very stable and are free from the motion artefacts. As for the online diagnosis and alarm, the result of the experiment is satisfying. The falls suffered by patients could be detected with high accuracy and sensitivity. Although several false medium alarms are raised, we are assured this problem will be solved if the high-quality ECG signals can be acquired. Finally, some additional areas that we would like to explore in the future: providing many more physiological parameters monitoring, such as galvanic skin response (GSR), blood pressure and saturation of pulse oximeter (SPO_2); carrying out a study on sleep apnoea syndrome.

7.3 Smart Clothes for Ambulatory Remote Monitoring

The prototype of communicating underclothes for medical remote monitoring was created. It delivers physiological information on the subject (cardiac frequency, breathing frequency surface and mid-temperature) as well as the environment and activity parameters (ambient temperature, fall detection). It also enables the automatic data transfer on event, with the localization of the subject. New information technologies offer today the possibility to ensure an access to the care for isolated people, people at risk or with reduced mobility (teleconsultation, teleassistance, remote support at home). They naturally include video, but ethical problems limit its development in the field of monitoring, the equipment of the environment, whose development is still in infancy and the wearing of sensors (Noury et al. 2004). Most of these solutions remain rather complex and/or intrusive and do not seem to ensure all the guarantees in terms of respect of the intimacy, even if the acceptability level is improving in the field of care. Studies have shown that while helping the patients at risk, the systems must be discrete and reliable (sensitive and specific).

They must be applicable the usual conditions of life and in the society people live in. In this context, the development of micronanotechnologies and their use in the world of health find a natural field of application. The integration and the miniaturization of microelectronics allows the development of microsensors (accelerometer, magnetometer) which can be inserted into the garments of individuals to analyse their physiological parameters (heart rate, breathe, temperature) Vivometrics®, SenSatex®) and activity (movements). Thus, the use of microsensors worn by the people seems to be a particularly smart and acceptable solution for individuals, their family and home-care professionals. The systems of telealarm, a model of worn equipment, are gradually developing and the simple system of call alarms is now completed by functionalities such as activity recording (Vivago®) and detection of abnormal events such as falls with the help of sensors such as accelerometers. The VTAMN project (Vêtement de Télé Assistance Médicale Nomade – Underclothe for Nomad Medical Tele Assistance) aims at measuring the activities and physiological parameters of subjects in their daily life with an original setup of sensors networked in a garment, distributed algorithms and the possibility of launching alarms through a cell-phone and to rescue the person after a localization with a worn GPS. The technological challenges of the VTAMN (Figure 7.6) were to identify the technological limits (elasticity) when mixing textiles (ITECH) with connectors, microsensors for the measurement of temperature (LPM-INSA), activity (TIMC-IMAG), respiratory (RBI) and cardiac frequency (MEDIAG) and power management and integration (TAM).

A GUI was developed (SPIM), to easily handle the signals and for visualization of the data along with the localization of the patient. A real effort was made

VTAMN

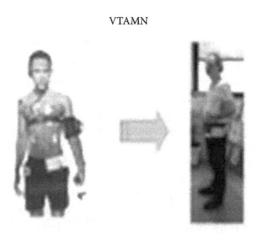

FIGURE 7.6
Technological break of the VTAMN project: wireless telemonitoring of the physiological signals in ambulatory conditions. (From Noury, N. et al., VTAMN – A smart clothe for ambulatory remote monitoring of physiological parameters and activity, *Proceedings of the 26th Annual International Conference of the IEEE EMBS*, San Francisco, CA, 2004, Vol. 2, pp. 3266–3269.)

FIGURE 7.7
VTAMN garment with the belt. (From Noury, N. et al., VTAMN – A smart clothe for ambulatory remote monitoring of physiological parameters and activity, *Proceedings of the 26th Annual International Conference of the IEEE EMBS*, San Francisco, CA, 2004, Vol. 2, pp. 3266–3269.)

to reduce the overall power consumption. As the batteries were too large to be integrated into the garment, it was decided to add a belt to the underclothes (Figure 7.7) in order to carry the batteries. For the same reasons, the main electronic board and the GPS/GSM modules were also placed on the belt. The overall weight of the VTAMN is 730 g, half of it is on the belt (battery life > 18 h).

7.3.1 Display Terminal Subsystem

The data from the VTAMN, transmitted through the GSM module placed onto the belt, is received on a PC in a *call centre*. The main screen displays the administrative data and physiological values. The signal base-band depends on the subject's activity but is still acceptable. The last screen points to the geographical localization of the person on the local map.

7.3.2 Testing and Evaluations

VTAMN project led to the implementation of three prototypes with the following characteristics: maximal integration in the textile, comfortable (370 g for the garment, 360 g for the belt), acceptable autonomy (>18 h), 3c leads ECG, breathing frequency, surface and mid-temperature, GPS localization and

GSM communication. The evaluation started with three healthy adults for 4 days. The signals obtained from the textile electrodes are significant. The movements of the thorax are correctly converted into a breathing frequency. The temperature is correct. The activity sensor is functional. The VTAMN is easy to dress. The thermal comfort is acceptable, even with high ambient temperatures (36°). The VTAMN results are quite encouraging. The ECG signals collected from the four *textile* electrodes are correctly transmitted to the call centre and the QRS signal is exploitable for cardiac arrhythmias detection. The thoracic–abdominal movements give the breathing frequency during normal activities (walking, etc.). The temperature sensors work correctly. An alarm is automatically launched from the fall sensor. The MMI was only prototyped in order to validate the other functionalities.

7.4 Development of Electronic Jerkin

Electronic jerkin (*E-Jerkin*) is a wearable electronics product made especially for a lifestyle object such as an MP3 jerkin and a Bluetooth connection to the mobile phones. Using this platform of the integration of electronics into textiles, the textile industry has the ability to quickly offer innovative products, to enter new markets and to strengthen their brands. The following modules are integrated into the jerkin: (a) audio module, (b) multimedia module and (c) flexible keypad module.

7.4.1 Module I: Audio Module

The Infineon programmable microchip is used for the digital music player in this jerkin; it consists of power management function and several general-purpose input/output lines (GPIOs). The 8051 microcontroller is used for controlling all functions, and the 16 bit DSP processor for signal processing. Input devices are the keypad and microphones. The microphones, earphones, storage media, keypads, displays and sensors are connected to this module. The built-in software offers several possibilities like this module that can be used as speakerphone, for speech compression and decompression, for music synthesis, for music decompression, the module also act as a software-modem, for speaker-independent and speaker-dependent recognition engines. This main module is designed especially for wearable electronics and smart clothes. The speech recognition offers hand-free for interaction with the electronics.

7.4.2 Module II: Multimedia Module

This module contains the battery and the multimedia card (MMC). The MMC can store 128 MB of memory and can be plugged into the computer. Also, it

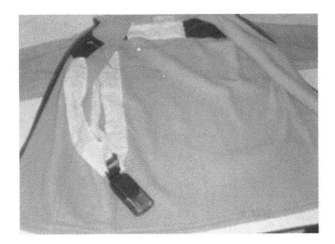

FIGURE 7.8
Multimedia module attached inside the E-Jerkin. (From Ashok Kumar, L. et al., *Asian Text. J.*, 6, 48, 2010b.)

can be recharged while being plugged into the computer and new music can be transferred to the MP3 player. This module is fixed to the jerkin by means of one piece connector as shown in Figure 7.8. This removable module is a Bluetooth gateway for connection to a mobile telephone, and a rechargeable battery which can withstand up to 8 h of battery life. The headphones can connect either to the MP3 player or to the phone.

7.4.3 Module III: Flexible Keypad Module

E-jerkin is equipped with a textile keyboard on the left arm and headphone in its collar as shown in Figure 7.9. The keyboard is lightweight and can be folded and easily fit into a pocket or bag. The flexible keypad module consists of sensor chip connected to the metal foils. The microphone and earphone modules are narrow fabrics connecting the audio module headphone and microphones. The keypad is a replaceable module and it can be fixed to any part of the jerkin. Conductive wire with flexible keypad is attached to the sleeve part of the jerkin, which is used to control the MP3-Bluetooth module. The pressure-sensitive five-button switch pad can be discreetly integrated into the sleeve of a garment. The buttons on the switch pad are used to control volume and simple functions such as play, pause, etc. The flexible key pad attached to sleeve is as shown in Figure 7.10.

7.4.4 Textile Integration

Embedding electronics into clothing is challenging. Beside washing and ironing problems, electronics may not function properly especially in demanding situations. Usability surveys have shown that consumers do not appreciate solutions that are uncomfortable with bulky devices and

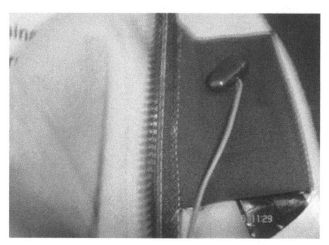

FIGURE 7.9
Microphone integrated in the collar of E-Jerkin. (From Ashok Kumar, L. et al., *Asian Text. J.*, 6, 48, 2010b.)

FIGURE 7.10
Flexible pressure-sensitive keypad attached in sleeve of E-Jerkin. (From Ashok Kumar, L. et al., *Asian Text. J.*, 6, 48, 2010b.)

visible cables. When integrating the electronics system into textiles, three major challenges should be overcome; they are (a) realization of the electronic components, (b) ensuring constant power delivery to the system and (c) providing a washable and flexible packaging, which does not cause any discomfort to the wearer.

Various types of natural user interfaces and connections have been developed for intelligent textile products, ranging from press buttons to Velcro tapes. Tapes, bindings and zippers can be used for integrating cables into

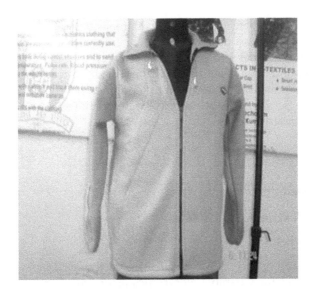

FIGURE 7.11
Full view of electronic jerkin. (From Ashok Kumar, L. et al., *Asian Text. J.*, 6, 48, 2010b.)

the garment. Conductive materials were sewn into the jerkin for connecting the textile keyboard, electronics module and to the collar headphone and microphone. It consists of intelligent docking, which is a special plastic material that allows the textile integration. The materials used are easy to sew/glue into garment. The weight of the material is very less and this minimal weight of the system aids for the whole integration of the kit. The docking station is used to plug in the MP3 Bluetooth module. The textile cable with microphone connector controls the whole system and is used for snap-in earphones in the collar. Snap-in microphone is waterproof and is integrated into collar of the Jerkin. The total system needs to be charged for its function and a travel charger is used to charge the MP3-Bluetooth module which keeps the system alive. The total operations are customizable through software modifications. Figure 7.11 shows the full view of E-Jerkin.

7.4.5 Power Supply

Power sources remain a key issue for the wearable electronics. The module is provided with LiPo battery with 780 mAh and the charging time of the LiPo battery is approximately 2.5 h. The playtime of the MP3 player is approximately 8 h and the talk time via Bluetooth of mobile phone will be more than 2 h. Two types of applications for wearable electronics are suited to generous apparel. One is communication and entertainment, and the other is location and position sensing. The wearable electronics applications described earlier are only a small selection of the possibilities it could offer. In the future, nearly, all the clothes or carpets could be made with integrated microelectronics. This microelectronics could improve our everyday life and offer a new way of interacting with microchips. The goal of this research has been to provide a better understanding of interactive electronic textiles. This research work identified the technologies, applications, opportunities and potential market appeal for interactive electronic textiles, while the survey captured expert perceptions and future insight to support the literature review.

7.5 Smart Textile Applications

7.5.1 MP3 Jacket

An example of a smart textile application is the MP3 jacket (Jung et al. 2002). Important for wearable electronics (especially for a lifestyle object such as an MP3 jacket) is the design: in this case, it has been made in cooperation with German Fashion School in Munich (Deutsche Meisterschule für Mode in München). One solution is to use the conductive ribbons as design element. Another important factor for this jacket is the robustness of the material. The jacket consists of several modules. All these modules are integrated into the jacket. One module is *the audio module*. It consists of an Infineon audio processor chip (size 7×7 mm²), an 8051 microcontroller, a 16 bit OAK DSP chip, program memory and a few auxiliary devices. The whole size of the module is $25 \times 25 \times 3$ mm³. The Infineon chip has a programmable microchip (and for this jacket, it is used for the digital music player), a power management function and several general-purpose input/output lines (GPIOs). The 8051 microcontroller is used for controlling all functions, and the 16-bit OAK DSP chip for signal processing. Input devices are the keypad and microphones. Microphones, earphones, storage media, keypads, displays and sensors are connected to this module. The built-in software offers several possibilities: the module can be used as speakerphone, for speech compression and decompression, for music synthesis, for music decompression (MP3), as a software-modem and for speaker-independent and speaker-dependent recognition engines. This main module is designed especially for wearable electronics and smart clothes. (Speech recognition offers hand-free for interaction with the electronics.) Another *module* contains *the battery* and the MMC. The MMC stores 128 Mb and can be plugged into the computer (it is recharged while being plugged into the computer and new music can be exchanged with the MP3 player). This module is fixed to the clothes by a one piece connector. The *keypad module* consists of sensor chip module connected to metal foils. These sensors are activated if a finger is close to a specific pad. One of these buttons is used for the activation of speech recognition. The *microphone and earphone modules* are narrow fabrics connecting the audio module and normal earpieces and microphones. Some modules could be added in the future to increase the usage of these jackets: for example fingerprint sensors for authentication, sensors for monitoring vital signs, wireless data receivers and many more. Jacket with the MP3 player is integrated – therefore, the jacket is very expensive. It would be better to prepare the jacket to be used with an external MP3 player (integration of the keypad only). Some sportswear manufacturers such as Burton sell jackets with integrated MP3 players.

7.5.2 Smart Carpet

The smart carpet is a grid of processing elements (PEs). The PEs could be equipped with sensors or actuators (Figure 7.12). At crossover points, the PEs are connected to the conductive fibres. A PE is a 32 bit microcontroller with four UARTs (*universal asynchronous receiver and transmitter*) used for communication with its neighbours. This PE is connected to the PC via a portal (*using the RS232 communication port*). The software used for the smart carpet is written in Java (*based on Java Swing Technology*) and is called smart carpet monitor (SCM). All algorithms are implemented as software simulator ADNOS (*Algorithmic Network Organization System*).

For this grid of PEs, a certain algorithm is needed: if a smart carpet is produced at a factory, there is no knowledge on where it will be used; therefore, it should be self-organizing. Self-organizing means, for example, the portal has some information; based upon this information and with communication

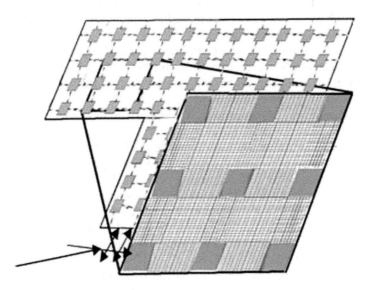

FIGURE 7.12
Embedded network of processing elements.

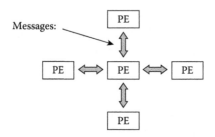

FIGURE 7.13
Phase elements and message integration.

between the nodes, each PE is able to determine, for example, its own position, orientation and distance to the portal (Figure 7.13). A smart carpet may be used at a room with no rectangular layout; therefore, it might be cut. If it is cut, several wires will be disconnected, maybe a PE is destroyed. So the algorithm has to detect these regions. The following description of the algorithm is based on the assumption that the PEs are placed in a grid, in a symmetrical way and that the portal has information about its position and orientation. Otherwise, the coherence and positioning phases are more difficult (one possible solution is finding coordinates based upon, e.g. triangulation). The smart carpet has two main phases: the organization phase, and afterwards the operation phase. The organization phase consists of eight sub-phases. The first phase is the *power-routing phase*. Like nearly all phases, it starts from the portal. Each PE checks its neighbours for leakages. (It can have a maximum number of four neighbours.) Defective PEs are disabled and if a neighbour is working, power is led through them. This workflow is used through the whole network until all PEs have power. During the *coherence phase*, messages are exchanged.

Starting from the portal (the portal knows the right direction), the first neighbour receives one direction. Due to the fact that there are only four communication lines, the PE knows the direction of the other three nodes. It sends the appropriate directions to its neighbours. After this phase, all PEs know its orientation. Afterwards each PE calculates its own temporary position (set to (0, 0) at the beginning) in the *positioning phase*. The portal starts giving the direct neighbours their absolute position value. Then, all neighbouring PEs check if they have to update their own position. (i, j) is the own calculated position and communicated position. If i > r, i is replaced by r, and if j > c, j is replaced by c. The new position is sent to the neighbours. The distance to the portal is also needed. It is calculated in the following way: starting from the portal, each PE sends its own distance to its neighbours. When receiving a message, the PE selects the neighbour with the smallest distance d, and sets its own distance d + 1. The new distance is transmitted to its neighbours. This is also referred to as the *distance phase*. Based upon the calculations made before, *routes* are *generated*. While using a regular backward organization algorithm, routes with minimal distance to the

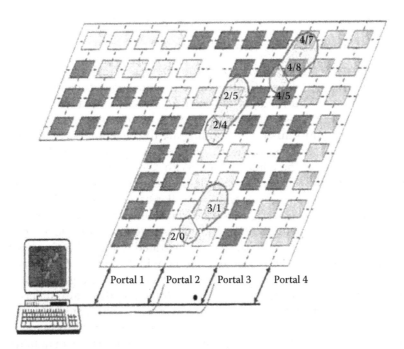

FIGURE 7.14
Smart carpet monitoring system. (From Jung, S. et al., A digital music player tailored for smart textiles: First results, Presented at the *Avantex Symposium*, Frankfurt, Germany, 2002.)

portal are generated. Then, the throughput calculated. The *throughput* is the number of routes leading through the grid. Afterwards each PE receives a unique id from the portal (Figure 7.14). This id is based on the established routes. Using these routes, data can be transferred through the carpet, and so the carpet is working. This smart carpet described earlier can be used in several ways: if the PEs have sensors that react on pressure, the smart carpet can be used as an alarm system. If a person steps on a sensor, the PE sends its number to the PC. The software on the PC may then check whether the person is allowed to step on this place. Otherwise, the software may set off an alarm. Another use case may be the detection of emergency cases. If smart carpets are used in homes of elderly people, they may detect when a person falls down and does not move any more. Then the software may call the ambulance.

The smart textiles applications described earlier are only a small selection of the possibilities smart textiles could offer. In the future, nearly, all of our clothes or carpets we use could be made with integrated microelectronics. This microelectronics could improve our everyday life (and offer a new way of interacting with microchips). But obstacles have to be overcome before smart textiles can begin their success: The biggest one is the cost: they have to get cheaper. Textile industry will not be interested in integrated microelectronics, if it is too expensive to integrate.

7.6 Development of Heating Gloves

Heating gloves is the development of the intelligent heating solutions to be integrated with the gloves. Customization of the systems enables the integration into nearly all kinds of gloves. Easy handling, maximum wearing comfort (size of the systems, low weight, etc.), high-power efficiency (intelligent system) and minimization of cabling are the benchmarks of this gloves. The automatic heating system of the gloves is very much needful in western countries during the winter season. With reference to the temperature in the atmosphere, the temperature inside the heating gloves will alter. In this research work, leather gloves and controller-based heating element were used to provide warm effects to the wearer. Commercially available leather material is purchased and heating coil is integrated with the gloves for heating system. Figure 7.15 shows the glove integrated with the heating system.

Heating gloves made up of leather material with heating system is used for cold-weather operations and pose unique problems with regard to maintenance of hand dexterity and performance. Research efforts are underway to find a way for soldiers to maintain dexterity during cold-weather operations. Cold-weather operations pose unique problems with regard to maintaining hand dexterity and performance. Currently, gloves/mittens that maintain comfort and warmth cause dexterity to degrade because of bulky material and loss of tactile sensitivity. If gloves are not used, hand and finger temperatures rapidly decrease during cold exposure, causing a reduction in hand function and manual dexterity. When the outside temperature is low or the wearer feels very cold, the user can manually press a button to switch on the heating system; once it reaches the maximum temperature, the user can switch off the system.

FIGURE 7.15
Heating gloves. (From Ashok Kumar et al., 2010.)

FIGURE 7.16
Inner and outer parts of the gloves. (From Ashok Kumar et al., 2010.)

Currently, there is no system or method to maintain dexterity during cold-weather field operations that has low-power requirements and is non-flammable. The system or method envisioned would employ technology using innovative engineering and/or a physiological approach that enables the resting warfighter to maintain dexterity at air temperatures at less than 32°F for more than 4 h. Figure 7.16 shows the inner and outer part of the gloves.

7.7 Development of Pneumatic Gloves

In recent decades, increase in the need for rehabilitation therapies has sparked interest in developing a clothing system using pneumatic system. The two trends, the use of adaptive pneumatic and the use of smart fabrics, have led to the development of the new flexible, pneumatic clothing. The main goal of this research work is to enable stroke-affected and physically challenged personnel to exercise for rehabilitation wearing pneumatic clothing in homes or workplaces without the assistance of physicians. The evaluation of the developed product has been conducted with physicians of various disciplines and patients with associated problems. The evaluated result shows the advantage, functionality and the need of the product. The diversification and technological advancement of textile sciences has been to such an extent that no area seems to be untouched by textiles. Products used for medical or surgery applications may at first sight seem either very simple or complex. In reality, however, in-depth research is required to engineer a textile for even a simplest cleaning wipe in order to meet stringent professional specifications. This research work reports the development of pneumatic gloves capable of supporting physically challenged people and helps them in hand rehabilitation. Textile-related technologies are once again having a significant impact on the future of the textile industry. The contribution of textile to pneumatic gloves is discussed.

7.7.1 Synthetic Rubber

The rubber material which is next to the skin will expand as there is an air supply. The Lycra layer will assist the gloves to expansion and make them fit the hand comfortably. The rexine material in the outer layer will not allow the gloves to blow up after the required amount of air is filled in the rubber material. It will create a reverse pressure in the opposite direction opposite of the rubber expansion, which will create a pressure on the fingers. This pressure will help the wearer to get the soothing effects of massage; when the air supply is released, the rubber layer will become empty and the Lycra and the rexine layers will relax and return to normal position. The Lycra layer will help in keep the gloves firmly on the fingers and brings the rubber material back to position. Repeating this operation continuously will provide a continuous massage for the wearer and help the patient to exercise for rehabilitation with the pneumatic gloves in their own homes or workplaces without the assistance of physicians. The air supply can be altered with the help of the pressure control valve. Depending upon the requirement, the pressure can be regulated with these valves.

7.7.2 Lycra

Spandex or elastane is a synthetic fibre known for its exceptional elasticity. It is stronger and more durable than rubber, its major non-synthetic competitor. Spandex fibres are produced in four different ways including melt extrusion, reaction spinning, solution dry spinning and solution wet spinning. Once the pre-polymer is formed, it can undergo further reactions in various ways and can be drawn out to produce a long fibre. The solution dry spinning method is used to produce over 90% of the world's spandex fibres. The basic characteristics of the Lycra material is shown in Table 7.1. For this

TABLE 7.1

Characteristics of Lycra Material

S. No.	Description	Specification
1.	Chemical composition	Segmented polyurethane urea
2.	Specific gravity	1.05 as a clear polymer, or 1.20 as pigmented polymer
3.	Moisture regain	1.5%
4.	Softening point	390°F/190°C
5.	Melting point	437°F/250°C
6.	Finish composition	Silicone oil with a metal stearate

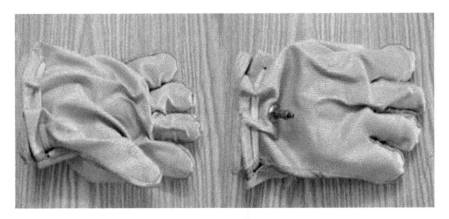

FIGURE 7.17
Front and back portions of the glove. (From Ashok Kumar et al., 2010.)

research work, 40^s cotton yarn count with 40 D Lycra Denier of 92% cotton + 8% Lycra composition of 200 fabric GSM were used. Figure 7.17 shows the front and back portion of the pneumatic gloves.

7.7.3 Rexine Material

Bonded fabrics are the fabrics that consist of at least one layer of textile material and the other layer of polymeric substrates bonded together by means of some additives or the adhesion property of any one of the layers of the fabric. Rexine fabric is one such bonded fabric having either cotton or any other textile layers, above which the rexine coating is provided. The rexine-coated material will have a little bit or no elongation which helps in designing of the pneumatic clothing. The water permeability of the rexine-coated material improves and the air permeability property is also improved.

7.7.4 Pressure Regulator Assembly

The pneumatic circuit for the rehabilitative arm was designed to control the rubber muscles that are attached to the device using two-way solenoid valves. Since solenoid valves are either on or off, two valves are required to control the pneumatic glove. One valve lets air in the muscle, causing it to contract, while the other valve releases the air out to the environment, causing it to extend and attain the resting position. A tank and an air filter were combined to provide the air supply to each muscle in the circuit. An example pneumatic circuit for two air muscles is illustrated in Figure 7.18. Figure 7.19 shows the pressure regulator assembly used for the research work.

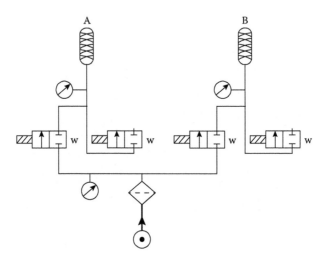

FIGURE 7.18
Pneumatic systems for pneumatic clothing. (From Ashok Kumar et al., 2010.)

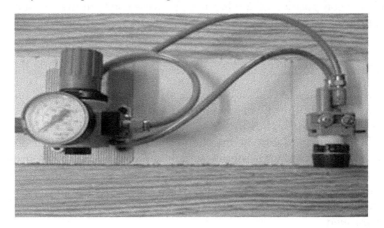

FIGURE 7.19
Pressure regulator assemblies. (From Ashok Kumar et al., 2010.)

7.7.5 Pneumatic Instruments

Pneumatic equipment provides the ideal solution for applications in the textile machine industry. It uses a natural medium that is unsurpassed in its environmental compatibility air. The technology ensures that, in the processing and manufacturing of materials and products which have to be reliably protected against grease, oil and other contaminants, such advantages are also combined with high productivity where excellence is always in vogue. The pneumatic components required are (a) pneumatic air supply unit, (b) pressure controller valve, (c) pressure regulator and (d) switches. Figure 7.20 shows the glove with pumping instruments.

FIGURE 7.20
Pneumatic gloves with pressure storage and pumping mechanism. (From Ashok Kumar et al., 2010.)

7.7.6 Pneumatic Gloves

The gloves consist of three layers. The primary layer is the cover, which is made up of artificial rubber, that acts as the conversion unit of the pressure applied to appropriate areas of the fingers, as shown in Figure 7.21. The secondary layer shown in Figure 7.22 of the gloves will be the Lycra that acts as the barrier for the air leakage and makes the gloves firmly fit the wearer's hand. Figure 7.23 shows the final layer made up of rexine material, forming the outer layer covering both the primary and secondary layers. The reason for choosing the rexine for this layer is to create the pressure on the fingers when there is air supply. When the air is pumped up to create pressure on

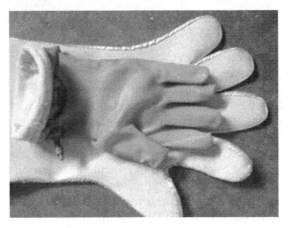

FIGURE 7.21
Primary layers made up of artificial rubber. (From Ashok Kumar et al., 2010.)

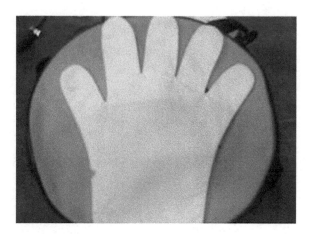

FIGURE 7.22
Secondary layers made up of Lycra material. (From Ashok Kumar et al., 2010.)

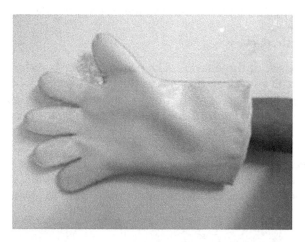

FIGURE 7.23
Final layers made up of rexine material. (From Ashok Kumar et al., 2010.)

the gloves, it compresses (squeezes on) the large artery in the fingers, stopping the blood flow for an instant The rubber material which is next to the skin will expand as there is an air supply. The Lycra layer will assist the gloves in the expansion and make the gloves to fit the hand.

The rexine material in the outer layer will not allow the gloves to blow up after required amount of air is filled in the rubber material. It will create a reverse pressure in the direction opposite to that of the rubber expansion which will create a pressure on the fingers. This pressure will help the wearer to get a massage effect. When the air supply is released, the rubber layer will become empty and the Lycra and the rexine layers will relax and come to normal position. The Lycra layer will help in the firmness of the gloves for the fingers and brings the rubber material back to position. Repeating this operation continuously will provide a continuous massage for the wearer and helps the patient to exercise for rehabilitation with the pneumatic gloves in their homes or workplaces without the assistance of physicians. The air supply can be altered with the help of the pressure control valve. Depending upon the requirement, the pressure can be regulated with these valves.

The pressure-regulating valve maintains an essentially constant output pressure (secondary side) independent of pressure fluctuations (primary side) and air consumption. Primary pressure at the connecting thread is reduced when air is exhausted from the push-on connection. The differential pressure regulator maintains a manually adjusted differential pressure between primary pressure at the thread connection and secondary pressure at the push-in connection. Pressure applied at the inlet end can be exhausted with no change in pressure at the thread connection end to the integrated valve. Most of the current rehabilitative devices for the hand use an external DC motor–controlled robot as the manipulator of the human hand. But these devices have many disadvantages: expensive, large, non-compliant, and they are not designed for rehabilitation at home, especially for the patients in their home recovery period. The solution to design a better rehabilitative device for the hand and forearm is to use a lightweight pneumatic actuator. Figure 7.24 shows the glove with pressure regulator assembly.

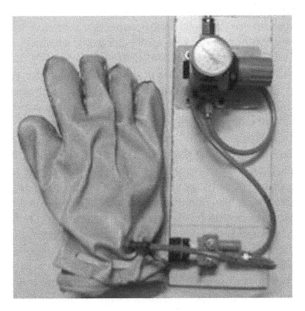

FIGURE 7.24
Glove with pressure regulator. (From Ashok Kumar et al., 2010.)

7.8 Gloves and Mobile Device: Product Design

An idea of a glove wirelessly connected to a mobile phone was developed (Berglin 2005). The idea has been developed with the aid of interviews and experimental product design. Interviewees were persons in professions who are exposed to difficult weather and whose tasks are physically demanding. The concept has been designed in a three-finger glove where the interaction takes place mainly by voice and hand gestures. The product design integrates the function in the construction of the glove. Wanted investigates the use of technology in a special context where working conditions are exposed to hard weather and converts it to a useful concept and product. Development in computing technology into smaller units has made it possible to integrate technology into a range of products like clothing. At the same time, textile material science has developed new textile materials, so-called intelligent textiles or smart textiles. Smart materials are materials that react to stimuli from its environment, react and adapt to the stimuli by changing their properties in response to the circumstances. Combining electronic devices and new textile materials could give an opportunity to develop new interactive products in textile and fashion with a new type of behaviour and use. Implementing technology into textile products changes both the use of the product and the user interface to the technology. Usually, a computer interface consists of a display and a keyboard and the design focuses on user tests, technical and semiotic functionality. When integrating interaction into products that we already have in our environment such as clothes, interaction design must consider the functional and aesthetical aspects connected to the product. In Hertzian tales, Anthony Dunne collects electronic products and ideas that supply aesthetical rather than practical or functional needs. It is interesting to see how aesthetical considerations often open the way for excellent functional products. The examples are results of experimental product design as a tool for exploring electronics objects and the questions that need to be given a thought are: What computer-related functions are of interest to implement in a certain product? How do we interact with such a product? How does the integration of technology affect the design of the product? Wanted is a project where those questions are explored in relatively extreme situations. In solving a problem for a user with special or extreme needs, the design is carried out to an extreme, which makes the issues clear. Wanted is a project where experimental product design has been used in order to find out how technology could be combined with a textile product for an extreme context regarding hard weather conditions. In this case, the product is a glove incorporated with mobile technology. A challenge in this project has been to develop a new type of interaction and to implement the functionality into the glove.

Wearable computing includes a wide a range of products such as mobile phones, PDA and smart clothing. The projects show the potential of wearable computing in different contexts for different uses. Most solutions in smart clothing are electronic products integrated into clothing, simply to illustrate the possibilities. In Finland, a group of researchers at Tampere University in co-operation with clothing manufacturer Reima has developed a survival kit for snowmobile users. New nomads is a project at Philips Electronics exploring the convergence between technology and fashion and the products; the project is divided into five distinct areas for different target groups. The interaction is mainly designed as textile keyboards and displays implemented in the clothing. Nomadic radio is a wearable audio platform developed at MIT Lab in Cambridge. The idea is to have a *hands-and-eye-free* product where the user gets access to remote information and communication. Instead of using keyboard and display as input and output device, nomadic radio uses speech and audio as interface. Some projects also show the potential of combining new textile materials with electronic components. Researchers at Infineon technologies have made an MP3 player integrated into a jacket. In this case, conductive fibres are used to lead signals between the different components like headphone and the audio module, for example. Conductive fibres are also used in the keyboard. Post, Orth, Russo and Gershenfeld present a type of keyboard in textiles, using conductive fibres and embroidery.

7.8.1 Concept Development

The arena for the concept and the experimental product design are situations of hard weather and working conditions. In order to explore the arena for design, it was decided to use interviews of persons from different professions whose working conditions are exposed to hard weather and physically demanding tasks. Since it was the first contact with users to identify some new concepts rather than fulfil user demands for one special professional group, four different professions with special working conditions and use of technology were chosen: Linemen, workers on ski-slopes, employees in the Swedish defence service, rescue workers and guides in Arctic area. In interviews, there are three kind of hidden information (e.g. Johnes), normal and abnormal activity, the reason for action and sources of information used by operators. Use of the result from the study of extreme working conditions gave the idea to facilitate the communication. Communication was central in handling required tasks, preventing accidents and in order to handle acute situations and accidents. In hard weather conditions, it is hard to reach and handle the mobile phone or radio and a headset is difficult to use. By wirelessly connecting the mobile phone to an item that is already used and that is easy to reach, the communication could be facilitated. It would also make sense to implement technology in a textile product. Using the glove as a mobile device allows one to make and receive calls without picking up the mobile phone. The connection to the mobile phone as a computer terminal also opens up the possibility of: getting access to GPS position and weather information through the glove. Another idea was to use an open function that identifies the user in order to allow access to different systems.

7.8.2 Product Design

The idea of communication was worked out in a first prototype where the interaction was developed. A three-finger Alpine glove, which is something between a glove and a mitten, was the basis for the design. It was important to develop a new type of user interface with no display or buttons that could be hard to handle with thick gloves. During the interview, the importance of just using one hand and the focus on few useful functions was pointed out. It was decided that phone calls should be made and received with the aid of the hand gesture and the voice. An interaction model was worked out where the call is activated by pressure on a part of the thumb, speaking in the hand and a possibility to change number by shaking hands (Figure 7.25).

The prototype and interaction was tested on the user group and it was easy to communicate the idea and the purpose of the product. Using only one hand during the act was appreciated and required gestures in the interaction did not seem to be any problem. One problem though was the placement of the headphone in the hand, since that is a position exposed to pressure. Alternative placements were discussed and the final choice was to place the headphone on the upper part of the index finger. By that change, there was a good posture and the headphone could be further pressed into the ear, which eliminated the problem of poor sound. The interaction model set three areas to design, the activation area on the thumb, the listening area on the index-finger and the speaking area just on the wrist. To identify the possibilities and the restrictions of the glove, the construction of the glove was studied. The construction was well adapted to the functionality and to the form of the hand. Every cut, seam and detail is connected to functionality and good fit. Construction solutions have their strength in using elements of good design and function. In this case, it was also possible to use the construction to lead conjunctions between the components. A general product design choice was to use the construction of the glove when implementing the functions into it. The seams were moved in the construction, creating a natural line between the areas, a line for visual affinity between the functions but also a possibility to lead conjunctions in (Figure 7.26).

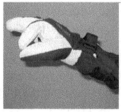

FIGURE 7.25
First prototype and interaction model. (From Lena, T. and Berglin, H., Wanted – A mobile phone interface integrated in a glove, In *Proceedings of Ambiene 2005*, Tampere, Finland, 2005.)

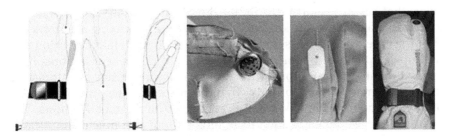

FIGURE 7.26
Final design solutions, sketch, prototyping and product. (From Lena, T. and Berglin, H., Wanted – A mobile phone interface integrated in a glove, In *Proceedings of Ambiene 2005*, Tampere, Finland, 2005.)

7.9 Future Research Scope

Use of experimental product design in order to find out how technology can be combined with a textile product for an extreme context with regard to climate and ergonomics has been studied by many researchers. This study shows how demands in extreme contexts can be used in experimental product design and how extreme circumstances make the arena for design very clear and focused. The questions to be addressed in research work have been: what kind of computer-related functions are of interest to implement in a textile product for such conditions and how that affects the interaction and product design. In this case, the computer-related functions are achieved from a mobile phone and implemented in a glove. The idea has been successful, since it has facilitated the problem with communication using the glove that is easy to reach and handle. The experimental product design has considered the ease of use, the embedded technology and the functional and aesthetical aspects of the product. Using the construction of the glove has preserved the function and aesthetics of the glove. At the same time, the solution has highlighted the new active parts of the glove and has also embedded the technology. When clothing is the user interface to computing technology, it is important to transform the traditional computer user interface to something that corresponds to the product. Future work in this project will continue in two directions. Both of them use the advantages and potential of letting the glove being wirelessly connected to the mobile phone. One direction is to implement an extending concept with GPS position, weather information and an open function. That will increase the interaction and amount of technology used and that will of course affect the product design. The other direction is to merge electronics and textile technology. What happens when a microphone is a part of textile instead being an electronic object that has to be embedded in the glove? How will the development of textile displays affect this kind of product? By just inserting electronics in clothing or textile products is not good enough. It is a total integration that will open up future opportunities. The new types of textile fibres have opened up new avenues for future applications, and it may be interesting to apply them in Wanted to study what this means for the product. Instead of being a mobile phone, Wanted is connected to the mobile phone. The result of that is: the embedded technology is not that much complicated which makes the merging goal more realistic.

7.10 Summary

This technology would then be beneficial to those living in remote areas where there is no supply of electricity from the grid, fuel may be scarce and expensive and maintenance of equipment is uncertain. Moreover, it could be used to reach power quickly to disaster-hit areas: those hit by earthquakes, hurricanes, floods or fire. Electronic Jerkin has been developed by integrating fleeced polypropylene fabric with Bluetooth and MP3 system. This microelectronics could improve our everyday life and offer a new way of interacting with microchips. The goal of this research has been to provide a better understanding of interactive electronic textiles. This research work identified the technologies, applications, opportunities and potential market appeal for interactive electronic textiles. Intelligent heating solutions developed eating gloves to be integrated with the gloves. The system's customization enables it to be integrated in nearly all kinds of gloves. The automatic heating system of the gloves is very much needed in western countries, especially during winter. The pneumatic glove was designed successfully with textile material: both the pneumatic and electronics components are found to be functioning well. The device can be used for rehabilitation at affordable rates and is convenient to operate at home just by pressing one button on the board. The developed product has been tested with physicians from various disciplines and patients with associated problems for validation. The evaluated result shows the advantage, functionality and the usefulness of the product. The various ratings given by the physicians of various disciplines for the afore-mentioned properties have been evaluated and charted in the graph given later. The average points for each property have been calculated and the final product is evaluated accordingly. The suggestion was encouraging and it can have wider scope in the medical field in India. The design of this prototype will lead to further research on developing a complete upper extremity stroke rehabilitation device. Such devices will hopefully become a reality in the near future to assist the overwhelming and increasing number of stroke sufferers.

Bibliography

Anliker, U., J.A. Ward, P. Lukowicz, G. Troster, F. Dolveck, M. Baer, F. Keita et al., Amon: A wearable multiparameter medical monitoring and alert system, *IEEE Trans. Inf. Technol. Biomed.*, 8(4), 415–427 (2004).

Ashok Kumar, L. and T. Ramachandran, Emerging trends in flexible wearable electronics for versatile applications, *Indian Text. J.*, 4, 62–66 (2010).

Ashok Kumar, L. and T. Ramachandran, Design and development of telemonitoring system using mobile network for medical applications, *Natl. J. Technol.*, 7(2), 1–7 (2011a).

Ashok Kumar, L. and T. Ramachandran, Teleintimation garment for personnel monitoring applications, *Melliand Int., Worldwide Text. J.*, 53–54, D45169–D45175 (2011b).

Ashok Kumar, L., J.C. Sakthivel, M. Sivanandam and T. Ramachandran, Development of self-assisting pneumatic clothing for physically challenged people, *Tech. Text., Germany*, 6, E98–E99 (2010a).

Ashok Kumar, L., P. Siva Kumar, K. Sunderasen and T. Ramachandran, Development of wearable solar jacket for remote and military applications, *Asian Text. J.*, 6, 48–52 (2010b).

Axisa, F., P.M. Schmitt, C. Gehin, G. Delhomme, E. McAdams and A. Dittmar, Flexible technologies and smart clothes for citizen medicine, home healthcare, and disease prevention, *IEEE Trans. Inf. Technol. Biomed.*, 9(3), 325–336 (2005).

Ayesta, U., L. Serran and I. Romero, Complexity measure revisited: A new algorithm for classifying cardiac arrhythmias, In *Proceedings of the 23rd Annual International Conference IEEE, EMBS*, Turkey, Vol. 2, pp. 1589–1591 (2001).

Banerjee, S., F. Steenkeste, P. Couturier, M. Debray and A. Franco, Telesurveillance of elderly patients by use of passive infra-red sensors in a smart room, *J. Telemed. Telecare*, 9, 23–29 (2003).

Christov, I., I. Dotsinsky and I. Daskalov, High pass filtering of ECG signals using QRS elimination, *Med. Biol. Eng. Comput.*, 30, 253–256 (1992).

Clarenbach, C.F., O. Senn, T. Brack, M. Kohler and K.E. Bloch, Monitoring of ventilation during exercise by a portable respiratory inductive plethysmograph, *Chest*, 128, 1282–1290 (2005).

Cohen, K.P., D. Panescu, J.H. Booske, J.G. Webster and W.J. Tompkins, Design of an inductive plethysmograph for ventilation measurement, *Physiol. Meas.*, 15(2), 217–229 (1994).

Carroll, D. and M. Duffy, *Modelling and Analysis of Planar Magnetic Devices*, Power Electronics Research Center, Electronic Engineering Department, NUI Galway, Ireland (2005).

Debray, M., P. Couturier, F. Greuillet, C. Hohn, S. Banerjee, G. Gavazzi and A. Franco, A preliminary study of the feasibility of wound telecare for the elderly, *J. Telemed. Telecare*, 7, 353–358 (2001).

Dittmar, A. and G. Delhomme, Microtechnologies et concepts Bio-inspirés: Les potentiels en Biomécanique, *Arch. Physiol. Biochem.*, 109, 24–34 (2001).

Dotsinsky, I. and T. Stoyanov, Power-line interference cancellation in ECG signals, *Biomed. Instrum. Technol.*, 39(2), 155–162 (2005).

Dunne, F. and A. Hertzian, *Tales: Electronic Products, Aesthetic Experience and Critical Design*, RCA CRD Research Publications, London, U.K. (1999).

Elton, S.F., Ten new developments for high-tech fabrics & garments invented or adapted by the research & technical group of the defence clothing and textile agency, In *Proceedings of the Avantex, International Symposium for High-Tech Apparel Textiles and Fashion Engineering with Innovation-Forum*, Frankfurt-am-Main, Germany, pp. 27–29 (2000).

Fensli, R., E. Gunnarson and T. Gundersen, A wearable ECG recording system for continuous arrhythmia monitoring in a wireless tele-home-care situation, In *Proceedings of the 18th Symposium on CBMS*, Dublin, Ireland, pp. 407–412 (2005).

Gorlick, M.M., Electric suspenders: A fabric power bus and data network for wearable digital devices, In *Proceedings of the Third International Symposium on Wearable Computers*, San Francisco, CA, IEEE, pp. 114–121 (1999).

Infineon Technologies – Products and product categories [Online]. Available: http://www.infineon com/products. Accessed 27 September, 2012.

Johnes, J.C., *Design Methods*, 2nd edn., John Wiley & Sons, New Jersey (2002).

Jones, M., T. Martin and Z. Nakad, A service backplane for e-textile applications, In *Workshop on Modeling, Analysis, and Middleware Support for Electronic Textiles*, MAMSET, San Jose, CA, pp. 15–22 (2002).

Jung, S., C. Lauterbach and W. Weber, A digital music player tailored for smart textiles: First results, In Presented at the *Avantex Symposium*, Frankfurt, Germany (2002).

Jung, S., Ch. Lauterbach, M. Strasser and W. Weber, Enabling technologies for disappearing electronics in smart textiles, In *IEEE International Solid-State Circuits Conference*, San Francisco, CA, pp. 9–13 (2003).

Kallmayer, C., New assembly technologies for textile transponder systems, In *Electronic Components and Technology Conference*, New Orleans, LA (2003).

Lagerholm, M. and C. Peterson, Clustering ECG complex using hermite function and self-organizing maps, *IEEE Trans. Biomed. Eng.*, 47(7), 838–848 (2000).

Leckenwalter, R., Innovations in new dimensions of functional clothing, In *Proceedings of the Avantex International Symposium for High-Tech Apparel Textiles and Fashion Engineering with Innovation-Forum*, Frankfurt am Main, Germany, pp. 27–29 (2000).

Lee, J.B. and V. Subramanian, Organic transistors on fiber: A first step toward electronic textiles, In *IEDM Technical Digest*, Washington, DC, pp. 199–202 (2003).

Lena, T. and H. Berglin, Wanted – A mobile phone interface integrated in a glove, In *Proceedings of Ambiene 2005*, Tampere, Finland (2005).

Levkov, C., G. Mihov, R. Ivanov, I. Daskalov, I. Christov and I. Dotsinsky, Removal of power-line interference from the ECG: A review of the subtraction procedure, *Biomed. Eng. Online*, 4, 50–57 (2005).

Lifeguard. Lifeguard jackets, http://www.lifeguard.stanford.edu. Accessed 23 May, 2012.

Linz, T., C. Kallmayer, R. Aschenbrenner and H. Reichl, Embroidering electrical interconnects with conductive yarn for the integration of flexible electronic modules into fabric, In *Proceedings of the Ninth IEEE International Symposium on Wearable Computers*, Osaka, Japan, pp. 86–89 (2005a).

Linz, T., C. Kallmayer, R. Aschenbrenner and H. Reichl, New interconnection technologies for the integration of electronics on textile substrates, In *Ambience 2005*, Tampere, Finland (2005b).

Mann, S., Smart clothing: The shift to wearable computing, *Commun. ACM*, 39(80), 23–24 (1996).

Marculescu, D., R. Marculescu and P.K. Khosla, Challenges and opportunities in electronic textiles modeling and optimization, In *Proceedings of the 39th Design Automation Conference*, New Orleans, LA, pp. 175–180 (2002).

Marculescu, D., R. Marculescu, N.H. Zamora, P. Stanley-Marbell, P.K. Khosla, S. Park, S. Jayaraman et al., Electronic textiles: A platform for pervasive computing, *Proc. IEEE*, 91(12), 1995–2018 (2003).

Martin, T., E. Jovanov and D. Raskovic, Issues in wearable computing for medical monitoring applications: A case study of a wearable ECG monitoring device, In *Proceedings of the Fourth IEEE International Symposium on Wearable Computers (ISWC'00)*, Atlanta, GA, pp. 43–49 (2000).

Mathie, M.J., J. Basilakis and B.G. Celler, A system for monitoring posture and physical activity using accelerometers, In *Proceedings of the 23rd Annual International Conference IEEE, EMBS*, Turkey, Vol. 4, pp. 3654–3657 (2001).

Mathie, M.J., N.H. Lovell, A.C.F. Coster and B.G. Celler, Determining activity using a triaxial accelerometer, In *Proceedings of the Second Joint Conference EMBS/BMES*, Vol. 3, pp. 2481–2482 (2002).

Mitcheson, P.D., T.C. Green, E.M. Yeatman and A.S. Holmes, Architectures for vibration-driven micropower generators, *J. Microelectromech. Syst.*, 13(3), 429–440 (2004).

Noury, N., P. Barralon, G. Virone, P. Boissy, M. Hamel and P. Rumeau, A smart sensor based on rules and its evaluation in daily routines, In *IEEE-EMBC 2003*, Cancun, Mexico, Vol. 4, pp. 3286–3289 (2003a).

Noury, N., A. Dittmar, C. Corroy, R. Baghai, J.L. Weber, D. Blanc, F. Klefstat, A. Blinovska, S. Vaysse and B. Comet, VTAMN – A smart clothe for ambulatory remote monitoring of physiological parameters and activity, In *Proceedings of the 26th Annual International Conference of the IEEE EMBS*, San Francisco, CA, Vol. 2, pp. 3266–3269 (2004).

Noury, N., G. Virone, P. Barralon, J. Ye, V. Rialle and J. Demongeot, New trends in health smart homes, In *Healthcom2003*, Santa-Monica, CA, pp. 111–117 (2003b).

Paradiso, J. and M. Feldmeier, A compact, wireless, self-powered pushbutton controller, In Tan, Y.K. (ed.), *Ubicomp 2001: Ubiquitous Computing, LNCS 2201*, Springer-Verlag, New York, pp. 299–304 (2005).

Paradiso, R., Wearable health care system for vital signs monitoring, In *Proceedings of the Fourth International Conference IEEE-EMBS on Information Technology Applications in Biomedicine*, Ischia, Italy, pp. 283–286 (2003).

Park, S. and S. Jayaraman, Adaptive and responsive textile structures, In Tao, X. (ed.), *Smart Fibers, Fabrics and Clothing: Fundamentals and Applications*, Woodhead, Cambridge, U.K., Ischia, Italy, pp. 226–245 (2001).

Park, S. and S. Jayaraman, Enhancing the quality of life through wearable technology, *IEEE Eng. Med. Biol. Mag.*, 22(3), 41–48 (2003).

Perng, J.K., B. Fisher, S. Hollar and K.S.J. Pister, Acceleration sensing glove, In *Proceedings of the Third International Symposium on Wearable Computers*, San Francisco, CA, IEEE, pp. 178–180 (1999).

Post, E.R., M. Orth, P.R. Russo and N. Gershenfeld, E-broidery: Design and fabrication of textilebased computing, *IBM Syst. J.*, 39(3–4), 840–860 (2000).

Post, R. and M. Orth, Smart fabric, wearable clothing, In *Proceedings of the First International Symposium on Wearable Computers*, Cambridge, MA, IEEE, pp. 167–168 (1997).

Rantanen, J., N. Alfthan, J. Impiö, T. Karinsalo, M. Malmivaara, R. Matala, M. Mäkinen, A. Reho, P. Talvenmaa, M. Tasanen and J. Vanhala, Smart clothing for the arctic environment, *IEEE Comput.*, 33(6), 15–23 (2000).

Rantanen, J. and M. Hännikäinen, Data transfer for smart clothing: Requirements and Potential technologies, In Tao, X. (ed.), *Wearable Electronics and Photonics*, Woodhead Publishing in Textiles, Cambridge, U.K., p. 198 (2005).

Rialle, V., F. Duchene, N. Noury, L. Bajolle and J. Demongeot, Health 'smart' home: Information technology to patients at home, *Telemed. J. E-Health*, 8(4), 395–409 (2002).

Riley, R., S. Thakkar, J. Czarnaski and B. Schott, Power-aware acoustic beamforming, In *Proceedings of the Military Sensing Symposium Specialty Group on Battlefield Acoustic and Seismic Sensing, Magnetic and Electric Field Sensors*, Laurel, MD (2002).

Rodriguez, J., A. Goni and A. Illarramendi, Real-time classification of ECGs on a PDA, *IEEE Trans. Inf. Technol. Biomed.*, 9(1), 23–34 (2005).

Russell, D.A., S.F. Elton, J. Squire, R. Staples and N. Wilson, First experience with shape memory material in functional clothing, In *Proceedings of the Avantex, International Symposium for High-Tech Apparel Textiles and Fashion Engineering with Innovation-Forum*, Frankfurt-am-Main, Germany, pp. 27–29 (2000).

Sawheny, N. and C. Schmandt, Speaking and Listening on the run: Design for wearable audio computing, *IEEE Comput.*, 31(7), 108 (1998).

Snyder, D., Piezoelectric reed power supply for use in abnormal tire condition warning systems, US patent 4,510,484, to Imperial Clevite, Inc., Patent and Trademark Office (1985).

So, H.H. and K.L. Chan, Development of QRS detection method for real-time ambulatory cardiac monitor, In *Proceedings of the 19th Annual International Conference IEEE-EMBS*, Chicago, IL, Vol. 1, pp. 289–292 (1997).

Spinelli, E.M., R. Pallas-Areny and M.A. Mayosky, AC-coupled front-end for biopotential measurements, *IEEE Trans. Biomed. Eng.*, 50(3), 391–395 (2003).

Starner, T. and J.A. Paradiso, Human-generated power for mobile electronics, In Piguet, C. (ed.), *Low-Power Electronics Design*, CRC Press, Boca Raton, FL, Vol. 45, pp. 1–35 (2004).

Sturm, T.F., S. Jung, G. Stromberg and A. Stöhr, A novel fault-tolerant architecture for self-organizing display and sensor arrays, In *SID Symposium Digest of Technical Papers*, San Jose, CA, Vol. XXXIII(II), pp. 1316–1319 (2002).

Troster, G., T. Kirstein and P. Lukowicz, Wearable computing: Packaging in textiles and clothes wearable computing lab, In *14th European Microelectronics and Packaging Conference & Exhibition*, Friedrichshafen, Germany (2003).

Van Langenhove, L. and C. Hertleer, Smart textiles: An overview, In *Proceedings from Autex Conference*, Poland (2003).

Wilhelm, F.H., E. Handke and W.T. Roth, Measurement of respiratory and cardiac function by the LifeShirt™: Initial assessment of usability and reliability during ambulatory sleep monitoring, *Biol. Psychol.*, 59, 250–251 (2002).

Zhang, Z.B., M.S. Yu, R.X. Li, T.H. Wu and J.L. Wu, Design of a wearable respiratory inductive plethysmograph and its applications, *Space Med. Med. Eng.* 19(5), 377–381 (2006).

Zheng, J.W., Z.B. Zhang, T.H. Wu and Y. Zhang, A wearable mobihealth care system supporting real-time diagnosis and alarm, *Med Biol. Eng. Comput.*, 45, 877–885 (2007).

Index